The
Horse
Training
Problem Solver

**YOUR QUESTIONS ANSWERED
ABOUT GROUND WORK, GAITS,
AND ATTITUDE IN THE ARENA
AND ON THE TRAIL**

JESSICA JAHIEL

Illustrations by Claudia Coleman

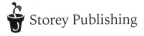
Storey Publishing

The mission of Storey Publishing is to serve our customers by publishing practical information that encourages personal independence in harmony with the environment.

Edited by Lisa Hiley
Art direction and text design by Cynthia McFarland
Text production by Liseann Karandisecky
Cover design by Kristy L. MacWilliams based on design by Kent Lew

Cover photograph by Dusty Perin

Indexed by Susan Olason/Indexes and Knowledge Maps

Printed in the United States by Versa Press
10 9 8 7 6 5 4 3 2 1

Library of Congress Cataloging-in-Publication Data

Jahiel, Jessica.
  The horse training problem solver / by Jessica Jahiel.
    p. cm.
  ISBN 978-1-58017-686-6 (pbk. : alk. paper); ISBN 978-1-58017-687-3
  (hardcover : alk. paper)
  1. Horses—Training—Miscellanea. I. Title.
SF287.J34 2007
636.1'0835—dc22

                                                    2007014565

DEDI

To three dear friends, now deceased, who understood horses and
humans and did a superb job of educating both:
Sally Graburn, Jill Hassler, and Dr. Reiner Klimke.

And as always, to my parents, who were born to teach
and have always done it so very, very well.

ACKNOWLEDGMENTS

No book is ever the creation of a single individual.
In the case of this book, several people (and one particular group
of people) stand out for their special contributions:

Madelyn Larsen, agent and friend, who found the ideal
publisher for this series of books.

Deb Burns, Lisa Hiley, and Ilona Sherratt, the Storey "dream team"
that gave the book its form.

Claudia Coleman, the extraordinary artist whose illustrations have
once again made horses come alive on the pages of a book.

Karen Fletcher, "Web Goddess,"
without whom the *Horse-Sense* mailing list and
Web site would never have become a reality.

Last but not least: the *Horse-Sense* readers who have
been asking questions and telling their stories for more than
10 years. They are all constantly struggling with the difficulty of
matching theory to practice and applying horse management
concepts and general training principles to the individual animals
that they love, care for, and try so hard to understand.
Without them, this book would not exist.

CONTENTS

FOREWORD

JESSICA JAHIEL'S WRITING IS INTELLIGENT and thoughtful, generous and measured, sound and systematic, clever and entertaining, clear and consistent, honest yet tactful, helpful and encouraging. Jessica is an educated, respected, experienced, and qualified instructor and teacher, who is joyfully optimistic about horses and their people yet realistic and practical with her advice. Her horse training approach is the classic time-honored, thorough method. She is more interested that her readers and their horses have a safe, positive experience than quick results.

As she answers many specific common questions, Jessica explains the difference between being considerate of your horse and spoiling him, between discipline and abuse. She describes how much is too much, and when to look for another way. She provides excellent detailed information about tack and techniques but she shows us that the best solution to a problem usually comes from a combination of looking at ourselves and seeing things from the horse's perspective.

She is a mirror of her horse training advice: "Be clear, consistent and kind."

As you read this book, you'll see that Jessica thoroughly enjoys horses, helping people, and writing. She makes us all feel that we and our horses are gold — worthy of all of the effort and time it takes to do our very best.

— CHERRY HILL
Trainer, instructor, and author of 30 books
and videos on horse training and care

PREFACE

THE CHALLENGES AND JOYS INVOLVED IN HORSE TRAINING are similar whether that training takes place in Boston or Botswana. The same issues affect horses and their trainers no matter where they live and what riding discipline they follow. Because the cost of a healthy, sound, fully trained horse is often prohibitive, many riders find themselves training their horses because they have acquired young horses, rescue or rehabilitation projects, or ex-racehorses that they intend to retrain for new careers. Because fewer riders are brought up in a horse-rich environment, many of the riders who find themselves training horses are doing so for the first time. They have never had the opportunity to watch the full process of horse training, much less the opportunity to become the apprentice of an expert in horse training.

Since the inception of the *Horse-Sense* newsletter more than 10 years ago, horse owners and riders around the world have sent hundreds of questions about horse training, how horses learn, how horses should be taught, and how their owners and riders can learn to teach them. As with the previous books in this series, *The Horse Behavior Problem Solver* and *The Rider's Problem Solver,* the people who ask these questions come from different backgrounds and different countries, follow different riding disciplines, own and ride different breeds of horses; and have different dreams and goals. And as with earlier books, all of the people who ask questions have three things in common. They share a passion for horses; they have open, questioning minds; and they wish very much to "do right" by the horses in their lives.

There are hundreds of books available on the subject of horse training. Many of these books are very helpful. But as so many readers have said over and over again, it's one thing to understand the basic principles of training, but quite another to achieve the practical, personal application of those principles. What they want and need is personal help: specific advice that applies to *their* situations and *their* horses.

Every question in this book came from a person who wanted help with an individual horse in a specific situation. By bringing together this collection of questions from riders and horse owners all over the world, and answering each one in its own specific context, I hope I can help readers analyze their own situations and find solutions to some of the problems they may encounter.

The aim of this book is to help people analyze their horse training in terms of both the "big picture" (philosophy, principles, and practices of good horse training) and the all-important smaller details that are sometimes overlooked. I would like readers to be able to take the solid fundamentals recommended in so many horse books and use the information in *this* book to help them apply those good fundamentals to their own horses.

Every interaction is training in the sense that every interaction teaches the horse something about the person who is handling or riding it. But education is a two-way street. If we pay attention, every single interaction we have with our horses will teach us something, too. Horsemen around the world share certain qualities: a willingness to listen to their horses, a willingness to look at the world around them and ask questions, and above all, a willingness to give their horses the benefit of the doubt at all times.

That's not all they share. When questions are about riding, nine — no, make that ten — out of ten situations are likely to require the rider to pay closer attention to her position and aids. When questions are about training, the common theme is once again the importance of the "basics," in this case, the rider's ability to observe her horse accurately, handle him kindly, and start at the beginning to develop the horse's minds and emotions — that is, his understanding and trust — along with his body. True training must address *all* aspects of the horse.

If there's a single "take-home message" in these pages, it is this: Listen, read, watch, learn, but don't just memorize lists and memorize the "right" answers. Think about what you learn, discuss it with your friends, relate it to your past experiences, apply it to your daily interactions with your horse — make it part of you. Keep your interest and your curiosity, and take joy in discovery.

My goal is to help you figure things out for yourself and to help you understand your horse so that you can teach him, train him, educate him in a way that will let him figure things out for himself, too. Most problems between humans and horses are the result of human actions — sometimes those of a human involved in the current situation, sometimes those of a human who was involved with the horse in the past. Either way, the person who is in a position to figure out and fix whatever's wrong is the current human in the horse's life — *you!*

JESSICA JAHIEL
Summerwood Farm

PART I

Basic Training

THOUGHTS ON TRAINING

TRAINING FROM THE GROUND

TRAINING FROM THE SADDLE: WHOA AND GO

TRAINING AT THE WALK

TRAINING AT THE TROT

TRAINING AT THE CANTER

TRAINING OVER FENCES

CHAPTER 1

Thoughts on Training

TRAINING YOUR OWN HORSE is one of the most enjoyable projects you can ever undertake. Dealing with the problems that arise is just part of the training process. No training program is ever going to proceed in a perfectly smooth, linear fashion, but by thinking clearly about your horse, your training methods, and your goals, you can achieve steady progress.

There will be times when your horse will not be sure that he can do something, and you will need to insist, pleasantly and politely. Most of the time, the magic words in training are "please" and "thank you"; occasionally the magic words are "now" and "thank you." If you are understanding, honest, fair, and appreciative, your horse will enjoy being trained by you.

Timetables, Watches, and Egos

Q You once gave a lecture at a conference called "Egos and Watches" or "No Time for Egos" or something like that. I've been asked to give a presentation on horse training at our local junior college, and I would love to borrow some of your material. That was the best lecture I ever heard on horse training, and I know these students could learn a lot from the ideas you presented.

A The lecture was called "Timetables, Watches, and Egos," but I must say that I like your "No Time for Egos" title just as much. Timetables, watches, and egos are three things that trainers must leave outside the stable door. The trainer's focus must be on the horse and in the moment, with total awareness of the "now" — but with perception, not with pressure. Good trainers understand that there are no shortcuts, that training is not a formula, a trick, or magic. It's not instant — it's a process that requires knowledge, understanding, patience, persistence, and time.

Good trainers know that things take time. If you take the time to train correctly, you won't need to invest additional time and money and effort later to correct your mistakes and fill the holes in the horse's training. Every horseman's motto should be *festina lente* — "make haste slowly." If you go slowly, you get there faster. More importantly, if you go slowly, *you get there*. If you're in a hurry, you won't get there at all.

A good trainer will let the horse's progress determine the training schedule.

Here are the keys to good training:

- The better the trainer, the easier the training process will be, and the less training time will be wasted on misunderstandings and miscommunications.
- A good trainer has long-term training goals in mind, understands how the short-term goals fit into the overall training plan, breaks down the short-term goals into tiny steps, and ensures that the horse masters each step before being asked to take the next one.
- Self-discipline before discipline: Good trainers use appropriate equipment for training; they don't use equipment to substitute for training. They know that riding and training effectively require hard work and self-discipline on the part of the human.
- Horses reflect their riders and even more so their trainers. If you want to change something about your horse, look closely at yourself and make the change in yourself first. Often, that's all you will have to do — the horse will reflect that change.
- Good trainers are always learning, always improving, and always aware that they don't (and can't) know it all.

A good trainer's goals for a horse must encompass more than a specific level of performance. If a trainer is exclusively preoccupied with developing a horse's obedience and mechanical skills, the horse will never become all that he could be. Goal setting must depend on the individual trainer and horse. If the horse is lucky, the trainer will have realistic expectations and the horse will not be hurt, frightened, or confused as he makes continual progress toward the trainer's goals.

Certain ongoing goals should shape and inform every training session and every trainer-horse interaction. One such goal is for a horse in training to develop and maintain a cheerful, enthusiastic attitude toward training and riding. Another ongoing goal is the improvement of the horse's appearance — over the course of training, a horse should become steadily more beautiful as his body develops correctly. The horse should also become steadily more enthusiastic and confident as his mind and understanding and trust develop correctly.

Trainers need to know their horses — not just what they see in front of them in the arena, but everything about their horses. If someone asks you to longe an unfamiliar horse, the horse's appearance and demeanor will give you some information about his conformation, condition, fitness, and training. His responses to your signals will give you much more information. But to be really effective, you need to know still more. Where does this horse live? Is he alone or does he have companions? Does he spend his time in a stall or a dirt enclosure or a field? How large is the field, what is the terrain like, how is the footing? How is he fed, what sort of hoof care does he receive, what sort of attention do his teeth receive? What has his life been up until now? What sort of work is he being asked to do? Can he do it? Will he be able to make progress at it? Does he enjoy it?

Good trainers realize that they, like the horses they train, are forever works in progress. They work hard to develop their powers of observation and analysis and their communication skills. They read and learn, watch others and learn, work and learn, study horses and learn, and strive to become the best trainers they can be.

Could I Train My Own Horse?

Q I am trying to decide whether it is feasible for me to keep a five-year-old gelding or if it would be better to find him a new home. I am very fond of him and want to do what is best for him as well as what is realistic for me. I am 43 years old, work full-time, and have all the usual family, work, and social life commitments. I am a green rider living in a sparsely populated area. As far as I know, the closest trainer is several hours away. The horse has had 30 days of basic training. I also took several lessons on him in an indoor arena.

In my estimation, as well as that of the trainer and my riding instructor, he is basically a good horse with potential but is still green, a bit spirited though not really hot, curious (and easily bored at times), can be willing and cooperative but needs a firm hand. He has bucked four people off, including me. I don't know the circumstances in the other cases; in my case I felt that it was a combination of his desperately not wanting to leave his pasture mate and my inexperience. It has taken me several months to come to terms with that incident. After looking into many options, I have come down to these:

A. Search diligently for a good home where he will get lots of time and attention and not be mistreated. I am willing to lose money if I could find the right person who is happy to bring along a green horse. The advantages are more time for me; less expense; less responsibility. The disadvantages are missing the horse and possibly missing out on the experience of solving the problem.

B. Take him back to the boarding facility with the indoor arena and keep him there for a few months, take more lessons on him, and hope that we gain confidence in each other. That situation was okay before, though I was a bit bored with going around and around the arena. Also, the facility is about 45 minutes away. Advantage: experienced people helping me every step of the way. Disadvantages: heavy cost in both time and money; somewhat boring after awhile; horse gains no experience on the trail where I plan to do most of my riding.

C. Keep him at home and try to work with him myself. I would have to have some major work done to create a riding arena. I would have to try to educate myself on horse training through books and videos. Advantages: a riding arena for future use; easier to spend time with him both in training and in just hanging out; could be a richly rewarding experience. Disadvantages: cost in time and money; my lack of experience could make everything worse; could be a horribly demoralizing experience.

Can you help me paint a realistic picture of training him myself? In your opinion, is it even possible for a green rider to do such a thing? I tend to believe that I need experts to do everything for me, whether it is building a rock wall or

Before you embark on a horse-training project, evaluate your own experience, knowledge, and skill.

making a dress. Maybe that is not necessarily true. After all, people have been training their horses for hundreds of years without professional trainers; surely they couldn't have all been disasters?

If you could comment on the joys and frustrations of the training process, including the time involved, and give your opinion on inexperienced people attempting this — also if there are particular qualities that you see in people who are good with training — I would be most appreciative.

A You've obviously put a lot of thought into this question. I hope you will not be disappointed that my advice is to go with your first scenario, in which you find a suitable home for this horse and look for the sort of horse you need now: a quiet, kind horse that enjoys trails and can help a novice rider learn in comfort and safety. You've described a horse that might possibly be a good second horse for you, and would probably be a good third horse, but is not a suitable first horse. Let me take your points one by one:

1. You want what's best for him: that would be consistent handling and training. This gelding is only five and has had very little training. Your instructor and the trainer say that he needs a "firm hand," and with only 30 days of training, he has already learned to buck people off. Assuming that this horse is a sound animal that is comfortable in his tack, this behavior indicates he has been pushed too hard and handled inconsistently. Bucking ought not to be part of the training process.

2. You want what's realistic for you: that would be an experienced, well-trained, kind, comfortable trail horse, probably 8 to 15 years old. You are a green rider with many demands on your time. You're in an isolated area without easy access to regular help. None of this will interfere much with your enjoyment of an older, quieter, well-trained horse, but all of it will interfere with your enjoyment, your progress, and your safety if you are trying to work alone with a young and green animal in your spare time.

3. When you were bucked off, you were shaken by the experience and needed several months to come to terms with it. That's nothing to be ashamed of — you're a novice, after all — but it means that you really do need to acquire that mileage before you take on a young horse.

4. Although you are a beginning rider, you were bored by what you were doing and thought that your horse was bored, too. This is usually a warning sign of poor instruction. Even if your entire lesson takes place at a walk, it should be interesting to both you and the horse, and it should be a pleasant experience for both of you. You need to know more than *what* to do, you need to know how, and why, and when, and for how long, and how to reward the horse.

5. Learning to ride is complicated and time-consuming enough; focus on that for a while, without adding the host of complications that arise when you also try to train a horse. Learn to ride, and learn about horses. And realize that many very competent, experienced riders never do "start" a young horse — it's not for everyone.

6. Good help matters. Even when you have an experienced horse, you will need some assistance and guidance. A green rider working alone with a green horse is a burning formula for disaster. Even when the horse and human are both good-hearted, misunderstanding will pile upon misunderstanding until utter mutual frustration sets in or until an injury occurs to one or both of the individuals.

7. Training a horse is like educating a child, requiring experience, knowledge, and skill. Because horses are large and powerful, and also surprisingly delicate in some ways, you have to factor in the physical risks to both student and teacher.

The trial-and-error method isn't appropriate for horse training. A key difference between horse training and making that rock wall (or dress) is that mistakes in horse training will make the horse suffer and can put you in real danger. Such results are far less likely when you are dealing with inert substances such as rocks or fabric. If you become horribly frustrated with your work-in-progress wall or your dress, you can scream at it and then abandon it for several days or a week or a month — it will be there, in the same condition and utterly unaffected by your outburst, when you return. That's not true of a horse. Horses are large and reactive and remember everything, and they will react to your behavior, not to your own belief in your good intentions.

Horses have reasons for the things they do. Once you understand horses thoroughly, you'll know what is natural for them, why they react as they do, what causes them to react, and how to encourage or discourage particular actions and reactions. Just as you can't evaluate a horse's illness or injury without understanding what would be normal for the horse, you can't evaluate a horse's reactions or actions unless you know which behaviors and reactions are normal.

It's not entirely true that people have been training horses without professional help for hundreds of years. There have been professional horse trainers for thousands of years. The goals of training may have been different (transportation, warfare, and agriculture rather than recreation and competition), and the equipment has changed over the centuries, but there have always been people who trained horses professionally, and the best trainers were in great demand. You're correct that many people have trained their own horses, but not without help. Right now, you need to focus on your own education. You can be involved with horses for the rest of your life, and that gives you quite a lot of time to learn and to develop experience and even expertise.

Good communication skills will help you. So will sensitivity and sympathy: You'll need both when you work with horses. These qualities are essential for trainers, but they must be accompanied by a good knowledge base. Horses must be dealt with *as horses*. Patience is vital — imagination perhaps less so, because unless you understand the nature of horses, your imagination may interfere rather than assist. Many kind, sensitive, imaginative humans are disasters as horse trainers, because they simply don't know enough about horses. They deal with horses according to their imaginations rather than according to horse nature, horse behavior, and horse logic.

Training a horse requires experience, knowledge, and skill. Horses are large and reactive and remember everything, and they will react to your behavior, not to your own belief in your good intentions.

A sane, sound, well-brought-up young horse of four or five, just beginning his formal training under saddle, should be able to become a nice riding horse after two years of consistent, competent training. It might take longer. It might not take as long. In either case, the horse would *not* be a fully trained, experienced animal at the end of that time — he would be ready for more training. A horse that does *not* get the best early training will take much longer to train correctly, because those all-important early years create a pattern of learning and a set of expectations in that horse, for good or for ill. If you were to attempt to train a young horse through the trial-and-error method, then even if you both survived intact, you would need to spend a couple of years retraining him later, with help. Horses started by novices usually require a great deal of retraining and time. It's better all around if you learn more before you begin.

Learn Your Horse, Then Train Your Horse

Q A friend of mine took her new mare to one of your clinics and you told her that she should learn her horse before she could train her horse. I loved that idea, and I have had it in the back of my head since then. Now I have a new horse that my instructor bought at an auction, and we don't know anything about him except that his conformation is good and he seems to have fairly good movement. We're basing this on watching him run in the arena, since nobody has yet been on his back.

My veterinarian thinks he is about eight years old. He was very thin at first, but now that he has been here for two months and gained a lot of weight, he still seems to have a nice quiet personality. My instructor thinks that he will be a good training project for me, and I am thrilled that she trusts me enough to let me try, but I know I need help. Given that we have just about no information about this horse, how can I "learn" him well enough to train him?

A What a wonderful question, and what an exciting project for you! Learning your horse means getting to know him. The better you know a horse, the better you'll be able to train him effectively. You've taken the first step just by watching him run in the arena. Watching a horse is an important part of training. You can

Watching your horse is the first step on the road to training him.

learn about your horse by watching him when he's loose, learn more by working him on the ground, learn even more by riding him — and then watch him again and learn still more. The secret is to watch carefully and make notes, asking yourself questions and watching until you have the answers. Everything you observe about your horse will help you understand him better once you're on his back.

Turn him out in the field. What does he do when he is first turned out? Does he rear and buck, run to the far end of the field, race over to the other horses, or walk to the nearest patch of grass and begin grazing? When other horses are present, how does he interact with them? When he plays, does he spend most of his time rearing or running? When other horses are *not* present, does he whinny for them, trot the fence line, or graze calmly? How does he react to windy days? What does he do when it rains? How does he react to snow? Does he like to roll? Does he roll often? Does he roll all the way over? How many times?

Turn him loose in the arena. Does he stand still, race around the edge, criss-cross the arena, or play in the center of the arena? Does he run or buck or roll? If there is a jump or a bucket in the arena, does he approach and investigate it, or does he avoid it?

Watch his movement. Is his walk slow and shuffling, or does he appear to be going somewhere? At the trot, how far forward do his legs reach? Are his front and hind legs parallel? Does his trot show suspension? Which direction does he prefer? How does he move from trot into canter — does he jump into canter, fall into it, or is it hard for you to see the moment when he changes gaits? Which canter lead does he prefer, and does he change his lead when he changes direction? When he stops suddenly from a canter, does he buck, rear, or crouch and slide to a stop? Is he visibly more flexible in one direction and stiffer in the other? How does he react to an audience? How does he react to music? Does he show off for other horses or for humans?

Put him on the longe line. How does he carry himself? Is his back lifted, does he raise his neck from the base and reach forward with his neck and head, does his tail lift and swing? How attentive is he to your signals? How quick are his responses? Does he bend easily on a 20-meter circle? What if you spiral him in until he is on an 18-meter circle, or a 15-meter circle? How does he react when you ask him to spiral out again? Is he more at ease in one direction than the other? How are his gaits — does he have clear footfalls at walk, trot, and canter? Does he move energetically, and will he lengthen his stride if you allow him to do so? When you ask him to stop and stand, does he stand quietly?

Now "learn" your horse by riding him. Is he tense or relaxed, fidgety or quiet, focused on you or focused on something else? How does he respond when you shift your weight, move the reins, give him a squeeze with your leg or a tap with your whip? Does he hold the bit quietly in his mouth, chew it, try to lean on it, or curl up behind it? Is his jaw relaxed or tense and tight? How about his neck? When he walks, does his back feel stiff and straight or flexible and round? When he begins to trot, does his back feel the same, does it seem to narrow and drop away from you, or does it seem to widen and lift? How does he respond to an opening rein, a direct rein, an indirect rein, a gentle half halt? When you turn him, is he moving from your leg and weight, or does he wait for the signal from the rein?

Never stop watching your horse. Even if longeing isn't part of your daily program, longe your horse from time to time and watch him carefully. Watch him as often as possible when he's turned out, and ask your instructor or another good rider to ride him occasionally while you watch. Have someone make a video of *you* riding him so that you'll be able to see how he moves and how he reacts when you are in the saddle.

Confused by Training Scale

Q I came to dressage rather late in life but am all the more fascinated because of approaching it as an intellectual challenge suitable for adults, or perhaps that's what I tell myself to compensate for being so much less supple and fit than the young people at the barn! I have been studying the list of the German training scale (rhythm, suppleness, contact, impulsion, straightness, collection), and I think I am clear about the principles and the order. What I am not clear about is the amount of overlap there may be between each two successive elements.

Can a horse become supple even if he is not yet confirmed in his rhythm? At the other end of the scale, can a horse that is not perfectly straight attain some degree of collection? Also, if a horse appears to be losing the element that the trainer is currently training, would it be correct for the trainer to return to the immediately previous element and practice that element before attempting the next one again?

A I think that what's confusing you is that these six concepts are generally referred to as "building blocks" or "steps," which seems to imply that each one is independent and that the trainer takes the horse from the first to the second, from the second to the third, and so on. Actually all six concepts are connected to each other in every way imaginable. The list is not strictly linear and it isn't a checklist; the trainer doesn't ever say "Ah, good, we have rhythm, cross it off the list and we'll move on to suppleness!" or "Hurrah, cross off contact, now we can focus on straightness!" Instead, each new item is added without the previous one losing any of its importance.

You're right: If a horse were losing — or struggling with — any of those "blocks," it would be correct for the trainer to go back, and possibly back and back, even to the very beginning. In fact, good trainers recapitulate the entire training scale on a daily basis — that's what they do when they are warming up their horses.

The building blocks concept is reasonably accurate in terms of the need to deal with them in a certain overall order and the importance of having each block securely in place before adding the next one. But there are many different ways of describing this logical progression. In very general terms, you could say that the first stage of training helps the horse retrieve his own balance and movement while carrying a rider on his back (rhythm and suppleness). The second stage builds the power of the horse's "engine" — developing the strength of his hindquarters and the bending and pushing power of his hind legs (contact and impulsion). Finally, the third stage involves an even greater development of hind-end power, promoting the horse's ability to carry the weight of his own forehand rather than pushing it (straightness and collection).

For any of the steps or elements to be achieved in a meaningful way, the previous elements have to be present as well. That's why I say it isn't a checklist — the list is cumulative, as is the effect of the training on the horse. The training scale is all about making progress and building on previous attainments. New elements can and should be added to keep training moving forward, but every previous element *must* be included *all of the time.*

Good training recapitulates the entire training scale on a daily basis.

Quick Comparison: Discipline and Punishment

Q For the past five months, I have been a working student for a well-known trainer. My riding has improved somewhat but I am not sure that I am learning anything useful about training horses, which is what I want to do with my life. I have never seen so much hitting of horses in my life as I have seen here in the last few months. I don't think of myself as soft or sentimental, but I have been accused of being both those things because I am not willing to pick fights with the horses that come here for training.

This trainer says it's important to start a fight with a horse during the first day he's here, so that he will immediately learn who is the boss. There are no exceptions. Young timid horses might get strong-minded later if they don't learn who the boss is right now. Older horses think they know it all (he says this about every horse older than two) and need to find out who is in charge.

Mr. X has a reputation for turning out horses that are very quiet and calm, even lazy. This was one of the reasons I wanted to work for him! Now that I know what he does, I don't understand how the horses turn out this way — the first few times I saw him beating up the horses, I really expected them to go crazy and trample him. Does this kind of training method really work? I don't have a problem with disciplining horses, I just don't see any reason to frighten them and beat them up. To me, that's not training, it's abuse. Can you give me a very short explanation of the difference between discipline and punishment and also explain how horses react when they are being disciplined or punished?

A I'm sorry that you're being exposed to these sorts of "training" methods. I agree that you're not going to learn anything useful there, and I strongly suggest that you leave as soon as you possibly can.

First let me address the issue of those "quiet, calm, lazy" horses. You've discovered one of the pathetic secrets of bad training methods. When horses — normal, average horses and also more sensitive horses — are systematically frightened, abused, and confused, they don't learn what the rules are or what is expected of them, they only learn that they are going to be hit again and again, apparently at random.

Since they can't make sense of their environment, have no way of controlling it, and are not able to escape it physically, they take the only possible escape route: They shut down. Imagine yourself surrounded by people who are shrieking in languages that you don't understand and beating you for no reason at all. If you reached a point at which you no longer were able to react to stimuli of any kind, you might appear to be a calm, quiet, lazy, insensitive person who just didn't notice what was going on around you. Actually you would probably be in some sort of dissociative fugue state in which you would still be moving and perhaps speaking, but you would be unable to remember who you were or what you had been doing.

Horses can have this reaction, and that seems to be the point of the "training" you have seen. You can see this on a much smaller scale at just about any boarding stable — there is always at least one rider who kicks or whips a horse constantly, claiming that "The horse is resistant" or "The horse isn't responding and needs a stronger aid." Horses in this position do eventually learn to ignore, or appear to ignore, the kicking, slapping, spurring, and whipping, and they accomplish this by shutting down and not registering any additional stimuli. Horses who are constantly pushed — all pressure, no release — and whose efforts are never rewarded or even acknowledged can also shut down like this.

Horse trainers need to be very clear in their own minds about the difference between discipline and punishment.

Discipline is fast, brief, and calm; punishment hurts and frightens the horse and has no place in training.

Discipline is a teaching tool with a clear purpose and a function. Discipline is understood by the horse, accepted by the horse, and makes the horse feel more secure. Discipline occurs when the trainer reminds the horse of the rules after (or just before) a minor infraction occurs; the horse learns a useful lesson and does not become upset or fearful.

Punishment is *not* a teaching tool — its only legitimate use is to severely discourage a dangerous, unprovoked aggressive behavior. Punishment has no place in training; it makes a horse afraid and insecure, and over time can create resentment as well as fear. Punishment occurs when the trainer has failed to notice that there was about to be a problem, and then failed to discipline the horse when it would have been appropriate to do so. Punishment involves rage, revenge, and the trainer's ego; the horse learns nothing useful and becomes upset and fearful. In other words, *punishment is discipline that failed.*

The difference is visible. When you hear someone yell at a horse or see someone hit a horse, *look at the horse.* His body and facial expression will tell the story. If you hear a slap and see a horse with a stiff back, tense neck, and tightly closed mouth, the horse has been punished. If you hear a slap and see a horse with a relaxed back, neck, and mouth, the horse has been disciplined.

When Is Punishment Appropriate?

Q I've heard and read a lot of conflicting information about when it is and isn't appropriate to punish a horse. I know that when a horse does something wrong, punishment has to be part of training, but I'm not clear on when it's a reasonable thing to do and when it's an abusive thing to do. Can you clarify when a trainer should punish a horse, and also what sort of punishment should be used?

A The quick and easy answer is that punishment is *not* part of training, and there is *never* a time when a horse should be punished for doing something wrong.

Think of your horse's training as being similar to a young child's education. If you had a little one in school, when would you find it acceptable for his teacher to punish him and what sort of punishment would you find acceptable?

The "punishment rule" is very simple.

If you're like most parents, you want your child to learn, and to learn how to learn, and to learn to enjoy learning so that he will continue to learn throughout his life. You want his teachers to help him learn, show him and tell him things, and make the learning experience enjoyable and pleasant. You want your child to look forward to going to school each morning — you don't want him to dislike it, dread it, and spend his time trying to find ways to avoid it.

All of these things apply to your horse in training. You want your horse to learn, you want him to learn to learn and to enjoy the process of learning. You want the process to be pleasant and enjoyable for him, and you want him to learn that *you* are a good person to be with. You want him to like you, trust you, and feel safe with you; you want him to have a good time when he is with you. You want him to whinny and come to the gate when he sees you approaching with the saddle — you don't want him racing to the opposite end of his field, or turning his back to you and trying to hide his head in the far corner of his stall.

It's quite possible to teach a horse using positive methods. There will be discipline, but *punishment* shouldn't come into the picture at all. Appropriate discipline, whether by voice, leg, or whip (never by reins), is a good trainer's correction in the form of an instant reaction that is neither violent ("I'll hurt you!") nor judgmental ("You're bad!"). Think of discipline as a course correction — "We need to turn left here," or, "Stop! Now straighten your wheels and back up another six inches, you're almost exactly lined up with the trailer." You're making a necessary correction, but there is no animosity and no hard feelings.

Positive training will make you, the trainer, work harder, but there's nothing wrong with that.

Disagree on Discipline

Q I have owned a four-year-old Quarter Horse gelding, my first horse, for about a year. He is intelligent, eager to please, and sane. Though I lack experience, I make sure never to go beyond my abilities, and I study books and videos and go to training workshops. I work at the stables where my horse lives, and I have received much guidance from the owner there, who breaks horses and does dressage. Before trying to work on technical riding skills, I concentrated on establishing relationships with the horses. All the horses there obey and respect me with the slightest gestures, and the owner has even expressed awe at how well I worked with them.

Unfortunately, a recent event made clear to me that he and I have some differences in thought. He offered to help me take a shoe off my horse, Blue. When Blue tried to pull his hoof away, the guy started yelling and hitting him. I watched the situation get out of hand quickly, becoming a contest of will between them, until I couldn't recognize my horse anymore! His eyes were rolling, he was moving all over the place, refusing to give his foot, and this guy was angry and cursing and kicking him. He grabbed a twitch and my horse stood still. After about two minutes (the shoe was difficult to remove) the fight between them started again.

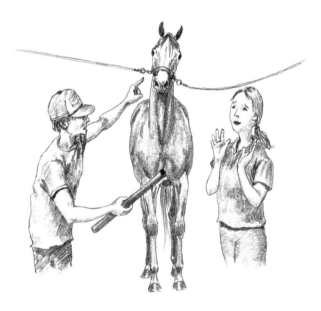

Hitting and yelling won't increase a horse's comfort or his understanding.

At the end, he was yelling at me, saying I am too gentle with my horse. My horse was quite simply nuts; I have never seen him behave like that before.

This man practices classical technique, using continued pressure. I have been using the Western method, releasing the pressure as soon as my horse complies with my demand. I didn't feel comfortable twitching him while he was standing still! Since he was doing what we wanted, I felt he needed a release to let him know that. Without any clear signs of what we wanted and whether he was doing it yet, he seemed confused and frustrated.

There is also the emotion thing. This guy often says you must "get mad" to scare the horse into submission. I am more a mental person than an emotional one. I act, I discipline, but never with emotion. He thinks that means I lack force.

My question is this: Can it cause behavioral problems when a horse is accustomed to one method and has someone work a different way with him? Or can the horse learn "This way with this person, that way with the other"? I am happy with my horse's discipline and work. I wonder if I need to insist that this man know and respect the method Blue is used to, so as not to screw up his head, confuse him, or make him lose his enthusiasm for work.

A It sounds to me as though you have a very good method of dealing with your horse, and I think you are also getting some good help from the owner of the stable. However, I agree with you that the situation you described was not good or appropriate.

Horses always have reasons for doing what they do. A cooperative horse will not pull its foot away just to be rude, or to "test" or "challenge" the farrier, but he may pull his foot away if he is uncomfortable, unbalanced, or frightened. Hitting and yelling will not help a horse become more comfortable, better balanced, or less frightened, and if the horse was not already nervous, he will be after someone has yelled at him and hit him. A twitch shouldn't be used as an alternative to good handling, but many farriers and vets want the owner to use a twitch if a horse becomes agitated or nervous, because they don't want to put themselves in a dangerous position. Handling a horse's feet puts a human in prime kicking range, and I'm sure you can understand why any farrier would be reluctant to work on a horse that was dancing and leaping.

Horses *can* learn different routines and associate them with different humans, or even with different tack. Stallions used for breeding, for example, know instantly whether they are being taken out for exercise or to go to the breeding shed, because they wear a different halter or bridle for each activity, and they quickly learn which is which. But horses appreciate consistent behavior, and I suggest that you accept help *only* from those people who are willing to treat your horse with the kindness and respect that you have taught him to expect.

It should never be necessary to frighten a horse into submission. You can be *assertive* like the herd leader, without being *aggressive* like a predator. Discipline must always be given without emotion, otherwise it becomes aggressive punishment rather than an assertive reminder of one's authority. You are right and the stable owner was wrong. He may well be correct in saying that you lack force, but since force has no place in horse training, I would take that statement as a compliment.

I suspect that the man was frightened, and was probably ashamed of being frightened, and that fear and embarrassment made him act aggressively toward your horse. He was probably also annoyed that your horse would calmly do whatever you wanted and would not cooperate with him. If he is a good man, he is probably already more than a little embarrassed about the incident and wishes it had not happened.

If you want him to use your techniques of horse handling, you may need to show him exactly what they are, and explain that *this* is the way you want your horse handled by everyone. If he is willing to comply with your request, then you can safely accept his help; if he is not, then you cannot. "Getting mad" is not a good training tool — I like yours much better: patience and calm. The next time this man offers to help you, unless you are certain he will use your preferred techniques, say, "Thank you, it's very kind of you to offer, but no thanks."

One final thought: Classical technique employs minimal and very brief pressure, with the emphasis always on the "yield," what the new Western trainers call the "release." Continuous pressure is *not* classical technique, it's just bad training.

Dealing with Resistances

Q What is the right way to deal with resistances? I have always been told that it's important to work through them, but I'm not sure I know what that means. Usually when I see other riders "working through resistances," they are beating or whipping or repeatedly kicking their horses, and I don't want to do that to my horse.

Is that really what working through resistances means? Is there some other way I could deal with a resistance? I worry that my horse might have trouble understanding my signals or might not be able to do what I'm asking, and I would punish him when he didn't deserve it.

I have had my horse for two years, and I would not say that he has been resistant yet. But others have told me that all horses are resistant by nature and that it's something that every rider has to deal with frequently, so it is just a matter of time before my horse shows resistance, too.

A   The way a rider or trainer deals with resistances often has less to do with the horse than with his or her personal beliefs about the nature of resistance. I generally see three rider responses to resistance:

- ▶ "How dare the horse disobey me!"
- ▶ "This horse doesn't want to cooperate, I must make my aids stronger!"
- ▶ "Oh dear, I must have done something wrong, now what *was* it?"

A good trainer won't push a tense, anxious horse, but will restore calm before asking again.

The first two interpretations tend to lead to more, bigger, and more frequent resistances, whereas the third, in which the *rider* accepts responsibility, tends to lead to fewer, smaller, and less frequent resistances.

When someone says, "My horse is resistant," I ask, "Resistant to *what*?" Horses on their own, without riders or trainers to tell them what to do, don't exhibit resistances. A resistance is a *reaction*. If my horse resists, the horse is reacting to *something I did*. My interest is in discovering the cause and putting the situation right, not in blaming the horse and certainly not in punishing the horse.

A resistance is the way a horse says, "What did you just say, I couldn't quite hear you?" or, "I don't understand what you just said," or, "I can't do what you just asked me to do." I have not yet met a horse whose resistance I could interpret in any other way. Unfortunately, I do meet many people who think that a resistance is typical horse behavior and that all resistances must be "pushed through," "worked through," or "punished" (and the three terms might as well be synonyms). Horses ridden by people who expect them to resist are likely to provide the expected resistance in some form. Horses generally do give us what we ask for, and some riders practically demand resistance by insisting that the horse *perform* rather than *try*, and by making a demand over and over until the horse finally gives in or gives up.

If someone like you is training a horse, things are different. If the horse offers a resistance, and the trainer's reaction is "Uh-oh, I pushed you too far, sorry, let's go back a step and relax and try again, and I'll take it easier this time," then I would expect that trainer to encounter very few resistances from horses in training. Your horse probably experiences occasional moments when he misunderstands an aid, or when he doesn't clearly understand what you've asked him to do, or when he understands but is not certain that he can do what you've asked him to do. In other words, he is a normal horse having normal experiences associated with being trained. The difference is that *you* aren't attaching the label "resistances" to these moments, because, luckily for your horse, you don't think that way.

Horses aren't devious; they don't spend their time devising clever methods to avoid work, although many riders and trainers seem to believe this is the case. To avoid resistances, pay attention to your horse at all times so that as soon as he becomes a little bit tired, sore, worried, or frustrated, you instantly redirect his energy into a positive action and then praise and reward him and give him a break. The more observant and understanding you are, the better trainer you will be, and the fewer resistances you will encounter.

"Natural" Trainer and Untrainable Horse?

Q I don't know if this was the right thing to do, but I took my horse home from the trainer's barn yesterday when she still had six weeks of training to go. I am hoping that you will be able to help me understand whether the trainer is right and I am ruining my horse by indulging her, or whether there is something wrong with the way he was training her.

Lia is a three-year-old mare, half Connemara and half Thoroughbred. I've owned her since she was eight months old, and I have trained her myself until now. I am 54 and have never backed a young horse myself, so I arranged to place Lia with a local trainer for three months, after which he assured me that she would be easy for me to ride. In retrospect I realize that I should have been a bit suspicious of this since three months is not long enough to train a horse thoroughly. In any case, I had heard good things about him from a neighbor (another mistake on my part, since I had not at that time ever seen the neighbor ride) and knew that he trained with the "Natural Horsemanship" pressure and release method.

I was allowed to come every other Thursday for half an hour to watch, otherwise I was to stay away because the owner's presence would supposedly distract the horse and disrupt the training. I agreed to this (yes, I know now that this was a huge error on my part) although I cried on the way home because I was unhappy about leaving her there. On each one of the three Thursdays that I watched Lia being ridden, I wanted to cry, and after the third one I returned home, talked with my husband, and took my trailer to go and bring Lia home.

First visit: The trainer pulled Lia's head to his knee and booted her sideways with his leg, saying, "The most important skill to have is the one-rein stop." She staggered and almost fell twice but he never let up. When I asked him if so much force was necessary, he said it *was* when a horse was spoiled and older than the normal age to begin training (two or younger).

Second visit: Lia was being taught to back, and apparently resisted when he first used firm pressure in her mouth, because she reared. When she came down, he did several very rapid one-rein stops, jerking her head to his knee and really shoving her hindquarters to one side. This time she did fall. I was horrified but also secretly glad that I had taken her to a trainer instead of having her rear and fall when I was riding her.

Third visit: Lia was still learning to back but had made great progress, he said. He had adjusted the reins to run through some sort of loop between her front legs (attached to the cinch I believe) so that she had to put her head down between her knees, and he then backed her across and then around and around the riding school. When he stopped her near me her whole body was shaking, and I could see blood at the corner of her mouth.

Forcing a horse into submission is an unacceptable training method.

He allowed her some rein and she immediately began to fling her head about, so he backed her all the way around again, did several more one-rein stops, and then put her away, saying that she had not yet learned to yield to pressure but that she would learn by the time I took her home.

I deliberately chose this trainer because I did not want my sweet mare to be harmed by traditional forceful breaking, but if this is Natural Horsemanship I want no part of it. Is there some other method I could use that would allow me to train Lia myself? Or is it hopeless — is it possible that she is just not trainable?

A There was nothing natural about this trainer, and as for horsemanship, *that* never came into the picture at all. You did the right thing in taking Lia home; I only wish that you had been able to take her away after or even before your first visit to the trainer's establishment.

"Pressure and release" should probably be written as "pressure and RELEASE," because that puts the emphasis where it belongs. The *release* is the essential element here. Pressure without release is not a training technique, because the horse learns not from the pressure but from the release.

When a horse shows you that you've applied too much pressure, or applied pressure for too long, or, as is very common, applied too much pressure for too long, you need to remove the pressure, pause, do something else, and then ask again, but this time more gently, more slowly, and more quietly. The horse must be given time to understand that something is wanted, time to figure out what is wanted, and time to offer the movement or posture that the trainer seems to want. For the trainer to sustain or, worse, escalate the pressure while waiting for the correct response only interferes with the training process.

A horse that is just learning to step back will do better if the trainer asks for a single step, then immediately releases, praises, and rewards the horse, pauses, and only then asks for another step.

Pressure and release is easy to overuse and misuse. Some trainers put heavy, unrelenting pressure on a horse and maintain the pressure until the horse performs the action they want. This works only if the horse is lucky enough to guess what's wanted and is physically able to comply. It's useless in the long term, because horses will learn to respond to pressure with one of three reactions, not one of which is productive: attempt to escape the pressure (flight), attempt to escape the trap in which they are caught (fight), and attempt to escape emotionally when physical flight and fight are both impossible (freeze or shut down).

Your mare was hurt, frightened, and confused, and was forced into submission instead of being given a chance to learn anything. Constant pressure teaches the horse only one thing, and that is to fear and dislike the person applying the pressure. Poor Lia had a bad time at the trainer's establishment, but *you* have learned several very important lessons that will stand you in good stead from now on.

First and foremost: It's the release that teaches.

Second: Trust your instincts. You cried on your way home, which means that at some level, you knew, or felt, that all was not well. Your instincts were wiser than you knew.

Third: You have a responsibility to your horse. Don't allow anyone to do anything that will hurt or frighten your horse. If something bad happens, you and your horse will live with the consequences for a long time.

Fourth: Be hesitant to accept recommendations from people whose own standards you don't know. If you had known more about your neighbor's definition of "good training," you would have taken Lia elsewhere.

Fifth: Beware of any trainer who does not want the horse's owner to see the horse being trained. This is never a good sign. The best trainers want the owner to see as much of the training as possible and want the owner to participate as much as possible. Even the best training is largely useless if a horse is sent home without its *owner* being given some training as well.

Sixth: Look at individuals, not at labels. This man was obviously neither a trainer nor a horseman: Fear, force, and violence are unacceptable training techniques. You don't have to give up on "natural" horsemanship — just look for a better trainer.

CHAPTER 2

Training from the Ground

SENSIBLE, HORSE-FRIENDLY TRAINING starts with work on the ground, because the training process begins as soon as there is contact between you and your horse. *When* you begin your horse's training matters much less than *how* you begin it. Training might begin just after your foal is born, just after you buy an ex-racehorse off the track, or just after you adopt a fully grown Mustang. Successful training can also begin when a horse is a mature adult. Whether you are using a round pen, a square pen, or no pen at all, the constructive work you do with your horse on the ground will prepare him for future work under saddle.

How Do I Teach Ground Manners?

Q As a first-time horse owner, I made the mistake of buying a horse that is much less well trained than I thought he was! I was told that he was a fully trained nine-year-old gelding who could be ridden both English and Western. In fact, Carlito is four years old, not nine. He knows very little about anything, though he can stop (sometimes) and steer (not very well), but is so nice that anyone can put an English or a Western saddle on him and ride him around. I have had him for six months and have fallen in love with him, and I want to keep him and work with him even though my veterinarian and my instructor both would prefer that I get another horse.

Because he is so sweet and quiet (my vet and instructor agree on that) and I feel safe around him, I want to train him myself. I will work with my instructor on everything involving riding, of course. But I would like some instructions on how to teach him good manners for everyday handling. I would like him to face me when I go into his stall to halter him, and then I would like him to wait for me to lead him out and for him to walk next to me. When I attach the lead rope to the halter, he immediately begins walking. Sometimes he goes ahead of me and sometimes he lags back. I think that if I can train Carlito to have nice ground manners, my instructor will take my training ambitions more seriously.

A You bought a horse that was advertised as mature and fully trained and found that he was young and basically untrained. Although you are a first-time horse owner, you want to keep the horse and train him yourself. I'm sure you understand why this makes your instructor nervous, and why she and your vet recommend that you replace him with a horse that *is* mature and trained. That said, Carlito is your horse now and from the sound of it he is not going anywhere. This isn't an ideal situation, but since Carlito is obviously sweet, calm, and tolerant, and your instructor is willing to work with you and your horse, then if you are prepared for a great deal of work and time, slow progress, and (inevitably) some frustration, then by all means, have a go at training him.

You are undoubtedly correct about your instructor — she is probably dubious about your ability to deal with the sheer scope of the project you've taken on. You are also correct about the importance of teaching Carlito good ground manners. He needs them, you need for him to have them, and yes, this *will* show your instructor that you are serious about your desire to train your horse. It will also show her that you understand and use the principles of good horse training and are ready to learn more from her.

If you want Carlito to face you in the stall, don't enter the stall until he is facing you. Unless your stalls are solid from floor to ceiling, he'll be able to see and hear you when you're standing outside his stall. Call him, praise him when he comes over, then open the door, pet him, put on his halter, and praise him again. If he turns away when you bring out the halter, wait for a moment and then call him again, praise him, pet him, and put on the halter. It won't take long for him to figure out what you want.

Horses are much more comfortable if their halters are put on correctly.

When you put on the halter, unfasten the crownpiece first. Then follow these steps:

1. Stand next to or slightly in front of his left shoulder (depending on his length of neck and your length of arm), put your right arm over his neck and grasp the crownpiece of the halter.

2. With your left hand, hold the halter in front of him, low enough that he will need to drop his head to put his nose into it. If you like, you can begin by holding a treat where he will have to reach through the halter to get it, which will quickly give him the right idea. If you use a verbal signal such as "head down" and treat only when his head is low, he will learn that dropping his head is comfortable for him and elicits praise and treats from you. You can phase out the treats over time. If you don't begin with treats, you'll need to allow more time, but in either case, patience is the real secret: If you hold the halter and wait, he will eventually drop his head and you can pet and praise him.

3. As he drops his head, pull the halter up on his head, using your left hand to keep the cheekpiece and buckle away from his left eye and using your right hand to bring the crownpiece over to the left side of his neck.

4. Buckle the halter, pet your horse, and praise him.

Note: If your halter has a "convenient" clip at the throat latch, ignore it. Those clips create unhappy, head-shy horses, because owners who use the clip invariably put on the halter by pushing the fastened crownpiece up the horse's forehead and over his ears, and then remove the halter by pulling it over his ears again in the opposite direction. Put on Carlito's halter properly each time so that he won't learn to associate the halter with the discomfort of having his ears squashed flat.

When you clip the lead rope to his halter, *just stand there*. If Carlito starts to walk off, ask him to stop and stand, or even to back a step or two and then stand. When you have his attention, position yourself at his shoulder, give a light tug on the lead rope, say "walk" or "walk on," and begin walking. If you do this every time you attach the lead rope to his halter, he will learn to focus on you and stand until he is asked to begin walking. For the first few weeks, practice stopping and starting several times whenever you lead him. This will help both of you develop the habit of focusing on each other.

For safe leading, it's best if your own shoulder is near the base of your horse's neck, roughly halfway between his shoulder and head. Keep Carlito in position next to you as you lead him and keep these points in mind:

- Hold the lead rope a foot or so below your horse's jaw. The lead rope will cross your body; the rest of its (folded) length will be held with your other hand. (Practice from both sides.)
- To get your horse's attention, speak his name and/or give a little tug on the lead rope.
- To stop, use his name, say "whoa," and stop walking. If necessary, give a little tug on the rope, but if he is paying close attention to you, he will stop when you stop.
- When you want him to move forward again, speak his name and tell him "walk on" as *you* begin to walk forward.
- Try to avoid making kissing or clucking sounds, because those are imprecise and generic. Carlito should pay attention to *you*, and you should give him precise commands.
- If Carlito forgets and begins to move ahead of you, give a brief tug on the lead rope to remind him to hold his position. If he becomes distracted and begins to slow his walk, speak his name and say "walk on." If you need to stop and regroup for any reason, do so, and take advantage of the opportunity to ask him to stop and then praise him for stopping and standing.

There should be a clear space between your side and your horse's body, and he should respect — that is, maintain — that space. If he crowds you, bend your elbow outward so that he will encounter it and remind himself to keep his distance. If you prefer, you can use the butt end of a whip instead: Hold it horizontally across your body and allow the horse's neck or shoulder to bump against it if he comes too close.

If Carlito is a typical horse, you will find that he enjoys these lessons and does his best to please you.

Leading an Excited Young Filly

Q My new weanling filly is six months old and she is the love of my life. Xena is tall and beautiful and very brave for a baby. I could sit and watch her for hours. I am having a little problem with her, however. She spends her nights in a stall, but goes out in a pasture with three other horses from just after sunrise to just before sundown. When I take her from the barn out to the pasture, she doesn't always come with me nicely. Sometimes she is so excited that she leaps and jumps around, and then other times she'll just stop dead and stand, and then she won't go forward.

I've tried three things that people have suggested. First, I tried pulling her forward, which didn't work at all, and then I was told this was wrong. Then I tried facing her and giving short pulls on the rope instead of a long pull, but she wouldn't come with me and at one point she even started to back up. Then I tried standing there waiting for her to move, which seems to work after a while, but then sometimes we'll walk another 10 feet and she will stop again. I want to train her right, but I need to be able to get her from the barn to the pasture every day. The distance between them is about 200 feet along a dirt path. I worry because there is no fence around the property, so if Xena got loose, she might go anywhere including out onto the road. What can I do to have her walk nicely from the barn to the pasture? The strange thing is, she knows how to lead! Before her mother went back to my instructor's farm, Xena was very obedient and would lead nicely to the pasture behind her mother, with no stopping.

A Your last sentence tells the whole story: Your filly doesn't actually know how to lead. When you led her behind her mother, she was just following her mother — the lead rope was incidental, and as far as she knew, you too were following her mother. Following mama provides considerable incentive for a foal to go forward. Now you're the proud owner of a large, six-month-old weanling filly who no longer has a mama to follow. When you lead her by herself, she doesn't have that same incentive to go forward, so you are going to have to teach her to lead.

For now, since you are absolutely right about the risks of an excited foal running around a property with no perimeter fence, you might want to have someone lead another horse to the pasture in front of your filly so that she can follow. You can do a little leading practice on your way to the pasture, but plan to hold most of your leading lessons in an enclosed space, either a pasture or an arena.

Since your filly already accepts the halter and lead rope and just gets "stuck" sometimes, you won't be starting from the very beginning, and it shouldn't be too difficult to teach her the rest of what she needs to know. Here are some suggestions for your leading lessons.

Pressure on the lead rope should be brief, not steady. You want her to come with you; you aren't going to drag her anywhere. Teach yourself to reward your filly quickly by creating a sag in the rope. She needs to learn that you will remove all pressure from the lead rope as soon as she even tips her nose in your direction.

A simple step sideways can be enough to unstick a horse that is reluctant to move forward.

Whatever you do, don't turn and face her. Staring at her will not give her *any* incentive to move toward you. You'll be exhibiting predator behavior, and that's more likely to frighten her than to elicit her cooperation. Look in the direction that you want her to go.

Carry a buggy whip or a long dressage whip. Whenever your filly stops — assuming that you haven't asked her to stop — the whip will help you remind her to go forward. While you are facing in the direction you want to go, with the horse at your right shoulder, holding the whip in your left hand will allow you to reach around behind your back with it and tap-tap-tap (not hit!) the horse on her hindquarters or hocks. The length of the whip lets you use it to enforce your verbal "walk on" cue and your facing-forward, ready-to-go body language. A few taps are generally all that's needed to "unstick" a horse, and then you can praise your filly as she walks forward. If you haven't used this technique before, practice your coordination on another horse before you work with your filly.

Move her sideways, then forward. If you have taught your filly to follow her nose, even just for a step or two, then you may have good results if you try to move her sideways as well as forward. When she gets "stuck," encourage her to come with you to the left or the right: walk her on a curve. Horses who become "stuck" on straight lines or when they feel lead rope pressure will often become "unstuck" and begin walking forward again if they are led in a circle instead of a straight line.

With the aid of your tap-tapping whip, you may be able to go from pasture to stall in a series of loops. If it's winter, you may create some funny-looking tracks in the snow, but all that matters is that when you ask her to move forward she answers, "Yes, okay, I'll do that." Once she develops the habit of moving when you ask her to move, and coming with you when you ask, she will learn to lead in a straight line. The circle stage won't last forever; as time goes by, she will become "stuck" less and less often.

Teaching to Tie

Q I have a new rescue horse that is almost a year old and, as far as I know, has never been tied. I've read horror stories about horses breaking their necks when they panic and pull back, and other horses who survive but aren't ever normal again or who have neck injuries that show up later in life. My other horses already knew how to tie when I got them. This will be my first time training a horse to tie. Everyone gives me advice and the only thing that's consistent is that if one person says, "Always do this," the next person will say, "Never do this" — and they're talking about the same thing!

I've been told to tie the horse high, to tie him low; to tie him to an inner tube, to never use an inner tube because it will teach him to pull; to use a loop of twine because it will break, to never use a loop of twine because it will break. I've been told to tie with a rope, to never tie with a rope but always with a bungee cord; to never tie with a bungee cord — you get the idea!

A Tying does not actually have to be one of the first things a horse learns. I much prefer *first* teaching horses to be connected to a person by a lead rope, *then* teaching them to lead (walk, turn right, turn left, stop, back, and so on), and only *then* teaching them to tie.

Whether you're working with a young foal or an older horse, it's a good idea to teach the horse to stand quietly while he is connected to you by a lead rope. Begin by attaching the lead rope to the halter while the horse is in the stall and holding the other end of the lead. You can groom the horse, scratch him, pet him, do whatever you like. There's no need to put constant pressure on the lead rope — occasionally put just enough pressure on to tip the horse's nose this way or that, and then relax

*Teach your horse to lead reliably
before you teach him to tie.*

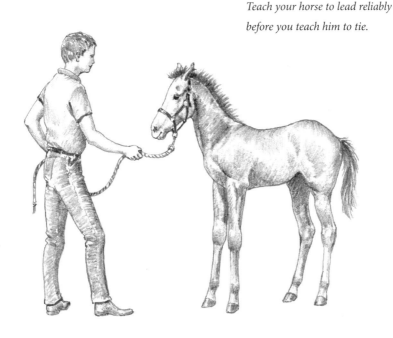

the pressure and praise and pet the horse as soon as he begins to move his nose in the desired direction. Do enough of this and you'll be well on your way to teaching your new horse to lead.

Change the venue to a larger pen or arena or turnout paddock, then to a field. If possible, "pony" your horse from another horse, which will teach him to follow when the pressure on the rope is coming from above. Give the horse plenty of time to discover that (a) being connected to someone or something by a rope is not a bad thing, and that (b) when pressure is put on the rope, he can make the pressure disappear by moving in the direction of the pressure. These are two lessons that horses need to learn early on but that are often neglected.

If the horse learns them well, the first time you tie him and stand back to watch, you will probably see a nonevent; that is, when he reaches the end of the rope and feels pressure on his head, he will take a step or two in the direction of the pressure, thus releasing the pressure, and then stand quietly. Just in case, it's still best to tie the horse high, to something tall, and to tie the lead rope to an inner tube instead of directly to the post or tree. But if a horse has spent a lot of time being held and led by a rope, he is unlikely to panic and harm himself.

Some trainers like to take a completely untrained horse, tie him securely, and let him fight until he realizes he can't possibly win. I don't like this method, even when an inner tube is used to diminish the potential damage. First, it's unfair to the horse. I believe in preparing a horse, physically and mentally and emotionally, for each new demand, not in making a huge demand ("You must submit!") of a horse that is completely unprepared. Leading is a lesson that leads logically to tying, so why on earth tie the horse and allow or encourage him to pull back, panic, and fight until he gives up — and then, later, teach him to lead? Second, it's unnecessarily dangerous. Why deliberately put a horse in a position where all of his instincts will tell him that his life is being threatened and that he must fight to escape? The fact that the method is designed to set up the horse to lose the fight doesn't make either the method or the risks acceptable.

Teaching Horse to Lower Head on Cue

Q I am not very tall; my horse is very tall. I want him to lower his head on cue so I can put his halter or bridle on him without having a fight. I have been trying to train him to do this but it isn't working. Can you help me? Here is how I have been training Shadow. I use a pressure and release method that is natural, so I know he understands it.

I put one hand on the bottom of his halter, by the lead rope, and I pull down. I tell him "head down," and I put my other hand on the top of his head to push his head down. The problem is he doesn't put his head down. Sometimes when I push really hard on his head between his ears, he will put his head down just for a minute but then if I stop pushing he pops it right back up high.

I know he knows what I want, but he just doesn't want to cooperate with the training. By now he should put his head down just when I tell him "head down," but he doesn't. How can I get him to cooperate? I thought the point of gentleness training and pressure and release was to avoid a fight with your horse, but what if your horse just won't cooperate with the training? I need him to give to my hand so I can release the pressure, but he won't.

A   Don't be so sure your training isn't working — you have taught your horse to play a game where your job is to push his head down and his job is to pop it back up again. "Head down" only means "Put your head down" if you've taught the horse to associate hearing those words with putting his head down. Otherwise, it's just a noise that means nothing at all or, in this case, that means "Get ready to pop your head up as soon as I stop pushing it down."

In other words, you're not actually teaching him to drop his head on cue. Or, at least, you haven't taught him that *yet*. Since from *his* point of view, you haven't even tried to teach him to do that, you could begin right now, do it a different way, and be much more effective.

I'm not sure what "gentleness training" is, but the name sounds pleasant enough. The problem is that what you're doing with your horse right now isn't really gentle and it's not really training, either, at least, not in the sense of teaching your horse to do what you *want* him to do. Let's try it another way.

Pressure and release is a perfectly valid system of training, but you need to understand how it works. There's more to

Your horse should be eager to put his head into his halter.

pressure than pushing your horse's head down — pressure exists as soon as you begin asking your horse to do anything. Your hand on the lead rope, pulling the halter downward, is already creating pressure — you don't need to use your other hand to push your horse's head.

The release means that you *stop* applying pressure. The release tells the horse that he's on the right track, giving you the right answer, doing what you want him to do. You need to release as soon as your horse gives you the slightest hint of what you want. Really good trainers release as soon as the horse even begins to *think* about giving a hint of the desired behavior.

The pressure-and-release system can work very well, but it doesn't work at all if you keep applying pressure until your horse performs precisely the action you want. Steady, incessant pressure works against training for two reasons. First, instead of letting the horse *try*, it makes a constant, nagging demand that creates resistance rather than cheerful cooperation. Second, the pressure is released only if the horse performs perfectly, which means you're giving no hints at all about what you want. The horse has to guess and perform a very specific action. You're not helping your horse figure out what that action is — the horse literally doesn't have a clue. It's the trainer's job to provide clues and reward the horse for figuring them out.

> The pressure you need to use is an asking pressure, not a forcing pressure.

When you pull and push your horse's head down, *he* isn't putting his head down, *you* are, and as soon as you stop pulling and pushing, up goes his head. As far as he knows, he's done what you wanted — he went along with your actions, and didn't pop up until you stopped telling him to keep his head down. The problem is that you want him to learn something quite different. You want him to put his own head down and keep it down, not because you're pulling and pushing but because you've asked him to do something, he understands what you want, he can do it, and he's happy to do it.

The pressure you need to use is an *asking* pressure, not a *forcing* pressure. It should be light, and you should end it (create the release) as soon as he shows any sign of doing anything. Don't push his head down. Put pressure on the lead rope, but don't hang your full weight on it, just put a little tension on it and keep it there. Tell him "head down." When you get a response, release the pressure and tell him "thank you." A response would be *any* shift in his head position that shows he's relaxing his head and neck even the tiniest bit. The second you feel even a tiny lessening of the tension on that lead rope, let *go* and say "thank you." Don't hang on because it wasn't much of a movement or because there's still tension on the rope — let go if the horse drops his head even a tiny fraction of an inch.

If you hold on and wait for a bigger drop, you won't get it. If you release and praise the horse for a tiny, almost imperceptible drop, you'll get a bigger drop the next time you ask, and you'll get it sooner.

Your old way involved teaching him to resist; your new way means teaching him to relax. This will be much easier on both of you. Once he has the idea, he'll begin dropping his head more easily and quickly, dropping it lower, and keeping it lowered for a longer time. All you need to do is be clear and consistent, react *instantly* to even a tiny lowering motion on his part, and do this several hundred times.

Whenever you set out to teach your horse something, you need to give him time to get the idea of what you want, time to get it right, and time to get in the habit of giving you the desired response. This is where practice and repetition come in.

When he drops his head, praise him — this would be an ideal time to offer him a small treat. Continue to praise and talk to him. At some point, his head will lift — he'll see or smell or hear something distracting — and when that happens, freeze. No talking, no treats. Wait for him to drop his head, then praise and reward him. Let him discover that dropping his head makes pleasant talk and tasty treats appear, whereas lifting it makes those things disappear. Most horses figure this out quickly, as long as the handler doesn't try to force the issue by pulling the lead rope.

At first you may need to use treats to persuade him that "down" is a good place for his head to be; as he learns to relax and associate lowering his head with pleasant experiences, you can treat less often and reward only when he puts his head down and keeps it there. Again, most horses figure this out quickly.

When he'll drop his head and hold it low until you're ready for him to lift it again, you can teach him a "head up" cue if you like.

Be quick and generous with your releases. Your horse will eventually learn to associate the words "head down" with the action, and "thank you" with your release, but the release counts much more than the praise. Remember, in a pressure-and-release training system, the *release* is the reward — the praise is just something to remind you to be polite with your horse.

Picking Up My Mustang's Feet

Q My Mustang is three years old and supposedly trained with "basic manners." I bought him two months ago from a woman who adopted him as a yearling and hasn't had time to do anything with him. He acts like a horse that is not trained at all. He barely leads — he'll stop or he'll pull in another direction because he wants to go and look at something. He is a nice boy with a friendly, dog-like personality, but some days I really feel that I have a completely untrained 1,000-pound puppy at the end of a lead!

My main problem is getting him to pick up his feet. My farrier is very definite that it is my responsibility to train my horse to pick up his feet and hold them up. When I try to pick up one of his feet, he turns and puts his mouth on my back or shoulder, and I get nervous that he might bite me (he has not bitten anyone yet). Then he just stares at me as if he has no idea what I want, and I can't get his foot up or even get his knee bent; it's like trying to lift a tree! How can I train him to pick up his feet? I'd like him to learn it anyway because I won't be able to clean his feet if we fight every time I need to have him pick up a foot.

A Your new Mustang probably wasn't given much training at all — "basic manners" could easily mean "followed other horses into the stock trailer" and "will let you put on a halter and attach a lead rope to it." Instead of thinking of him as a horse with a few holes in his basic training, you'll be safer to assume that he has no training at all and then train him accordingly.

Since he doesn't lead well, teach him to lead. Teach him the basics of pressure and release, so that he learns to understand that a push or a tap means "move away, please," and a release means "thank you, that's what I wanted." He needs to learn those two things before you can begin asking him to pick up his feet and hold them for the farrier. We are all so accustomed to handling our horses' feet and legs that we sometimes forget that in nature, nobody does this, and it's something that horses must be taught to accept.

When you're ready to begin working with his feet, I'd like you to take two precautions: Wear your helmet, and bring a friend to help you with this part of the training. The helmet is for your safety, and the friend is for your safety, the horse's safety, and the sake of the horse's education. It's very helpful to have a competent, calm, horse-savvy adult holding your horse while you teach him how and when to pick up his feet.

Don't begin by trying to pick up a foot — you need to start much higher on his body. Begin by standing close to your horse and running your hand slowly and firmly (no tickling!) from his withers to his elbow and down each of his front legs to the fetlock and hoof. You'll go through a similar process with his hind legs, standing close up against him and running your hand slowly and firmly from his hindquarters all the way down his hind leg, past the hock and down to the fetlock and hoof. You need to be able to do this, over and over, on each leg *without your horse reacting* — and only *then* should you proceed to try to pick up one of his feet. Take your time with this part of the process — it's the foundation for all the rest. It shouldn't be hurried, but it should be repeated, and often. Make it part of your new horse's daily routine for at least a full year.

With time and repetition, your horse will learn to relax when his legs are handled.

There are two more things you should keep in mind, for safety's sake. First, being kicked *is* a real danger, but you can avoid most of the risks by staying close to your horse, making your movements slow and deliberate, and always being aware of his level of physical and mental comfort. If he starts to become uncomfortable, back off and give him a moment to relax.

Second, a horse's hoof doesn't have to kick you to cause damage. Have you ever had a horse step on your foot? Horses can't see their own feet and don't know when we foolishly place our own soft feet where their hard ones are about to step. When you are working with your horse's feet, your own feet should be *next to* your horse's body, never *under* it.

When you're ready to pick up one of your horse's feet, if you still feel as if you're trying to lift a tree, there is probably a good reason. Since your horse puts his mouth on your back when you try to lift his front hoof, he's obviously leaning toward you, and that means his knee is locked. Of course you can't pick it up — neither can he, in that position! But from that position he could bend his *other* knee and pick up *that* foot quite easily.

While you're on your horse's left side, standing by his shoulder, have your friend tip your horse's nose to the *right*. Look down at his left knee. If it's relaxed, as it now should be, you'll be able to pick up his left front foot, bending his knee as you do so, and then tip the sole of the foot up toward the barn roof.

Hold the foot for a few seconds, still tipped upward, and then announce "foot down" and *slowly* return his foot to the ground. Then praise him, pet him, give him a moment to think about it, and do it all again. Each time, add a few seconds to the amount of time you hold each foot in the air. When you can hold each foot up for a couple of minutes at a time (have your friend time you — a minute can seem like a very long time), begin moving the foot into different positions and holding it there — a little higher, a little out to the side, and so forth. For your own purposes, one minute in a low, close-to-the-body position may be enough, because you're not likely to do anything more with your horse's foot than clean it out with a hoof pick. But your farrier needs to do much more, and you don't want your horse to panic and pull back because the farrier is holding his foot for a longer time, or because he is holding it higher or more to the side.

Talk to your horse; let him know when you're happy with what he's doing. If you aren't happy — say, if he leans toward you and starts to rest his weight against you — stop talking, and use your knuckles or two stiff, extended fingers together to prod lightly (poke several times, don't use steady pressure) at the horse's shoulder or hip, depending on which hoof you are holding. As soon as he begins to shift his weight away, stop prodding and praise him.

For the hind hooves, the process is similar. Again, you'll want to keep your back straight. Stand next to your horse's left hip, with your back toward your friend. Horses typically stand with one hind leg locked and the other relaxed. Look at your horse's left hind leg — if he's carrying his weight on it, and the toe of his right hind leg is just resting gently on the ground, it won't be easy to pick up his left foot. Once again, you'll need to prod your horse, this time in the hip, to make him shift his weight.

His reaction, after a poke or two, should be to shift his weight onto the opposite hind foot, locking that hind leg and relaxing the one nearest you. Stop prodding, praise him, and run your hand down his body from just next to the tail all the way down his leg to the fetlock. When you reach the fetlock, use his name and tell him "foot" while you pick up his fetlock and tip the sole of his foot toward the barn roof. If he shifts toward you, putting his weight onto that foot again and relaxing the opposite hind leg again, just resume prodding until he shifts away and relaxes the leg again, then praise him and pick up the foot. At first, don't attempt to hold any foot for a specific number of seconds or minutes, or in a particular position. Take your time, remember to pick up all four feet in turn, and keep doing it until your horse easily and calmly picks up whatever foot you've indicated and holds it up without pulling his leg away, leaning on you, or showing any other sign of impatience, annoyance, or fear.

We are all so accustomed to handling our horses' feet and legs that we sometimes forget that in nature, nobody does this, and it's something that horses must be taught to accept.

Among the other variations in foot position, be sure to incorporate two that your farrier will always use: trimming position (sole facing up) and rasping position (foot and lower leg pulled forward, sole facing down). Even if your horse is kept barefoot, he will need to be able to hold his feet in both positions.

The more time you spend training your horse to accept all of this, the better off he will be, because farriers will find him easy to work with.

Young Horses and Longe Training

Q I will be training my yearling fillies to longe but not until they are at least eighteen months or even two years old. In the meantime, I'll work with them on ground driving and free longeing. They both already know the signals for walk, trot, and whoa. I know there are concerns about longeing young horses, so when I start them on the longe line, what are some guidelines that I ought to follow?

A You are wise to wait; if you can wait until your fillies are two, you will be wiser still. Working on a circle is too stressful for young legs and can cause permanent damage. When your fillies are ready, put boots on all four legs (young horses, especially on a circle, find it all too easy to hit themselves), use a longeing cavesson and a longe line that's at least 35 feet long (no side reins), and ask them for walk, trot, and whoa. Since they already know the words, that should be easy. If you longe them for 10 minutes (five each direction) every other day for a few months, they should be off to a fine start.

Longeing and Voice Commands

Q I will begin longeing my three-year-old filly in the spring, and I was wondering what you use for voice commands. With my older gelding, I have always used "waaalk," "trot," and "can-ter" for upward transitions, and "whoa" to do progressive downward transitions. Other people say "whoa" should be reserved for stopping, and they use those other commands for upward *and* downward transitions. I'm wondering if using the same command, for example for an upward trot transition and a downward trot transition, would confuse the horse. It would confuse me if I were a horse!

My coach uses noises for walk, trot, and canter cues because she says that when you take CEF (Canadian Equestrian Federation) level one coaching certification, and you do longe-line testing, it is in a big indoor arena with three or four other horses, all longeing at the same time. The horses get confused and don't differentiate between their neighbor's commands and their longer's commands. She uses unique noises so that her horses never get mixed up.

A   You're sensible to wait until your filly is more mature before starting her longe work. Voice commands, as I'm sure you've discovered, are as much a matter of *tone* as of the specific words used. Your horse will be able to distinguish quite easily among "waaalk," "trot" "can-*ter*," and "whoa." I don't use "whoa" for downward transitions, for several reasons.

First, "whoa" has a specific meaning: *stop*. I want to keep that association strong and undiluted, so that when my horse hears me say "whoa!" he knows that I mean *stop* and not *slow down* or *change gait*. If your horse is ever tangled in its own longe line (or in someone else's), puts a foot over a side rein, or steps into wire or something else dangerous, or if a piece of equipment breaks or another horse is suddenly loose in the arena — I've seen all of these things happen — you need to be able to stop your horse instantly, *on voice*. If "whoa" always means *stop* you'll be able to do this effectively; your horse won't have to wonder what it means *this* time.

Second, when you ask your horse for a trot, "trot" or "tee-rot" should mean *trot* whether the horse is standing still, walking, or cantering. It doesn't matter whether the transition is upward or downward; when you ask for a trot, that's what your horse should understand, and that's what your horse should do. Similarly, "walk" means *walk* whether your horse is standing still, trotting, or cantering. If a horse associates the word with the gait, he won't be confused — what *would* confuse him would be your saying "whoa" when you mean "go from canter to trot" or "go from trot to walk."

Use either his name or "And . . ." as a preparatory command *before* you tell him "walk" or "trot" or "canter" or "whoa." A preparatory command ("Silver, waaalk" or "And waaalk") is like a verbal half halt. It lets the horse know that something is coming and allows him to rebalance himself and prepare to *do* something. You'll find that your horse's transitions are smoother and that he can remain relaxed if he knows you will warn him that a command is coming.

Third, I don't use "whoa" for transitions because I want prompt, accurate, energetic, *forward* transitions whether they are upward or downward ones. When I'm longeing a horse at canter and I ask for a trot, I want him to move smoothly *forward* into trot, using his joints and muscles — I don't want him to canter more and more slowly and *fall* into a trot. Similarly, I don't want a horse trotting, then jogging, then collapsing into a slow walk; I want a trot-to-walk transition that is forward and prompt and active.

If I want to modify the horse's gait or activity level on the longe while he's working in a particular gait, I'll ask him to move more slowly by saying "easy trot" or "easy canter" or "easy walk." "Easy" is a word that horses figure out almost instantly, especially if you say it on a long exhale and match it to your body language (lift your rein hand slightly, stand taller, and allow the tip of the whip to drop). If I want the horse to move along more actively, I'll say "walk on," or "trot on" — and match this to my body language by lifting the tip of the whip so that it points at the horse's hocks rather than at his hind heels and by stepping slightly back toward his hip, putting myself in a position to drive him forward.

Don't worry about longeing in a large arena with several other horses. If your horse is focused on you (which is a matter of training, habit, and *your* focus), he will listen to you, not to anyone else. Even if another person is cracking a whip every few seconds, your horse will still be attentive to you and responsive to your commands.

I wouldn't bother with trying to make "unique noises." For one thing, most people come up with the same noises. For another, if you have your horse's attention, he will be working from your voice and your body language, both of which will be quite clear to him. A well-trained horse, with one ear and one eye on his handler, is not easily distracted by another person's commands to another horse.

If your horse is *not* paying attention to you, shouting, whistling, or using funny sounds won't do a lot of good. But he'll watch your body, and he'll listen for your voice. Calling a child in from the playground doesn't involve using a special noise so that he won't get confused — little Tommy will hear *his* mother's voice and go to her even if 10 other mothers are calling their own children (and even if some of those children are also named Tommy). Your horse has better hearing than any child, and the longe line itself provides you with a constant source of communication. Just longe normally, and use conventional commands. After all, what will happen if you are asked to longe another horse as part of the exam, or if your own horse (heaven forbid) is ill or lame when you go for your test, and you have to take another horse with you?

You can teach a horse to respond to almost any signal, but using simple, common signals will help you when you work with other horses and will also help your horse when someone else works with him. People who teach their horses "distinctive" or "unique" signals aren't doing them a favor — other handlers, who don't know that the horse has been taught this way, may punish the horse for not responding or for responding "wrong" to a conventional signal.

Why Would Circles Be Stressful?

Q I just bought a 16-month-old filly. I plan to train her in the round pen at my boarding barn. The round pen is 45 feet across. I know not to longe her because that could be stressful to her legs, but the person who teaches lessons at my boarding barn told me that round-pen work can be stressful, too. Why would it be stressful for the horse to work free in a round pen? She won't have anybody on her back and she won't be wearing a bridle or a longe bridle or side reins, so where will the stress come from?

A The problem with round-pen work is that, like longeing, it puts stress on the horse's body. It's the circle work that causes the stress, and the smaller the circle, the more stress there will be. Small-circle work of any kind (on longe, long-lines, or mounted) puts strain — torque, actually — on the leg joints, especially on hocks and fetlocks. It puts even *more* torque on the joints of a young/green/unbalanced horse, which will be putting more weight on his inside legs anyway. It's possible to break down a horse of any age by overdoing circle work.

Even running "free" in a round pen can be stressful to a horse's joints.

- Condition matters, of course — bone and tendon and ligament strength, plus integrity of joint cartilage, all play a part.
- Footing matters a good deal — *any* horse will eventually become lame if circled endlessly, but it will happen much more quickly and obviously if the longeing is done on a hard/rocky/uneven surface.
- Speed matters too, since it affects balance and impact. Many people don't know how to longe, and do it far too fast and without understanding that the horse needs to learn to balance on the circle.

Circles stress the horse's inside leg joints — this is why vets want to see horses longed on a circle during a pre-purchase exam or soundness check: they know that the circle will bring out subtle lameness that might not have been apparent on the straightaway. It's also why those vets will *stop* the circling the second they see any unevenness; they know that asking the horse to keep moving under those conditions will make the problem worse.

If you want to work your filly in circles, use the 45-foot round pen only for short sessions at walk and jog, and only if the footing is good. She's young and you have plenty of time. Teach her good ground manners, let her learn to like and trust you, and take as many lessons as you can so that when she's old enough to start real work, you'll be able to bring her along gently and slowly and keep her happy and sound.

Round-Pen Size for Training

Q The round pen where I board and train is approximately 30 feet in diameter. I work my wonderful 16-hand, six-year-old Arab-National Show Horse gelding in the round pen at walk, trot, and canter for about 15 minutes before we ride (no longe line). My question: Is the pen too small? (I wouldn't ask if I didn't have doubts!) If so, what's the minimum?

A Round pens, like arenas, come in different sizes for different purposes. One size does *not* fit all. There are breaking pens, working pens, and training pens — and those are just the three basic types. Your worries are legitimate. You're trying to use a breaking pen as a training pen, and it just won't work.

A lot of older facilities have tiny round pens, known as "breaking pens" or "bull pens." These are often only 35 feet in diameter. These tiny pens can be useful for slow unmounted work such as teaching a horse to tie or lead, sacking out a horse, and introducing a horse to tack. They can also be useful for teaching horses and new owners the basics of round-penning at a walk and jog, and they can be reasonable places to turn out a stallion or a foal to get some air and more freedom and exercise than a stall provides, but that's about the extent of their usefulness.

Small pens should not be used for fast exercise or serious longeing. Unless the horse is highly trained and superbly developed physically, it won't be able to move correctly or even carry itself correctly bent on such a small circle. And small pens are not suitable for ridden work. If the horse is carrying a rider, there are even more demands placed on his legs and balance. Even if the footing is perfect and your horse is warmed up, the tiny circles are too stressful for his leg bones and joints.

A "working pen" might be 50 feet or more in diameter. This would allow you to work a horse at a walk, jog, and trot, but probably not at a canter unless the horse is compact, balanced, experienced, and well trained. It would definitely *not* be the place to introduce a horse to the idea of cantering under a rider.

A "training pen" should be at least 70 feet in diameter. This permits the horse to work on a 20-meter circle (66 feet), which is the basic working circle for a low-level dressage horse in the first years of his training and a good *minimum* size of circle for any horse beginning training. Since horses don't usually work up against the side of the pen, but on a track to the inside of the pen wall, any round pen used for riding should be at least 70 feet in diameter, and if you're going to do a lot of cantering in the pen, it wouldn't hurt to make it even larger.

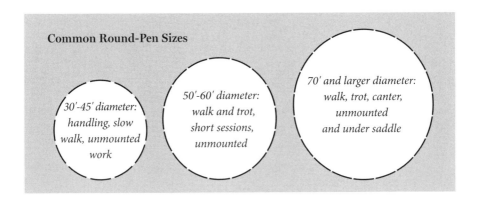

Common Round-Pen Sizes

30'-45' diameter: handling, slow walk, unmounted work

50'-60' diameter: walk and trot, short sessions, unmounted

70' and larger diameter: walk, trot, canter, unmounted and under saddle

If a larger pen isn't an option, then your question must be "What can I safely do with my horse in the pen I have now?" Perhaps you could invest in some longeing equipment and some good longeing lessons instead. If good longeing lessons aren't available, don't risk doing without them — longeing done badly is worse than no longeing at all. Your horse will be safer if you warm him up by walking him around the arena once or twice, then mounting and riding him at a walk for 15 minutes or so, then continuing your warm-up under saddle. This will be easier on him than dealing with the small circles imposed by a round pen that is too small to work in.

Trailer-Training

Q My mare is very difficult to load into a trailer. We can sometimes get her in a stock trailer without a lot of fuss, but she doesn't like straight-load two-horse trailers with partitions, and she flat-out refuses to step into a slant load. I took her to a clinic where a well-known cowboy clinician was going to train her to get into the trailer by making her run in circles on a lead rope until she ran into the trailer. I made him quit after about 15 minutes because she was dripping with sweat, you could see the whites of her eyes, and she was still not going to run into that trailer. The clinician got mad and said I obviously wasn't serious about training my horse, and that she had a real respect issue and he could see why.

I didn't have any trouble getting her home in the stock trailer; I think she just wanted to get away from that place and I don't blame her. But she was lame for a long time after that and my vet said she had strained a suspensory ligament, probably when she was racing around on that short rope. Is this a good way of training a horse to get in a trailer?

I have heard about leaving a trailer in the pasture so that horses can become used to it, and also about putting the horse's feed in the trailer in the pasture so that the horse has to step into the trailer to eat. Are those good ways to train?

My instructor wants me to use a method in which you stand behind the horse and keep tapping it with a whip until she gets into the trailer. I don't understand why that would work, and in any case my mare probably wouldn't get in because she is afraid of the sound of her hooves on the ramp. I am considering trading in my trailer for a step-up model unless you can help me get her over the fear.

A No, racing a horse in a circle on a 12-foot rope and attempting to run her into a trailer is definitely *not* a good way to train a horse to get into a trailer. The horse usually *will* leap into the trailer at some point because it seems potentially less terrifying than the scary, exhausting world outside the trailer, but what is the point of creating a state of panic and fatigue in a horse just to make her leap into a trailer to escape a frightening environment?

There are many ways of teaching horses about trailers. Here is another version of the one you experienced at the clinic but that can be done without damaging your horse's body or her trust in you. The setup is similar: You line up your truck and trailer in or just outside of a large enclosure with safe footing ("large" doesn't mean "riding size" — even a 30' x 30' pen would do, because you aren't going to ask the horse to work at speed). With the trailer open so that the horse could step in if he wanted to, you turn the horse loose and ask him to move around the enclosure, but not at a canter or even at an active trot; a walk or slow trot would be ideal, just as long as the horse is in motion.

Horses are clever animals and, if not pressured into fear, will both experiment and extrapolate.

My method of doing this involves a longe whip used as an extension of your arm: it is not whirled over the horse's back or cracked or slapped against the ground or used to hit the horse, it just indicates direction and the need to keep moving. Holding the whip, I turn as the horse moves around the enclosure, keeping the whip pointing behind the horse at a level midway between fetlock and hock. I will lift the whip nearer hock level to ask for more energy or speed, and drop it nearer fetlock level to ask the horse to trot or walk more slowly. For the horse to learn, slowly is excellent — and very slowly is ideal. A horse moving slowly is able to think and consider; a horse being chased at high speed can't think of anything beyond the fact that he's in danger and needs to escape.

Although most horses teach themselves what to do in half an hour or less, be prepared to allow your horse as much time as he needs.

Most important, I never take my eyes off the horse, not even for a second, because whenever the horse shows any interest in the trailer, I reward him by relaxing my body and letting the whip trail on the ground — this tells him "Okay, you can stop moving, you don't have to go forward now." I do this in response to *any* sign of interest, whether the horse veers toward the trailer, slows down slightly while passing the trailer, or even looks at or flicks an ear at the trailer.

Eventually, the horse begins to associate a feeling of comfort with looking at the trailer, then with slowing down near the trailer, then with standing by the trailer, and finally with going *into* the trailer. Eventually the horse will learn to load himself calmly into the trailer and stand there quietly. My job is simply to tell the horse, "Move gently, but keep moving — except when you're near the trailer, then you can rest," until the horse has an "Aha!" moment and realizes that "near the trailer" is a peaceful and restful place to be. Horses are clever animals and, if not pressured into fear, will both experiment and extrapolate; every horse I've ever taught in this way has quickly figured out that walking into the trailer would be a good idea.

So, that's the first step: Let the whole idea of getting near, inspecting, and getting into the trailer be the *horse's* idea. The horse gradually teaches himself to feel comfortable looking at, then approaching, then getting into the trailer, with no pressure from you, because you deliberately remove all pressure whenever the horse goes near the trailer.

The second step is more difficult psychologically — for humans, that is. When the horse gets onto the trailer, you cannot rush forward. You must continue to stand quietly, whip trailing, doing *nothing*. Wait. The horse will probably get off

Some trailers are designed
to let you load the horse in . . .

the trailer, possibly scrambling and a little excited, and again, you must stand quietly until the horse is out, and then gently and clearly send the horse forward again. You aren't punishing her for leaving the trailer, you're just reminding her that the trailer is the "standing" zone and the rest of the enclosure is the "keep moving" zone. The horse will shortly walk into the trailer again, stay on it a little longer, and get off a little more slowly and calmly. Once again, make yourself quiet and immobile until the horse unloads himself, and then quietly send the horse forward.

Your goal is to help the horse teach himself that he is comfortable near and inside the trailer, and that the trailer represents a chance to relax with no pressure of any kind. The more often he teaches himself that, the more comfortable he will be with the trailer. This lesson is particularly effective on a hot, sunny day, as horses are quick to recognize that standing in the trailer means a chance to enjoy the cool shade.

The third step is one you'll have to continue forever, and that is to ensure that your horse can remain calm and comfortable on the trailer. Make the drive as easy and pleasant as you possibly can, with minimal use of brakes and accelerator, and as few turns as possible. Trailer your horse with a familiar companion, and make your destination a place that is pleasant for horses.

. . . but it's always useful to teach a horse to load himself.

Leaving a trailer in the pasture isn't actually as simple as it sounds, and it's often not such a good idea anyway. Pastures aren't very good places for trailer tires — there's often too much moisture in the grass. Also, it's not just the trailer you would need to leave there, but the truck as well. *Never* let a horse enter a trailer that is not securely hitched to a truck — and be sure that the truck is in park, not in gear or in neutral. If there's one thing that really *will* make a horse terrified of trailers for life, it's starting to enter one and feeling it begin to tip and roll away.

The method your instructor prefers is useful and involves teaching the horse a simple cue that means "go forward." When the horse is lined up behind the trailer, perhaps with someone in the front of the trailer holding (but never putting pressure on) the lead rope, you can use your cue: the tap-tap-tapping of a whip on the back of the horse's hind legs or on the hip — whichever is easier for you, since of course you'll need to stand to one side, never directly behind the horse. Tap-tap-tap, gently and annoyingly, over and over, until the horse takes a step forward or picks up a foot (don't worry if he puts it down again in exactly the same place) or even shifts his weight forward or stretches his neck forward. This method, too, involves paying constant, close attention to the horse so that you can stop giving the "go forward" cue as soon as the horse even *thinks* about going forward.

When your mare eventually loads herself, you'll need to stay out of the way (toss the lead rope over her neck so she can't step on it) and allow her to unload herself, which she may do in a hurry. Neither the person in back of the trailer nor the person in the front of the trailer should grab or pull the lead rope. Allow the mare to leave as quickly or as slowly as she likes, and then walk her around for a minute and begin again. This is hard for most humans because we want to use our hands and pull things and control things, but you have to stand back and leave the horse alone. She must get off the trailer by herself before you touch the lead rope.

She may need to take herself off the trailer 10 or 20 times. Don't worry about it — she's teaching herself that when she gets on the trailer, she isn't trapped and doesn't have to panic. If she is like most horses, she will first rush off the trailer, then leave the trailer in a more leisurely way, each time staying a little longer and leaving a little more slowly. But this is a lesson that horses must teach themselves. Resist the temptation to use your hands and the lead rope; your job is to wait, then to pick up the lead rope calmly when she is safely out of the trailer, bring her back to the trailer, and ask her to get into the trailer again.

Many tasks are easier if you can break them down into small, manageable segments. Since your mare is afraid of the sound of her hooves on the ramp, you can deal with this issue separately by placing a sheet of ¾" plywood flat on the ground. Lead your mare onto the plywood, let her stand on it, and then back her off it again. It may be exciting at first, but persevere until the sound no longer bothers her. You want her to become calm and then bored with the whole idea before you present her with an actual trailer.

If you have difficulty leading her onto the plywood, then you'll know that part of the problem is that she hasn't been trained to lead. You'll need to teach her that lesson before you go back to working with a trailer.

Once your mare is loading easily and reliably, make it part of her regular routine. If possible, use different trailers from time to time. Move the trailer to a different location once in a while, because sometimes a horse that has always loaded calmly just outside the barn will balk if asked to load into a trailer that is 50 or 100 feet away from the barn. Invest your time now when you don't actually have a deadline or an urgent need to get your mare on the trailer.

Always train for tomorrow — remember that whatever you do with a horse today will affect the way that horse responds tomorrow and the next day, so be careful to think in the long term. Horses who are hurting or afraid can't think effectively. Horses who are confused and then punished will typically become resistant or even aggressive; horses who are confused but are handled calmly so that they can think and figure out the situation will tend to respond calmly, and that's why the slow, quiet approach to trailer-training is best. And speaking of slow, quiet approaches, there are still more ways to teach a horse about trailering. You might enjoy clicker training and target training, both of which are effective.

Clicker and Target Training

Q I'm interested in different ways of communicating with my horse. What can you tell me about clicker training and target training?

A Clicker training is an easy, enjoyable method to teach a horse just about anything simple that you would like him to learn. "Step forward," for instance, will be useful whether you are asking your horse to step forward onto a trailer, into veterinary stocks, or into an unfamiliar stall at a competition. Clicker training is applied operant conditioning. It was first developed for use with dolphins and quickly became popular with many dog trainers. It's also quite useful with horses, especially horses that belong to owners/riders who are new to horses and not very fluent in "horse" language. There are books and videotapes, classes and clinics that can help you learn this method in detail, so I'll just give you a brief summary here.

Using a small plastic-and-metal clicker, you make a clicking sound as a "bridge" between your horse's action and your praise or reward. You accomplish this by clicking instantly when your horse shows any sign of doing what you want him to do. In the case of a horse getting into a trailer, for example, you would not wait for the horse to go into the trailer and then reward him, you would click when the horse looked at the trailer and whenever the horse leaned or reached forward or even shifted his weight forward, *whether his feet moved or not*. All of those would be signs of moving forward — or at least signs of thinking about moving forward — and should be rewarded.

The usefulness of the clicker is twofold: For the horse, it gives an instant and clear, consistent, signal that says, "Yes, *that's* what I wanted you to do, well done!" It's always much easier for the horse to understand which of its actions you are praising and rewarding if the click comes *during* the action instead of afterward. This is also good for you. Clicker training can help you develop two very important habits: paying close attention to your horse and giving him a clear, instant reward for even the tiniest "try."

You won't have to use a clicker forever — you'll eventually be able to replace the clicker sound with a word or a click of your tongue — but clicker training is a lovely, simple, clear, and consistent way to teach your horse.

A logical follow-up to clicker training is target training, which involves teaching your horse to focus on some object (the target) and perform an action toward that object. An example would be a visual target — say an orange cone, a ball or some other toy, or a plastic sign with an obvious, easy-to-see mark on it, like a bull's-eye. Even a plastic flyswatter can be a target. You use target training to teach your horse to go to the target and put his nose on it; you could put the target in the trailer and send the horse to put his nose on the target (thus loading himself into the trailer); you could move the target from one side of the trailer to the other to teach your horse to load himself into a specific side of the trailer.

Target training isn't difficult or mysterious. Do you have a dog? If you have taught your dog to come to you when you call him, go into his crate when you tell him "kennel," and lie down on the dog bed in the corner of the living room when you say "cushion," you've already done some target training using stationary targets. If you've taught your dog to stay at your side when you go for walks, turning when you turn and stopping when you stop, you've already done some target training using a moving target. You can teach your horse to go to a stationary target (e.g., to the target or cone in the trailer) or to follow a moving target — that's what the lady in the circus is doing when she waves her whips and her horses come to her; move to one side or the other; and rear, kneel, and bow on her signal.

Like all good training, target training is gradual. You begin with nearby targets and small actions: You put a target near your horse, the horse looks at the target or puts his nose toward the target, and even if you know that he's doing it out of simple curiosity, you reward and praise him (your clicker will be useful here). As the horse becomes more accomplished at reaching for and touching the target, you make the exercise more demanding by placing the target at a greater distance, where the horse must take a few — and eventually many — steps to reach it, or where reaching the target will mean entering a stall or a trailer.

To teach your horse to follow a moving target, you can signal him to put his nose on the target and then move the target (slowly, of course) as he approaches so that he has to follow it to touch it. Ask for just a step or two at first, then a few more steps, and eventually you'll be able to move the target all around the arena or field and have the horse follow it. Once your horse understands the basic idea of tracking and touching the target, you'll be able to use it in many different ways. Here's one example: A horse that is anxious about being handled by the farrier or veterinarian can be reassured by allowing him to see and touch his target, and by praising and rewarding him for doing so. Saying, "Look, here's your target, aren't you clever for touching it, here's your treat," is much more effective than jerking the lead rope and saying, "He isn't going to hurt you, just stand still!"

Clicker training and target training promote positive feelings on both sides, as well as better communication. Your horse can relax and enjoy the training process, and you can become more aware of your horse and more clear in your requests and rewards. Your improved communication will continue to serve you well even if you prefer to continue your horse's training in other ways. Experiment with a clicker; try a bit of target training! You'll both enjoy it.

CHAPTER 3

Training from the Saddle: Whoa and Go

RIDERS OFTEN WORRY TOO MUCH about those first few rides. As long as your horse has been well prepared and you can keep the environment as familiar as possible, your first ride should be enjoyable, not nerve wracking. It will still be exciting: It's thrilling to see the world from the back of a horse that you've started and are now riding for the very first time. But it shouldn't be exciting in a dramatic way — there should be no bucking, rearing, or other pyrotechnics. A well-prepared young horse will inevitably become confused at some point, but is much more likely to stand and ponder than to explode.

Mounting for the First Time — Too Soon?

Q My problem horse is a three-year-old Paint gelding. I am a beginner/intermediate rider, and Max is the first foal I've raised. The problem is that Max is *so* smart that he is resistant in his training. He has never injured me, but he has been difficult from the day he was born. He is very independent, throws fits if he doesn't get what he wants, and was nearly impossible to halter train. He dragged us all over the paddock, ignoring the pressure and pulling away many times. He goes over fences and even through hot wires.

He is impatient and paws while tied, in a stall, or in a trailer. It took forever to train him to load. We tried when he was younger, but he bounced around a lot and pulled back and refused to go forward. After he grew up emotionally a bit, at the age of three, we ended up ponying him and he went right in.

Max opens gates and turns light switches on and off. One time he broke a light bulb in his stall, trying to mimic me pulling the chain switch. When I'm fixing fences, he will pick up the hammer by the handle and move his head up and down repeatedly. Then he will pick up the fence cutters and lightly tap a strand of wire. Amazing, huh?

I know that Max is not the kind of horse who could go to just any trainer. I intend to keep him for the rest of his life. I want him as well trained as possible, with good ground manners and nice trail skills; nothing fancy, just a great pleasure riding horse. I found someone who seemed like a good fit and for the first three weeks, everything went well. At first the trainer did a lot of ground work with Max, because he confirmed that Max is very intelligent and very impatient and needed the extra time.

When the trainer (who is into "resistance-free" training) felt that Max was ready for mounting, they went three or four times around the round pen, and suddenly out of the blue Max started bucking like a rodeo champion. The trainer, who says he can ride a buck, went flying and was injured. So, he went back to the ground work for another week, but the same thing happened the next time.

Now the trainer is unwilling to train Max, and I can't say that I blame him. This colt has been worked nearly every day for six weeks now. His ground manners are a dream. He is willing to do many things for me, but not for others. The trainer says that Max "just won't give him his mind." I will not give up on Max, but I do not want to get hurt. I know that he needs an experienced rider for a while. I want to find the right solution.

A   You have a very intelligent and observant horse that sounds like a lot of fun. What I want you to keep in mind is that even the most intelligent horse is still a *horse* and needs to be trained accordingly. The message I'm getting from your letter is that Max is being pushed too fast and is not ready for what is being done with him right now.

I see this often. An intelligent, active young horse goes to a trainer who puts him in the round pen and teaches him basic round-pen work, and the horse learns everything very quickly. It takes an exceptional trainer to look at a horse that responds to every movement made by the handler, and say, "He's not quite with us yet; he's responding fast but he's not relaxed and he's not really acting out of *understanding*." This is your horse's situation: His actions show that he is not ready to move out of kindergarten into first grade.

It's easy to push an intelligent and active young horse too far, and at that point, the horse's reaction will *seem* sudden and out of proportion, but it won't be. It will be perfectly appropriate for that horse and his level of understanding — and for the amount of pressure the horse can take. Intelligence and the ability to learn quickly do *not* necessarily indicate an exceptional ability to handle pressure.

Accepting what happens is one thing; skills acquisition and mastery are something else. When a young horse learns to go, turn, and stop in the round pen, that's one thing; when that horse learns to *understand* what he is doing and feel relaxed about it, that's something else. The horse that is tense and not entirely sure of himself will typically accept what is done until he simply can't handle the pressure any more, at which point he will react strongly, perhaps even explosively. Bucking, kicking, and running away are reactions that say, "I'm not comfortable, I don't like this, I don't want to be here, this is too much for me."

The best trainers have the ability to pay very close attention to the horse, to notice when the horse is feeling too much pressure, and to back off just *before* (not *after*) the horse becomes upset. Less accomplished trainers are less focused and less observant, don't read the horse as accurately, and don't react as quickly or as appropriately when they should back off and let the horse settle down. You've described a horse that has been pushed too far, too fast, by a trainer who tried to do too much at once.

When a horse is comfortable with his ground work and the trainer decides that it is time to get on the horse, there should be no fireworks. The trainer should get on, praise the horse, get off again, and give the horse some peace. Tomorrow is soon enough to get on again. The trainer who gets on and off 50 times the first day may *think* that he's teaching the horse "This is what you have to get used to," but the horse may well be hearing "I want something from you and you're not giving it to me, which is why I keep asking the same question over and over."

When the trainer has gotten on and off for several days in a row, using a mounting block to minimize the pulling sensation and torque on the horse's back, and *if* the horse is relaxed and confident and balanced, it will be time for the next step, which is sitting quietly in the saddle, using a familiar sound cue to ask the horse to move off, and then simply going with the horse wherever he chooses to walk, jog, or stop, allowing the horse to figure out how to move and balance. The good trainer won't use reins or legs; he'll just sit quietly and make occasional reassuring sounds while the horse explores his options.

This is *not* the time to try to make the horse go at a certain speed or in specific directions, and it is *not* the time to make the horse go around and around the pen or the arena. This is the time to let the horse teach himself that he can move and balance in spite of the weight on his back, and it's the time to talk to and praise the horse so that he can be reassured that he is doing well.

At first, just lie across your horse's back, praise him, dismount, and praise him again.

Bucking is not one of the options a well-started horse will typically explore, because when everything is done pleasantly, quietly, and *gradually,* the horse can learn to make his own adjustments and feel physically and mentally comfortable at each stage. But if the trainer decides to push on that first day, the horse is likely to be pushed far beyond his comfort level, and *that's* when the horse, unable to answer or even understand the trainer's questions, will typically begin to buck.

My advice to you would be that you go back several steps with Max. He needs to be confirmed in his ground work, not just doing it but understanding it, so that he will have confidence in you. Then he should be introduced to under-saddle work in a way that will connect to and build on that ground work. Verbal cues are excellent because they can carry over in a seamless way from ground work to mounted work.

It's important to keep your intelligent horse from being bored. A clever horse standing around in a stall with nothing to do will have many ideas, some of which you may not like (e.g., that light bulb). Turning Max out in a field or large paddock *all* of the time would help quite a lot. A horse that has put in 15 or 20 miles of daily walking in pasture is much less likely to "blow up" during training. Max needs to interact with his environment and with other horses. He's intelligent, sensitive, and quick to learn, but he's no more mature than any other horse of his age. He needs stimulation from turnout, interactions with peers, and well-spent time with you.

If you want to keep Max happy and entertained while he learns, I think that you and he would both enjoy and benefit from clicker training. Get a copy of Alexandra Kurland's book, *Clicker Training for Your Horse*, and a clicker, and start having fun. Clicker training is a form of operant conditioning that will help you with your timing and will carry over nicely to your under-saddle work. You can easily shift the click from an actual clicker sound to a verbal equivalent of a click, which can be any word or phrase you choose. I recommend "thank you"; it's a useful phrase with horses, just as it is with humans.

And, by the way, you cannot *make* a horse "give you his mind." You can teach the horse to trust you and believe you, and you can work with the horse's nature and his way of thinking, and then he will learn the pleasure of focusing on you and dancing with you, so to speak. But you cannot go into the ring and say, "Okay horse, today I want your mind" — and you cannot get the horse's mind just by making him move his feet. There's much more to good round-pen work than that.

Max is obviously intelligent and needs to be kept interested, but that's true of any horse in training. Don't be so focused on his intelligence that you forget he's still a horse with horse needs, horse motivations, horse thought processes, and horse reactions. When you work with him, please be careful, just as you should be with any horse, and don't skip any steps. A very intelligent horse, like a very intelligent child, may be able to master each step quickly and easily, but you'll need to be sure that each step is actually mastered and not just skipped. Skipped steps invariably come back to haunt you later, usually at the most inconvenient moment possible.

Max is a baby — he's just three, his body is changing, he's teething, and he's literally a different horse from week to week, just in terms of his development and balance. Give him the time he needs. Even the most intelligent horse in the world won't be fully developed until he's at least six or seven years old (and with larger breeds, eight or nine is often more realistic). Intelligence doesn't dictate maturity or development. Think of it this way — even if a woman were the most brilliant genius the world had ever seen, she would *still* need nine months to make a full-term baby. Some things just can't be rushed.

Mounting with a Helper

Q I'm getting ready to mount my new horse for the first time. I'm very excited and also a little afraid. My friend Carrie is going to help me on the big day. Here's my plan: I want to longe Blackie first, then when he is warmed up and relaxed I will stop him and take the mounting block out to him (he is used to seeing it in the corner of the arena). Then Carrie will hold him, I will lie over his back on my stomach, and she will lead us for a few steps. Then I will sit in the saddle, and she will lead us in a circle. Then she will go back to the center of the circle and longe Blackie in a circle for a while with me in the saddle. Is this a good plan, a bad plan, too much, or not enough? If it's okay to do this, what should I do when I'm in the saddle — how much can I ask without being unfair?

A You have a clear plan, and you ask good questions, so I suspect that you have brought Blackie along carefully. Good for you. I don't think this is a bad plan, but it may be slightly too ambitious for that very first day, and although you thought of almost everything, there are a few other things that you should consider. After you read this, you may want to wait another week or so while you put some of these thoughts into practice — they should make the big day more peaceful and enjoyable for everyone concerned.

Longeing Blackie to warm him up and relax him is an excellent idea — you should do whatever is involved in his normal longeing session. The one change I would suggest is that you and Carrie might want to work together for the next week or so, with you longeing him for the first 10 minutes (assuming that you longe him for 20 minutes) and Carrie longeing him under your supervision for the second 10 minutes. When you finally sit on his back, he will feel more secure if he is already accustomed to Carrie being at the other end of the longe line.

Taking the mounting block out to him is something else you might want to do for the next week or so. I'm sure that he has seen it in the corner for so long that he won't be bothered by it *as long as it stays in the corner*. Think like a horse: What sort of thing would sit quietly in a corner and then suddenly appear in, or move to, a different part of the arena? That's predator behavior, and it might cause him to become anxious.

If you want Blackie to be completely relaxed about the mounting block, practice bringing it out to him while Carrie holds the longe line; you can put the mounting block next to your horse, pet him, climb it and stand on it, and pet him again. This will get him used to three things:

1. The sight of the mounting block right next to him in the arena
2. The sound of you climbing the mounting block
3. The sight and feel of you towering over him as you stand on the mounting block

Make him familiar with all of those things before you actually get on him.

Finally, plan a less ambitious first day. If you lie over his back, pet him, praise him, and Carrie leads him a few steps, you can slide off, pet him again, and put him away. He'll have plenty to think about!

The next day, you can do it all again. This time you can lie forward on his neck and slide your right leg over so that you're sitting in the saddle, keeping your torso low until he relaxes. If you sit up suddenly like a jack-in-the-box he may become fearful. Take it slowly. Talk to him, pet him, sit up slowly and gradually, and let Carrie lead you for a short way, then stop, praise him, and pet him, get off as smoothly as you can, and then praise him, pet him, and put him away for the day.

The day after that, you can let Carrie lead you on a circle, and if he is calm about it, she can retreat to her end of the longe line and let him walk another full circle or two before stopping him so that you can dismount.

Those first sessions with you in the saddle will form Blackie's opinion of having a rider on board, so make them pleasant and brief. If you stay in the saddle until he has time to become tired and a little bit sore in the back, he'll form an entirely different opinion of the whole procedure. Ideally, you will leave him thinking "Why is it over? That was fun!" each time. Remember, you can always speed up your training later if you want to — but you will *never* get a chance to go back and do something more gently or slowly.

When you reach the point at which Carrie can longe your horse with you in the saddle, you'll be able to do some important work. Just remember that Carrie should be the one controlling the horse — yes, he's your horse, but he has no understanding of the rider's aids yet, and he'll need the security of understandable signals from Carrie. You should add your own light aids to *reinforce* the signals that Carrie gives, so that he can learn to associate his actions with your aids. Eventually, when he is getting most of his signals from you rather than from Carrie — and this may take several weeks or longer — you will be able to ride him off the longe. Even then, he'll feel safer, at least for the first few times, if you continue to circle Carrie at first.

In traditional training, the trainer would be the person holding the longe line and the trainer's *assistant* would be put up on the horse and told what to do and when to do it. It's trickier when you are the horse's owner, trainer, and rider, and you want to be the one on the horse's back. It's good that you are able to hand the longe line to a friend you can trust.

The big day should be peaceful and enjoyable for everyone concerned.

Worried about Our First Solo Ride

Q My first solo ride on my homebred horse is just a few days away, and I am incredibly nervous. I think I've done everything right — he is three years old and he learned to longe at two, then I did long-lining for three months to get him used to getting signals from someone who wasn't always visible. I backed him on the longe line with my instructor's help (she longed, I sat in the saddle and tried to project calm confidence).

We've walked and trotted (and even done three steps of canter with me holding my breath) on the longe line with the circle getting bigger and bigger. Now it's getting to be time for that first ride *off* the longe. I know this has to happen, and I'm really not worried about going on to do the rest of his training from the saddle. He isn't the first horse I've trained, but he is the first homebred horse I've brought up from a foal and trained, and boy does that ever make a difference!

I'm the kind of person who feels more secure with lists of things to check off. Could you possibly give me a list — maybe things that Caruso should know before Emily unclips that (longe line) umbilical cord and turns us loose?

Your very first moment off the longe line will be a thrilling one.

A You're right, that longe line really can feel like an umbilical cord — or a lifeline. Still, the day is bound to arrive when you and the horse are ready to work on your own. It shouldn't be too terrifying as you'll be in your same familiar arena, and your instructor will be right there.

Before I "fly solo" on a horse that has recently been backed on the longe, I want to be sure that the horse:

- Is trusting, confident, cheerful, and accepts me on his back
- Has been physically and mentally developed on the longe line and on long lines so that he is strong enough and secure enough to remain relaxed with a lifted back and balance himself *and* a rider
- Has been taught voice commands and has begun to make a connection between those and the very basic aids I'll be giving him from the saddle

That's my list. It's not very long, but if you can check off those items, you're ready for that first solo flight. There *is* something special about a homebred, home-trained horse — perhaps it's the fact that we bear full responsibility for everything they know, because there's no previous owner to praise or blame. Focus on the positive: Even if you're in a familiar indoor arena with your instructor present, and you're in the saddle for only two or three minutes, nothing compares with that first ride on your very own horse.

Rein-back Frustration: "Pressure and Release" Isn't Working!

Q I've been trying to learn "pressure and release" training, but I think I must be doing it wrong. Here's an example. When I do a rein-back, I put steady pressure on my horse with my legs and reins, and when he takes a step back I release the pressure. Then I use pressure again and he takes another step back. I have been doing this for about two weeks and it works okay; he will take a step back whenever I do it, but the reason I think I must be doing it wrong is that the clinician who taught me this said that it should get easier and easier until I would just have to think about my horse stepping back and he would do it. That isn't happening. I have to use the same pressure every time with my reins and legs, and today I had to use more pressure to get him to step back. I am always careful to release right after he steps back, so I don't understand why this isn't working right. Am I misunderstanding it, or is there some "trick" to this release idea?

A You're on the right track, and your attitude shows good horsemanship — you're asking what you are doing wrong, not what your horse is doing wrong. Congratulations — that shows a good understanding of training. I see only two problems with your "pressure and release," both of which you'll find easy to correct. First, you are right: There *is* a "trick" to the release idea. Second, you need to understand the idea of the release, because once you understand that, you'll be able to do the trick.

The trick is this: If you want your horse to learn to yield to pressure, you can't use a steady pressure, and you can't wait until the horse has already done what you want and then release. You have to use a light, brief pressure, and release it as soon as your horse even *thinks* about doing what you want. Here's how that might work in practice: You want your horse to step back. You give a brief squeeze with legs and fingers, and just before your horse takes that step back, he tucks his nose slightly and softens and shifts his body a little. *That* is the moment when you must release, so that your release says, "Yes, you're doing it!" instead of waiting until "Yes, you've done it."

It may seem odd to release before your horse actually takes that first step back, but this is the heart and soul of the whole idea of the release! If you're using pressure and release to train a horse, the release must be used immediately, as soon as the horse *begins* to yield to the pressure.

Your timing will have to be very good; the success of this training depends entirely on the trainer's ability to perceive the horse beginning to yield. You know your horse and when you're in the saddle you're in an ideal position to feel every tiny movement he makes. Those tiny movements are exactly what you need to notice and reward. If your legs and hands say, "Take a step back, please," and your horse tucks his nose in slightly but no other part of his body moves, he's not yielding; but if his nose moves only a fraction of an inch and you can feel his body move or even his weight shift backward, he's thinking about backing, he's trying, and *that try* must be rewarded by your instant release.

When you release during or even just before the step you asked for, you're sending the message "Yes, good boy, you're on the right track, I like what you're doing right now!" Your horse will respond by giving you a bigger "try" the next time you ask, and the next time, and the time after that. With each new "ask" from you, and each yield from him and release from you, he is figuring out just what it is that you want, and he is learning without being forced or becoming fearful. *He* can stay relaxed because *you* are relaxed. By giving him time to think and rewarding him as soon as he begins to think in the right direction, you're encouraging him to figure out the correct answer to your question. Good training is easier on him, and it's also easier on you: Instead of escalating your aids, you'll be making them softer, quieter, and subtler. Meanwhile, his responses will become stronger, more immediate, and more enthusiastic — that first tiny weight shift backward will become a lean, then a shuffle, and finally a step back, and you'll have gotten what you wanted in a correct and horse-friendly way.

Warning: Don't get so enthusiastic that you try to back your horse across or around the arena. Focus on quality, not quantity. Ask for one step, release, and if you get the step, relax a moment and then ask for a second step, release, and if you get that step, relax for a moment and then ask for something else — walk forward or trot forward — and do that for a few minutes to oxygenate and stretch and relax both of you before you stop (somewhere else in the arena, not the place where you stopped the first time) and ask for another step back. When your horse is able to offer several smooth steps back with pauses between them, you can begin to refine the rein-back by asking for the first step, then releasing as he begins to step back, then asking again before he has completed the step back — this will get you the

Too much "ask" and not enough "release" make both horse and rider stiff and anxious.

second step, and asking again before he completes the second step will get you the third step, and so forth. As long as you are releasing while he is stepping back, each "ask" will clearly be a separate question relating to the next step.

Over time, as you and your horse learn to read each other quickly and clearly, your horse will begin responding to such subtle signals that you'll be able to get a rein-back just by thinking about it.

Slack Rein, Gentle Contact?

Q I need to know what my horse is trying to tell me. Galaxy is a four-year-old Quarter Horse-Arabian cross. I started riding him last year very lightly, and we now go out on the trail three days a week and work in the arena on the other days, alternating. We take Mondays off. My problem is with the contact. I use a very gentle snaffle, a French-link that fits him well, and I try to work with him always very quietly according to Natural Horsemanship methods. His teeth are fine; they were just floated last month by a very good vet/dentist.

Here is the problem. I try to ride him with total slack in my reins at all times, except when I want something from him, and then I use one rein at a time. That is the way I was taught to be gentle and natural. My reins are very thin and light, and Galaxy's bit is very light. But Galaxy is getting more and more strange about his bridle; he throws his head up whenever I move my hand, even if I just move it randomly with the reins still very slack and hanging. Then when I use the rein, he keeps throwing his head even after I release it, which I do almost instantly after giving him a signal with the rein. And I use only one rein at a time. So what could be the problem?

I am leaving so much slack in the reins they are hanging on his neck, so I cannot be putting too much pressure on his mouth.

Soft, steady contact can be a security blanket for your horse.

I would understand his reaction if I was hanging onto his mouth with tight reins like our local Dressage Queen (sorry, I know that was catty, but I hate to see her horse suffer), but I am leaving the reins completely slack and just using them one at a time, very quickly, when I have to. What is going on and what can I do? Am I doing something wrong and hurting my horse? He is so young — if I have done something really bad, how can I fix it and will he learn to forgive me?

A First, sit down, breathe, relax, and stop worrying so much. Horses are very forgiving animals, and I'm sure that Galaxy won't hold your mistakes against you. We all have an amazing ability to notice the faults of others while overlooking our own — that's human nature. I think I can explain what your horse is trying to tell you, but if this information is going to help you, you'll need to stand back and look at your riding very objectively, in *horse* terms rather than in human terms.

You ride with a loose/slack rein, and then give a quick pull and release on one rein when you want something from the horse. You believe this practice to be gentle and natural, and you look at the local "DQ" with horror because she has constant contact with her horse's mouth, something you believe to be forceful and unnatural. You can't understand why your horse flips his head whenever you use a rein, and (I'll guess) you are confused by the fact that the DQ's horse doesn't do this.

"Natural" has become a buzzword, but it has no meaning when it's used in the context of bridles and bits and riding. It is not natural for a horse to carry a human or to wear a bridle. It is not natural for a horse to take directions from someone who is sitting on his back with hands connected, however loosely, to a piece of metal in his mouth. Natural, to a horse, would mean complete freedom to wander around at will, walking here and there, nibbling grass constantly, rolling whenever rolling seemed like a good idea, and interacting with other horses whenever that felt right. Humans don't fit into this picture, even if they only offer pats and carrots. From the horse's point of view, there is no such thing as natural horsemanship, because nothing about being ridden is natural.

Instead of trying to do things you have been told are natural, just relax, listen to your horse, and do what is *right*. Natural is nothing more or less than *one* version of "not natural" (out of the many different "not natural" options available). Once you get rid of loaded terms like "natural" and "dressage queen," you can start to look objectively at horses and riders.

It's a mistake to assume that contact is bad, and that no contact must therefore be good. Harsh, heavy, pulling contact is bad — there's no question about that! But "less is more" only if the horse is happy and comfortable. Most horses are happy and comfortable when they are allowed to determine the amount of contact, and when the rider is careful to (a) find out what makes the horse comfortable, and then (b) maintain that level of contact until the horse asks for a different level.

If contact is gentle, steady, and communicative, as it should be, horses generally regard it as a source of security and constant communication between horse and rider.

You will never take "too much" contact as long as the *horse* is allowed to decide how much contact is appropriate. Sometimes what the horse prefers seems like "not enough" to the rider, and then the rider needs to keep the rein as light as the horse wants it to keep the horse happy. Sometimes, though, the contact the horse prefers seems like "too much" to the rider, and then it's equally the rider's obligation to keep the contact at a level that makes the horse comfortable and happy. Contact should be measured in ounces, not pounds, but rider and horse have to be "on the same page." Otherwise, a horse that finds half an ounce of contact to be comfortable will be unhappy with a rider who thinks that three ounces are ideal; and equally, a horse that finds three or four ounces of contact comfortable will be unhappy if the rider insists on saying, "No, no, lighter is better, so you can only have half an ounce, I don't *care* how you feel about this, I'm being *kind!*"

I think that your horse is trying to tell you that he is unsure of your hands and afraid of the sudden movements of the bit in his mouth. The no-contact, one-rein-at-a-time school of riding is a Western one, based on a very old Spanish/Moorish approach, that assumes several things about horse and rider. One assumption is that the horse is fully trained. Another is that the rider is fully able to communicate with the horse through shifts in weight and position. Still another — this is key — is that the horse is wearing a *curb* bit, and that the horse is educated to the curb, and that the bit and reins are *heavy*, so that every tiny movement of the rider's hand shifts the balance of the bit in the horse's mouth and means something specific to the horse.

This kind of riding was never meant to be done in a snaffle or with lightweight reins. It's not meant for the beginning of training; it's a way to finish the horse's training. In the early stages of training, a horse would wear first a bosal, then a snaffle, and would be ridden on contact with constant communication between the horse's mouth and the rider's hands.

Keeping the reins loose most of the time and using a quick take-and-give (or, as the horse perceives it, a quick snatch-and-drop) with one rein whenever you want something has an effect on your horse that isn't good. The horse is not confident, relaxed, attentive, and able to respond instantly and easily to the slightest movement of the bit in his mouth, and to the slightest lift or drop of the rein. He hasn't reached that stage of his training and you aren't using that kind of bit or reins.

A horse in Galaxy's situation can't relax and move forward confidently with a relaxed jaw and closed mouth because he is constantly worried that he is about to experience a hard jerk from the bit in his sensitive mouth. I know you're thinking, "But I would never jerk the bit!" but from your horse's point of view, and from *his* end of the reins, that's exactly what happens whenever you take all the slack out of one rein and apply sudden pressure and release. Because the bit is a light snaffle, and because the reins are light and hanging looped, there is no early-warning system that tells Galaxy that you are about to pull a rein. When you do, what feels to you like a very light pull feels to Galaxy like a sudden jerk with no warning. For a horse, that movement is not just painful but frightening: What you think of as a signal, he perceives as an attack.

So yes, you're doing something that's hurting your horse. You're doing it out of goodwill and a wish to be kind and natural, but none of that makes sense to Galaxy. Change your style. If you want to ride without direct contact and with your reins always looped, you'll need a nice heavy ported curb and heavier reins. If you want to use a snaffle, you'll need to ride with consistent, light, reassuring contact, so that you'll be able to *speak* to your horse by tightening your fingers for an instant and then relaxing them again, never pulling, but never losing contact with your horse either.

Whatever provides your horse with constant gentle communication and reassurance, and enables you to give him direction without surprising or hurting him will be right. Listen to Galaxy. He'll tell you what is right for him, and that's what should matter most.

Now, back to my first point. *Horses are forgiving.* Your horse will forgive you very quickly. He probably already gives you credit for good intentions. Make the changes you need to make, listen to your horse, and stop making yourself miserable about past errors. You did the best you could based on what you knew at the time. Now you know better, so you can *do* better. Move on! That's what your horse will do.

CHAPTER 4

Training at the Walk

Do you think about training your horse at the walk, or have you gotten into the habit of "working" only at trot and canter? For many riders and for far too many horses, the walk is a neglected gait. Some riders avoid working at the walk for fear of ruining it; some just prefer the trot and canter. Don't be afraid to work your horse at the walk. It's true that a good walk is born, not made, but many horses that seem to have stiff, short-strided walks have simply never been worked correctly at that gait. Help your horse develop and enjoy the best walk that his conformation will allow — a good, swinging, powerful natural walk is a great asset to a horse.

What Ruins the Walk?

Q My instructor is always yelling at us to get out of our horses' faces at the walk. We all know that the walk can be ruined, because we've heard it ten thousand times. Yesterday she got a new horse in for retraining and she made us all come and watch him work. He is two-and-a-half hands taller than the horse I usually ride, but he can't keep up with him at the walk! My instructor says that this big expensive horse has a standout example of a ruined walk, and we should all pay attention and resolve never to ruin a horse's walk. It looks awful, like it's the wrong size walk for the horse — a little fat pony walk on a big tall Warmblood! So, what ruins the walk? Whatever it is, I really want not to do it. And if it hurts the horse (my instructor says it does), why wouldn't the horse just rear or buck or run away?

A Walks are usually ruined by the trainer and/or the rider. I've never seen a horse in the field ruin his own walk or ruin any other horse's walk. As for exactly how the trainer/rider ruins the walk, that depends.

I would guess that somewhere in this horse's history someone attempted to force collection by using draw reins or tight side reins or other restrictive equipment. Trying to put a horse "on the bit" with the reins can also ruin the walk. When riders force more contact (shorter rein, heavier pressure) than the horse can accept, or insist on the horse maintaining the same head position for longer than the horse can physically maintain it, the walk will suffer.

If a horse is uncomfortable and looking for an escape route, his walk may become crooked, unbalanced, or pacey (lateral). He may give up and lean hard on the rider's hand, or he may drop his back and become inverted and hollow. If you watch closely, you may notice that an uncomfortable horse will try to ease the pressure or achieve a tiny stretch here and there by popping his shoulder outward or leaning his haunches in.

Most of the horses I see with ruined walks arrived at that point very slowly and with no drama. They didn't rear or buck or run away because there was no sudden moment of severe pain to provoke a violent reaction. Their walks were ruined by perfectly nice, ordinary riders who were thinking about other things, focusing their *work* efforts on the horse's trot and canter, and never treating the walk as a *real* gait.

If you want to be sure that you never ruin a horse's walk, teach yourself to *ride* the walk actively every time you're on a horse; not by pushing and pulling and pumping, but by using your legs to ask the horse to step out actively and energetically with his hind legs, and relaxing your seat to invite the horse to lift and stretch his back.

When you walk on contact, allow the horse to move your hands and arms as far forward as necessary; don't limit his motion by blocking your own motion. When you work *off* contact — on a loose rein — be sure that you are still sending the horse actively forward with your legs and allowing with your seat. "Loose rein" means that your horse may put his nose on the ground if he likes, but it doesn't mean that he can slow down, take short steps, stop moving from behind, and drop his back.

Laziness is one way to ruin a walk; focusing on the horse's head position is another. Keep your focus on the energy you create in the part of the horse that's *behind* you, and allow the horse to use his head and neck freely — on contact or on a loose rein — in *front* of you. Teach yourself to be a good rider and trainer who always works horses from back to front, and whose question is, "What's happening with the hindquarters?" and not, "Where is the horse's nose?"

Reins that are too tight and short will block your horse's forward movement.

Can a Short Walk Stride Be Improved?

Q I recently moved my horse into a real boarding stable that has a lot of acreage including some beautiful trails. I'm so happy to have the chance to interact with other riders and give my horse a chance to enjoy the company of other horses. So far everything is great, and my nine-year-old grade gelding is mellow about it all. But now that I am watching other horses and riders, and most of all now that I can go on trails with other riders, I've discovered that my horse has a very poky walk and takes short steps. I guess I never noticed this before because I always rode alone and Leo's walk was just his regular walk, so I didn't think about his stride length at all.

Now it is starting to cause me some frustration because I would like to be able to walk and talk with my new friends on the trails, but Leo just can't keep up with the other horses. We'll start out walking with the others and then we fall behind and finally have to trot to catch up. I can't ask the others to make their horses walk "smaller," but is there anything I can do to make Leo walk "bigger"? According to my veterinarian and the owner of the boarding stable (she teaches hunter/jumper lessons), he has good conformation and there's no reason he should have such a short walk stride.

A Turn Leo out into an arena or put him on the longe line, and take a good look at his walk. You know what it's like when you are riding him, but what is it like when he has no rider and no saddle? If his stride is just as short and choppy when he's on his own, then there may be some physical problem preventing him from striding out. But if his stride is noticeably longer when he's on his own, you have two possibilities to consider.

Many horses move with short strides under saddle because something about the saddle or the rider makes it uncomfortable for them to move with longer strides. If your saddle is interfering with the free movement of Leo's shoulders, a very normal reaction would be for him to walk with short steps, so check the saddle fit and the position of the saddle on Leo's back. Sometimes a walk will improve instantly when a saddle that's positioned too far forward is pushed back into the correct position.

If your saddle fit and position are fine, then have someone, perhaps your barn owner, watch you ride. If your lower back and hips are stiff, if your arms are tense, or if your elbows are locked and can't follow Leo's head and neck movements at the walk, moving with shorter strides would be a normal reaction on his part.

If your riding isn't causing Leo to take shorter strides, then try riding in a way that encourages him to take longer strides. If you do that day after day, he will eventually get into the habit of taking longer strides at the walk.

The easiest and most natural way to ask your horse to take longer steps at the walk is to give him a slight nudge with the calves of your legs, one at a time, timing your leg aids to match the swing of his belly. Ride him at a walk for a few minutes with your legs stretched softly against his sides — just resting, not acting. You'll feel his belly swing from left to right and back again with each stride. Use your left leg whenever you feel his belly swing to the right, because that's the moment when his left hind leg is lifted and moving forward. When his belly swings back to the left, use your right leg, because that's when his right hind leg is lifted and moving forward. That moment, when the leg is in the air and coming forward, is the best time for you to say, "Reach a little more forward with that leg, please."

Training Leo to take longer walk strides means training yourself to ask Leo to reach forward more strongly with his hind leg each time he takes a step. It's going to be hard work for you as well as Leo, so don't try to make a dramatic change in his way of going in one day or over the weekend. Ask him for a few longer steps at a time — six or eight, say — and then allow him to relax and walk his normal stride for a few steps before you ask him to take longer steps. Asking for too much at once will make him sore and unhappy, but what you want is to show him that he can *comfortably* take longer steps and cover more ground at the walk. Incorporate this exercise into your arena warm-up, beginning with just a few minutes at a time and taking a week or so to build up to a 10-minute session.

Always monitor his reactions and responses so that you'll know immediately if he is becoming tired, sore, or both. This exercise is much more demanding than you might think, so if your horse becomes unhappy or cranky, don't blame him, just give him a break and ask for less when you return to the exercise. Better yet, pay such

You can use trail rides to improve your horse's walk.

close attention to your horse that he won't have a chance to become unhappy. Don't wait for an overt refusal to cooperate — notice if he begins to move more slowly, to hesitate a little, or to make smaller movements with his head and neck. Those signs may mean boredom, muscle tension, or fatigue, and you can avoid those problems by taking a moment to do something else that involves a stretch. When your horse needs a break, send him forward into an active trot on a large circle, rise lightly or go into your two-point position so that he can lift and stretch his back comfortably, and invite him to stretch as much as he likes, even if that means putting his nose on the ground. Then bring him up again slowly and go back to your walking exercise.

Do a little more each time, even if it's just a step or two, but never do so much that your horse becomes actively uncomfortable. When you've gotten into the habit of asking him for longer steps, and he has become more comfortable offering you longer steps, ask a friend to go out on the trail with you again for half an hour or so. Don't ask someone whose horse has a huge, loose, long-strided walk; choose someone whose horse's walk is slightly longer strided than Leo's. Ride out together and use your new exercise to try to keep Leo walking next to the other horse at all times.

Being out with the other horse will give Leo more incentive to increase his stride length, especially now that he knows he can do it. Take breaks: When he begins to tire, you can halt and chat, or trot and chat — either option will give his muscles a chance to relax. Be careful not to push him too hard or for too long. For one thing, he needs to enjoy this; for another, if you ask for more than he can comfortably give you at the walk, his efforts to comply are likely to result in a breaking up of the walk rhythm. Horses who are pushed and rushed at the walk often shift into a lateral walk where the rhythm is no longer four beats but two beats, pause, two beats.

Now that you know how to build the "reach" into Leo's walk, don't fall into old habits and accept short walk strides (at least, not once he's thoroughly warmed up). Many horses lose their lovely natural walk stride under saddle, partly because of discomfort but mostly because the rider accepts the short strides and the horse doesn't mind putting out less effort to give the rider what she seems to want. Remember that the walk is a gait, not a "time out," and ride Leo accordingly. He may never achieve the long, loose stride of a tall Thoroughbred or a Tennessee Walking Horse, but you should be able to develop and lengthen his walk stride enough to allow him to walk comfortably next to other horses on the trail.

Mare's Walk Isn't Four-Beat!

Q I took my mare, Sukey, to a clinic last week and heard some bad news about her walk: instead of 1-2-3-4, 1-2-3-4, the rhythm is more like 1-2, pause, 3-4. There's not a very big pause, but it's there. I thought at first that Sukey might be reacting to the long drive to the clinic but then realized that the clinician was right because even after Sukey was rested and warmed up she was still moving the same way. I had to really focus to feel that she was doing it, I guess because I'm so used to the way she walks and I don't ride any other horses. I don't attend many clinics, maybe one every couple of years, and there are no instructors in my area at all — there aren't even many horses around here.

Sukey was sold to me as a Welsh-Connemara cross, but she didn't come with papers and she could be just about anything — the clinician guessed Quarter Horse-Arabian. She's seven years old. I have two questions. How can I train her to do a regular walk that goes 1-2-3-4? And second, I'm fascinated by dressage; that's what I want to do with Sukey even though we're on our own and it's really the blind leading the blind (or is it the lame?). Please tell me that I can still do dressage even if Sukey's walk isn't exactly right!

A You can certainly still do dressage, although if you were planning to participate in competitions, you could expect low marks for an impure walk. Properly done, dressage will improve Sukey's strength and flexibility and responsiveness, so I would encourage you to follow your dream. Plus, dressage can create remarkable changes in horses, and it's entirely possible that her walk will improve considerably.

Begin by identifying Sukey's best tempo at the walk — the one at which her walk is nearest that 1-2-3-4 rhythm. Experiment: Ask her to walk a little faster, then a little more slowly, looking for the tempo that corresponds to the purest rhythm (and the smallest pause between 1-2 and 3-4). Make this your default walk — the one you start with, work in, and return to whenever you catch yourself walking her at a faster or slower tempo. You may be lucky enough to discover that at a certain tempo, she *has* an even, four-beat walk rhythm!

There are many exercises you can do to help regulate the walk. Let's say that you've found her best walk, and that even that walk is still very slightly lateral (1-2, pause, 3-4). Take her outdoors and walk her on trails where she can be both relaxed and forward at the same time. Put her on a loose rein and notice how much she moves her neck and head when she walks (forward and down and up and back). Ride her on a long rein — still with her neck stretched, but this time with a light feel of her mouth — and notice if there is any change in her head and neck movement.

If there is, then watch yourself very carefully, because you may be restricting her walk with your body or the reins. The wrong contact can have a very damaging effect on a walk, and "wrong" can mean no contact, too much contact, uneven contact, or off-and-on, unpredictable contact. It's easy to fall into bad habits, especially when you work alone, without help from an instructor or a regular clinician. It's also easy to put more pressure on one rein than the other, or use more pressure with one leg than the other, and both of these uneven rider actions will be reflected by the horse's uneven movements.

To help Sukey, spend some time during every ride focusing on yourself, developing a tall, relaxed, upright posture with steady, relaxed legs and arms that move from the shoulders to follow your mare's head and neck movements — hands steady, fingers closed on the reins, and elbows opening and closing as she walks. Let *her* do the walking — you may use your legs to keep her up to tempo or to ask for a slight increase in stride length, but your seat should follow her back and your hands should follow her mouth.

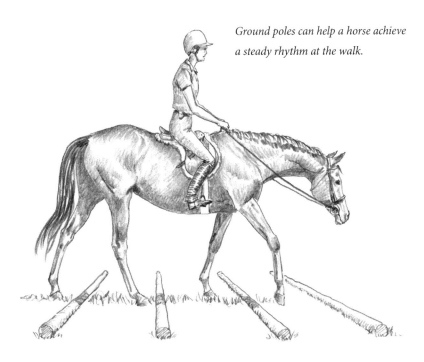

Ground poles can help a horse achieve a steady rhythm at the walk.

I'm going to suggest two items of equipment and two exercises. The first item is a metronome — not a big expensive one, but a small one that clips to your clothing and provides a soft (or loud — it's usually adjustable) ticking sound to accompany your work. Set it for the tempo that you've identified as matching Sukey's best walk, and then do all of your walk exercises at that tempo. If there's no one supervising you, it's very easy to slow the walk whenever you get to a corner, make a turn, or perform a lateral movement. The metronome will help you do all of your walk work at the same speed.

The second item of equipment is a set of cavalletti, or even ground poles (blocked so that they don't move out of position). A line of cavalletti or ground poles set at a comfortable distance for Sukey to walk through (every horse is different, so you'll need to experiment with the distances) will help her improve and purify her rhythm, especially if you pay attention to your metronome and use your legs (a pulsing aid from your calf muscle, one leg at a time, always on the side of the hind leg that's moving forward) to keep her going forward at her best tempo. In addition, having to pick up her feet even a little bit higher will encourage Sukey to use her joints more actively. Stepping over objects on the ground will encourage her to reach forward and down with her neck, and if you are balanced and using your legs to encourage her forward as needed, she will eventually begin to lift her back as she walks.

The two exercises I'll suggest for you are leg-yielding and shoulder-fore. (See chapter 9 for more detail on these.) Both will help her hind-leg reach and flexibility. In your everyday work, incorporate lots of walk-trot transitions, which will help build Sukey's strength and also give you the opportunity to add "trot breaks" to her walk-work (just don't try to trot through poles that are set for walk).

Event Prospect Has Hurried Walk

Q I am searching for an event prospect and recently went to look at an eight-year-old Hanoverian-Thoroughbred mare. She was bought as an unbroken three year old by a very young teenager who rides/shows pony hunters. The mare has been ridden hunt seat only, largely on trails as she really is not a show ring hunter type. Her arena training is limited. She rushes forward if you even touch her with a leg or attempt to sit the trot or canter. None of this concerns me, as I know it is merely a matter of training.

However, this mare has a very lateral walk that is quite forward and somewhat hurried. I tried slowing her walk to see if she would change to an even four-beat walk and it improved but still felt lateral. Is this correctable? I suspect she may have a sore back from the close contact saddle on her medium-wide back. Her owner uses a bounce pad, which of course narrows the saddle more, because she says the horse is cold-backed. When I attempted to sit her canter and trot she hollowed her back, threw her head in the air, and rushed forward. Could this lateral walk be caused from back discomfort?

If a sore back isn't the cause of the lateral gait, what are the possible causes and can she be retrained? This is a lovely, well-built mare just within my price range with a very nice trot. Her canter and jump work were really poor, but if I'm right about the sore back then I suspect they would improve with a well-fitted saddle and correct training. What are your thoughts on a lateral walk?

A You're right about the retraining. If you buy this mare, you will need to start from the ground up and take her through all the stages of training. If you want to event her, plan to wait at least a year, maybe two — it will certainly take a year or more to get her to that point, especially if she is sore-backed and needs some time off. Having said that, she still sounds like a nice mare and is probably worth the trouble it will take to retrain her.

Comfortable tack and a good rider will help a horse "use herself" better at the walk.

If you like the way the mare moves without a rider, and if her other gaits are good, the bad walk is probably the product of bad riding with unsuitable tack. I suspect that her young owner rushed her training to get her jumping quickly and used the walk as little as possible. Other causes are the same ones that cause uneven walks in humans: soreness somewhere, or a tight muscle, or a loose tendon or ligament, or habit. It could also be uneven development (quite common) or a rider who sits with her weight unevenly distributed or uses one leg more strongly than the other (also quite common).

A rushed walk is usually a lateral walk, and it's the walk of an unbalanced horse, usually one whose rider has maintained a tight grip on the reins. This horse has been falling away from her pinching saddle, which is not conducive to a good walk, and she has probably also taught herself to take little quick steps at the walk, so as to avoid using her head and neck that were being held in by the rider.

There are many ways to work on the walk. Some involve doing specific exercises at the walk, and some involve working at the trot and canter to gymnasticize the horse, thereby improving the walk as well. Begin by getting a wider saddle, put a thick pad or a gel pad under it (this is *just* for this one ride), and try her again. If she has a sore back, as she probably does, she won't be perfect but she should be more comfortable, and you should be able to tell immediately. Don't try to sit the trot — she's too sore for that, and in any case you should never sit to the trot until the horse's back comes up!

If they have a round pen or even a small arena, or a large arena, or a paddock — any enclosure large enough for the mare to make at least a 20-meter circle — remove the tack and ask if you can free longe her. It doesn't matter if she doesn't know how; just stand in the center with the longe whip and keep her moving. Watch her. If she looks infinitely better *without* a rider, she may be the horse for you — if she passes the vet check.

Great Trot, Comfy Canter, Bad Walk

Q My horse Laramie is almost four years old. He has a great trot and his canter is very comfortable, but his walk doesn't match his other gaits for quality. I have heard that the walk is the easiest gait to wreck, so I've deliberately avoided working him very much at the walk just in case I do the wrong thing and mess it up. He is part Thoroughbred and part Shire, so maybe he is just going to take longer to grow up than most horses? If I ask him to take bigger steps at the walk, he gets pacey.

Can he grow out of this problem, or should I just figure that he will always have a good trot and canter but a lousy walk? I would be willing to do walk exercises. I would like to take him out on the trails, not necessarily big all-day rocky ones, just some of the easier two-hour trails near our farm, but I don't know if I should do that since we would have to walk so much of the time. I just hope I haven't messed him up already. I've had Laramie since he was a weanling so if he has any problems, I caused them.

A Relax, you have plenty of time — he's a young horse! By all means, take him out on those trails. Sometimes young horses that are ridden primarily in arenas can develop short, uneven walks because they aren't used to striding out at the walk. They're not used to doing any real work at the walk, either, because riders often become focused on working at the trot and the canter, and use the walk only as a sort of rest period between trot and canter efforts.

Asking your horse to leg-yield on the trail is a helpful exercise.

The walk can and should be worked like any other gait. It is easier to ruin than the trot because it requires that the rider either allow the horse a completely loose rein or be able to maintain soft contact and follow his head and neck movements. If anything is wrong — say the reins are too short or the rider's elbows are locked — the horse will try to minimize the movement of his head and neck and will end up taking short walk steps. If this goes on for too long, the horse may forget that he ever *had* the ability to do a forward, reaching, ground-covering walk.

One reason I like longeing horses as part of their warm-up is that longeing allows the rider to watch the horse in action and see whether he is having any problems at any gait. I'm going to suggest that you longe your horse briefly, *at the trot,* before you ride him. Whenever a horse is having problems with the walk, it's best to let him warm up, loosen up, and stretch out at the *trot* before doing any real walk work. Just longe as you normally would, on the largest possible circle, on good footing, using a longeing cavesson, a long (35' or longer) longe line, and no side reins. Even 10 or 15 minutes of trotting will help your horse walk much more comfortably and easily afterward, and it would be quite appropriate to trot him on the longe and then take him out and walk him on the trails.

Don't hold onto his head, but do ask him to step up and stride out from behind — you want him to walk as if he had a purpose, as if he were going somewhere.

Walking on trails on a long rein would be a wonderful therapeutic exercise for your horse. Anything that develops the power of his hindquarters and helps him lift and stretch his back will help his walk. On the trail, he's likely to be a bit more forward because he'll be interested in seeing what's around the next turn — in the arena, he knows *exactly* what's around every turn. There's an easy exercise you can do to help develop the strength of his hindquarters and the flexibility of his hind leg joints, stretch his topline, and send him forward into his bridle: Leg-yield from one side of the path to the other, or on and off the path if there's room.

Keep him moving forward and keep him interested — his walk will improve. When you work him at the walk in the arena, give him a reason to reach forward and a reason to pick up his feet — ground poles set for his normal walk stride will do that, and if or when he becomes bored, add a couple of inches to the distance between poles and let him figure out what he needs to do to get through the line easily. If you do these ground pole exercises, ride him on a loose rein — not just long, but *loose*, so that he'll have every reason to reach and stretch and no reason at all to restrict his head and neck movement and shorten his stride. Keep him marching actively forward whenever you're walking — on the trail, in the arena, on the road, anywhere. Don't hold onto his head, but do ask him to step up and stride out from behind — you want him to walk as if he had a purpose, as if he were going somewhere.

Contrary to what you may have read, a horse's walk often can be improved all out of recognition. Your horse is large, young, and still growing, and he will continue to grow for at least another three or four years. A lot can change in that much time, and chances are good that when he's mature, his genetic heritage will come through for him — a short, choppy walk wouldn't be normal for a Thoroughbred *or* a Shire.

By the way, it's a sign of real horsemanship that you're willing to take responsibility for your horse's gaits and development — well done, you!

CHAPTER 5

Training
at the Trot

Most riders spend a lot of time working at the trot. Many riders find their horses' trots frustrating — too short, too high, too fast, too slow, or too much on the forehand. If you are one of these riders, you should take comfort in knowing that the trot is by far the easiest gait to improve. With patience, persistence, and systematic, good training over time, you should be able to create a dramatic improvement in your horse's trot.

Young Horse with Jackhammer Trot

Q My horse has a very uncomfortable gait when you first ask him to trot. He walks well and has a smooth canter, but his trot is like a jackhammer until after about an hour of work, at which point it becomes smoother. He does have fairly short pasterns, but all the same he *can* produce a better trot. He is an Arabian-Quarter Horse cross, four years old and 14.1 hands. He was broken in when he was around two-and-a-half years old and was used by the Mennonites to haul a buggy, but he refused to pull heavy loads so he was sold and retrained as a riding horse. I bought him four months ago. He has a very sensitive mouth and seemed uncomfortable in even the mildest bit I tried, a French-link eggbutt snaffle, so I use a Dr. Cook's Bitless Bridle, in which he is much happier.

He used to go hollow-backed with high head carriage, but he has been undergoing training for the last six weeks (not by me) and his head is now coming down and his outline is improving. When he was working in the buggy, the collar caused some muscle wastage on his neck, which is now beginning to recover, and he is generally muscling up all over, so I'm mildly hopeful that as he strengthens up his trot might improve. He lives out, and when he comes to my place full time he will be kept out, too. My Australian saddle appears to fit him and because his gaits do improve as he warms up, I don't think this is the problem. He's been ridden in a Western saddle with the same results.

How can I get him to offer a smoother trot when first asked? I live in the jungle and do all my riding on trails — there's no arena or round pen (we don't even have any level ground), so all training will have to be done on the trail.

A The fact that your horse's trot smoothes out after an hour tells me that while he is capable of doing a good trot, he may *not* be capable of doing it until he is thoroughly warmed up. In the long term, whether you can get a better trot when you first ask for it will probably depend on *when* you ask for it and on what you have asked him to do *before* you ask for the trot.

Here's something you can try: Do his initial warm-up at the walk as usual, and after 20 minutes or so of active walking, allow him to canter on a long or even a loose rein. If he wants to put his nose near the ground, let him do it; it's not a problem so long as his hindquarters are engaged and his back is lifted. After the canter, trot him and see what sort of trot you have. If it's not noticeably better than his usual trot, I will be very surprised.

This is a technique that I use most often with older horses (say age 16 and up) that don't trot well, even after a long, thorough walk warm-up, until they've had the opportunity to stretch their legs, backs, and necks at a canter. I generally warm up these horses at walk and then allow them to choose the next gait. They invariably prefer to canter. After they have cantered and stretched their toplines, they offer a much-improved trot, and they also seem to find the trot more comfortable to perform.

Given your horse's history and your description of his posture and movement when you purchased him, it's probable that until you began working with him he was never asked to use his hindquarters or his belly muscles properly, and he was never asked or expected to lift his back. Your horse is very young, but this walk-canter-trot warm-up might suit him because of his background; he may well be stiff and sore in his hindquarters and back. Even at his young age, he has been through a lot, and may have many of the same issues — including, perhaps, arthritis — that typically afflict horses in their teens. If a warm-up routine devised for the comfort of older horses can help him relax, stretch, and offer a better trot, incorporate it into your training.

Some horses need to canter on a long rein and stretch their muscles before they trot.

How Long to Improve a Rough Trot?

Q My horse's trot is very rough, difficult to ride, and impossible to sit. I would love to train him to have a longer, smoother trot, but how? My veterinarian and my instructor have both evaluated his conformation, and we all agree that he is a well-built horse with normal shoulders and pasterns, not hugely long and sloping, but not short and upright either. In other words, there is no good reason why his trot should be so horrid. He is three years old and was backed just six months ago. Is there a way to train him to create a better trot, and if there is, how much time will I need to make the change?

A You don't mention what you did with the horse before he was backed, or how his trot appears when he is loose in the field or being free-schooled or longed. If his trot is longer and smoother under any or all of those conditions, then the trot you are getting under saddle is definitely *not* the best trot he can do. If his overall conformation is good, his shoulders are free-moving, and his hooves are correctly trimmed and balanced, then the question to ask is: *Why* is he not able to offer his best trot under saddle? Could the saddle itself be interfering with his shoulder? If it fits badly or is positioned too far forward on his back, that's a possibility. Be sure to check the saddle fit with a rider on board — many saddles appear to fit well until riders are added, and then everything changes for the worse.

Your horse is so young that you should keep another possibility in mind: His apparent difficulty may simply reflect the fact that he is still learning to carry himself properly under the weight of a rider. Slow, gradual strengthening work should take care of that.

Good riding over time is the best way to help a horse improve his musculature and his gaits. Your horse has not been under saddle for very long and is still very much a work in progress. Developing a horse is a matter of making incremental changes. If you ride your horse gently and well for the next six months, at the end of that time I would expect to see him using his hind legs a little more actively and lifting his back and the base of his neck a little higher. I would hope to see him offering a slight flexion at the poll, not from being pulled in by the rider's hands but because his overall posture would make it easier for his head to hang relaxed from his more developed and stronger neck. At this point, your horse's face would not be on or anywhere near the vertical, but somewhere in the vicinity of 30 degrees in front of the vertical.

In six months he will be able to accept regular training but will not be ready for heavy work. If you can help him strengthen and work from behind, his trot should improve enough by the time he is four that he will be ready to continue a *slightly* more demanding version of the same work for another year.

Walk-trot transitions will help him, as will transitions within the trot. You can also begin teaching him to lengthen and shorten his stride. Anything that helps improve his balance will help improve his trot, because as he learns to carry more weight behind and to reach more with his front legs he will become more flexible in the shoulder. You can use ground poles to help him learn to lengthen his trot stride — just be sure that he continues to use and develop his hindquarters while he's learning to reach in front. Use a metronome set for the tempo of his best trot, and monitor him constantly so that you can be absolutely certain that the changes you're asking him to make aren't having a bad effect on the pure "one-two" rhythm of his trot.

Here is an exercise to try: Make a line of three or four ground poles, set for his normal trot stride. For many horses, this would mean setting them three and a half feet apart, but you should use *your* horse's most comfortable distance as a starting point, whether it's nearer three feet or four feet. Riding in a half-seat, ask the horse to trot down the line on a long rein. As he learns to trot through the poles in stride, working from behind and extending his neck, you can begin to increase the difficulty of the exercise by gradually increasing the distance between the poles and by adding more poles to the line. Use your legs to encourage him to

Sometimes improving a horse's trot is as easy as pushing the saddle back into its correct position.

reach for the poles, and allow him to lengthen his neck and stretch as far down as he likes, provided that he is still working off his hindquarters and you can feel his back lifting underneath you.

I can't promise that you will improve your youngster's trot in a specific number of weeks or months or years. I can promise that if you work him carefully and build up his strength and flexibility while keeping him sound and happy in his work, his trot *will* improve over time.

Tossing Head on Circle

Q My seven-year-old Quarter Horse-Oldenburg gelding has me stymied. I need an exercise or something to fix this problem. I should probably say that I can take only one or two lessons a year, if I'm lucky, and I do most of my training myself. He was a green horse when I bought him three years ago. He was quiet with a nice personality and could do walk-trot-canter but that was about it. Since then I have worked a lot on basic dressage with some books and videos and a lesson once in a while.

When we do trot circles, there are two moments on every circle when Tornado suddenly tosses his head, and I can feel his back stiffen. It doesn't last long at all, just one or two steps, then his head goes back down and his back comes back up and he's fine until we get to the other side of the circle and it happens again exactly the same way. He does this every time, but only in one direction. He doesn't do it when we walk or canter the exact same circle!

A You've probably already considered the obvious possibilities and had your vet and perhaps a chiropractor have a look at your horse. You have probably already considered and dismissed saddle fit as a possible problem, since your horse lifts his back and moves smoothly the rest of the time. Still, it would be a good idea to have the saddle checked, because sometimes this sort of "only occasionally, only on a circle, only in one direction" problem is due to a slightly twisted or warped saddle tree or a slight unevenness in the stuffing of the saddle. Sometimes the imbalance is so minor that it's almost imperceptible, but it can still be enough to cause the horse to feel his shoulder being pinched at one or two points on the circle. All it takes is one slightly larger step by the horse, one pinch from the saddle, and there's your head toss.

Riders can also cause head tossing. If there's a point — or two points — on the circle where *you* do something different, including shifting your leg or your fingers, looking down, moving your shoulder slightly, tightening your seat, or even rising slightly higher, your horse's head toss may be a direct result of your action. Pay close attention to your own posture and movement on those circles; you may find the answer in yourself.

If the answer is definitely with the horse, then do many, many transitions within the trot and between the trot and the walk and canter. Also do many half halts — and here again, pay close attention to your own posture and movement. Sometimes a head toss means that a rider has attempted a half halt with insufficient leg and either a blocking seat or a too strong or non-yielding hand.

An always-useful exercise in this situation is a trot spiral. Using your legs and weight (but as little hand as possible), shift your horse out onto a larger circle, then a larger one, and then begin to shift onto smaller circles. Just before your horse becomes frustrated or uncomfortable with the smaller circle, begin to shift him outward again. Keep your horse's attention on you while keeping him bent, balanced, and forward, and the head toss should disappear.

An occasional head toss may be a reaction to the rider's hands or position.

Repairing a "Draw Reins Trot"

Q One of my riding students has returned from a holiday with a new horse, much to my dismay. He is pretty, but if one looks past his height (16.2 hands) and his shiny coat, there is very little muscling to be seen and what there is, is in the wrong places. My student offered to purchase the horse after taking a lesson on it because she liked the horse and felt sorry for it, a compelling combination I must say. His owner insisted that the horse's curled-up posture was "collection." My student knew better but was horrified when the owner then said that it was a game that the horse wanted to play. She was told that the horse "always tries to lean on the rider" and "refuses to carry himself, he wants to trick the rider into carrying him." That's when she decided to purchase him and save him from his situation.

Even at a walk on a loose rein, this horse is overbent. At the trot, he is horribly overbent. My student has good position for a beginner (six months of lessons) and knows better than to pull a horse's head in like that, but this horse has quite obviously been overbent for quite some time, and one is tempted to conclude that he may have been ridden in tight draw reins since he was first backed! Every step he takes appears to be going tippy-toe down a steep hill (he takes tiny steps with the hind legs). What am I to do?

My student realizes that she can't fix this horse but refuses to send him back to the previous owner. She has offered to pay me to school the horse daily for several months. That's all well and good, but I have never reschooled a horse with such an extreme overbending problem and am not sure how to go about it.

I was trained to do most remedial work at the trot, and I am willing to do that, but this horse's trot is atrocious; from the shoulder forward he curls up like a frightened hedgehog, and from the withers back he is stiff as a board, with the hind feet not even stepping up under the stifles! How can I repair this poor horse's "draw reins trot"?

A You have a difficult project ahead of you, but perhaps not an impossible one. It's a pity that your student didn't consult you before making the purchase, but surely you can understand why she felt that she had to rescue this animal from someone who would pull a horse into such a painful physical position and then insist that the horse was deliberately moving badly. We could debate — probably endlessly — whether the horse's poor musculature causes it to move badly or whether having been forced to move badly that has ruined his musculature. I suspect the latter, and I think you should proceed as if that were the case, because then you can *do* something about the problem.

Plan to enlist some professionals to help you, and warn your student that she should expect to invest in some massage and some chiropractic treatments for this horse over the next few months. If I were you, I would begin with a veterinarian's visit and a referral to a chiropractor, and I would begin *now* doing as much massage as possible.

Explain to your student that the process of rehabilitating her horse will take time. Tell her that when a horse has moved incorrectly for many months, its muscles have developed to support the incorrect movement. Her horse will need to use these muscles in different ways and develop their strength and length in order to move well. Changing the horse's posture and movement will need to be done simultaneously and gradually. A simple change of tack and rider won't be enough.

Your student's instincts were good. This horse is not leaning on your hands or tucking his head to his chest out of perversity or a desire to play a game. He is doing what he has learned to do, and what his muscles now support — he is used to this, and although I doubt it feels good to him, it probably feels normal.

Your goal will be to eventually ride on light contact while the horse carries his own head and neck and you carry your hands. Walk-work over ground poles would be a good place to start, along with walk-work outdoors on trails or through fields.

It takes time and patience to rehabilitate a badly trained horse.

Horses who have been developed properly through correct training develop the ability to "sit" with their strong hindquarters, tighten their belly muscles, lift and stretch their backs and necks, lift their necks from the base, and carry their heads suspended gently from their necks. Horses who have been forced into knots with draw reins have no such abilities. Their hindquarters and bellies are weak, their backs are tight and stiff, and they aren't able to carry themselves and move in balance. Their incorrectly developed muscles don't allow them to move correctly.

If you were to turn this horse out into a field today, it probably could not move like a normal horse or assume the postures of a normal horse. This horse *would* benefit from turnout, but I suggest that you do a few months of remedial training first, *then* turn the horse out for a few months, and then begin again with longeing followed by a great deal of walk-trot work on long reins over the course of several more months.

This could work out nicely, since your student wants you to work with the horse daily for the first few months. Explain that the horse will then need a few months in a field, and after that, when it goes back under saddle, you will work with her and the horse simultaneously. By that time your student will have had another six months of lessons and should be ready to take an active (supervised) role in the retraining of her horse.

This horse may never be an outstanding athlete, but it can certainly move more correctly, be more comfortable, and have a chance to enjoy his life as a riding horse. He deserves to be given that chance.

Will Rhythm Beads Help My Horse's Trot?

Q This may not sound like a serious training question, but it is. My horse is not reliable or steady at the trot — she never seems to be sure how she should be trotting, and she goes from fast and short to longer (and much more comfortable for me) and back again without warning. She is also very nervous about sounds. If she hears a strange sound, she loses her focus easily and it's hard for me to get it back.

I was riding her on the trail and met a woman whose horse was wearing a necklace with little bells and rattles on it. I was positive that Mona would flip when she heard the noise, but she just looked at the other horse for a minute and then she was fine. I rode along with the other rider to ask about the beads, and then we trotted for part of the trail. After a while I noticed two things. One: Mona doesn't trot at the same speed for very long, even when she is following a horse that is dead steady, but she was *totally* steady following the horse with the bells. Two: Mona is usually not a lot of fun on trails because she is so spooky and reactive; she doesn't relax even when she has another horse in front of her. I thought she must really like the woman's horse, but the woman said she had seen this happen a bunch of times before and it was the sound of the little bells. I am seriously thinking about ordering some rhythm beads for Mona. Do you think they could really help her trot?

Most horses seem to enjoy the sound of rhythm beads.

A I see more and more "horses with necklaces" these days, and yes, the beads and bells do seem to have a calming effect on quite a few horses. Why not try some rhythm beads? They're inexpensive and they're fun. Horses and riders both seem to enjoy them quite a lot. This wouldn't come as a surprise to anyone who has driven a horse — think of sleigh bells! Horses are very responsive to rhythmic sounds, and many horses will deliberately nod or shake their heads if they have a jingly curb chain or a few bells attached to the bridle.

Rhythm beads can be a pleasant and relaxing addition to a horse's tack, particularly on trails. They make it easy to track the rhythm and tempo of the horse's gait, and they seem to help riders as well as horses. Your mare will trot more steadily if you rise to the trot rhythmically at a very steady tempo, and rhythm beads might help you with that.

Some riders ride with tiny metronomes that click or beep at a certain tempo. Others listen to music while riding: some through earpieces, and some through the sound system that pipes music into the riding arena. Riding to music is fun, and music and rhythm make other activities more enjoyable. Can you imagine an aerobics or dance class without music, or a silent skating rink?

If you acquire rhythm beads for your mare, take some time to show them to her and let her hear their sound. If you are at all worried that she might be nervous at first, you might want to carry them or hang the beads from the saddle at the start of your first ride, so you can drop them easily.

CHAPTER 6

Training at the Canter

FOR SOME RIDERS, the canter is the most enjoyable gait; for others, it's the most intimidating gait. Very often, a horse with a "canter problem" is actually having a problem with strength or balance, or simply doesn't understand what the rider wants. Whether your horse is a youngster just learning to carry a rider at the canter or an older horse with strong, established habits, it can be frustrating to think that your horse's canter problem may have started on your side of the saddle. On the other hand, you should find it encouraging to realize that you can probably solve the problem and begin a new phase of canter training just by correcting yourself.

First Canter

Q I have trained my five-year-old Arabian mare by myself since she was a yearling. We are doing great. I haven't cantered her yet, but we have gone on short trail rides alone (at the walk with a bit of trotting), and she has pretty good balance at the trot on circles. I'm wondering what preparations I should make with her prior to cantering. I've read that you should teach the turn on the forehand first, which I have started to do. How do I know when she is ready to canter with me on her back? How do I teach the leg commands for canter, and what's the best, safest way to go into a canter for the first time? If she doesn't understand something right away, she gets angry so I want to do everything I can to make it easy for her. I'm always very gentle and patient with her. She knows her voice commands really well, and I talk to her almost constantly.

A That first canter is always exciting, especially when you've brought up and trained a horse yourself. Horses enjoy cantering, and your first canter will be much more fun for both of you if your mare is strong and balanced enough to be able to canter comfortably without cantering at great speed. Horses become more excited as they go faster. Unbalanced canters tend to be too fast, and a horse that lacks the muscular strength to canter easily will tend to fall into a fast, unbalanced canter.

If your mare has a good walk and trot, can balance well on circles and through changes of direction at the trot, and can offer a longer or shorter walk and longer or shorter trot in response to your aids, the canter shouldn't present any problems.

The best and safest way to canter for the first time is to do it in a place where you feel safe and the horse feels secure. A familiar arena with good footing is ideal: the fact that it's enclosed means that you'll both be able to stay relaxed and unworried. However, as your mare becomes stronger and more balanced, she may offer a canter all by herself somewhere — when you come to a slight upward slope on a trail, for instance. If this happens, and you're balanced and ready to go with her, just relax, keep your legs *on*, and tell her what she's doing ("can-*ter*") in rhythm with her stride. This will help her associate the word with the action and understand that it's something you like and want her to do.

It's always easier if the horse has done enough work on the longe to understand and respond reliably to your voice. A horse that rocks into a smooth canter on the longe whenever you say "and can-*ter*" will be able to offer the same transition under saddle. Your mare knows your voice and knows the words, and this will make that first transition much easier.

There are two ways a horse can pick up a canter: one is to trot faster and faster until she falls into a canter; the other is to use the hindquarters to "power" or "jump" into the canter. The second way is the way you want to do it, but while your mare is learning to canter under saddle, accept *any* canter and tell her how wonderful she is and how pleased you are.

If your mare makes the transition to canter but doesn't offer the correct lead, that's not cause for concern. Let her canter on — don't bring her back to a trot, and don't say "no." At this point, the lead isn't as important as the gait itself. When you ask for a canter and she canters, praise her and allow her to canter for at least 10 or 12 strides, so that she'll understand that you really *did* want her to canter. If you are too concerned with the lead and you try to correct that within a few strides, she won't think, "Oh dear, I cantered on the wrong lead," she'll think, "Oops, I guess I wasn't supposed to canter." So let her canter down the long side of the arena, tell her she's a good girl, and worry about leads later.

If she's balanced and supple, leads won't be much of a problem anyway, especially if, at first, you always ask for the canter in the corner on the short side of the arena, and then canter down the long side. On a straight line, it won't challenge her balance. You can trot before the corner, which is where being on the

Your horse's first canter will be easy if both of you are relaxed and balanced.

wrong lead could cause her to scramble around the short side of the arena. Like everything else in riding and training, it's all about preparation. Some riders get unduly focused on execution — they would find it all much easier if they would remember to set it up and let it happen. That advice, by the way — "Set it up and let it happen" — is something you will hear from both classical trainers and the best "natural horsemanship" trainers. There's a reason for that: It applies to all horses and all training, all the time.

Put your mare in the best position to pick up the canter easily and calmly from a round, energetic, balanced trot. As you come to the second corner of the short side of the arena, sit a couple of strides, look *up*, put your outside leg back a hand's width, and *relax* your inside rein as you say "and can-*ter*!" Ask, and give her a chance to respond. If she trots faster instead of cantering, don't keep pushing until she falls into the canter. Instead, circle at the trot, regroup, and when she's balanced and energetic and round, set her up for the canter and ask again.

Let her canter on. At this point, the lead isn't as important as the gait itself.

When she canters, sit up, stay tall, let your arms follow the movement of her head and neck, and talk to your mare. If you can talk to her in the rhythm of the gait, so much the better — praise her, and tell her what she's doing, just the way you did when she first cantered on the longe. Tell her "can-*ter*, can-*ter*, can-*ter*" and let your lower back move with the saddle. Encourage her forward into a soft elastic contact, not into a fixed or a pulling hand.

Turn on the forehand is a good exercise to teach a horse to move away from the leg, and it's useful for opening and closing gates from horseback. But you do *not* want your horse to get into the habit of shifting its haunches every time you put your leg back. A cantering horse should be straight, not carrying its haunches to the inside. Too much turn on the forehand can cause problems later when you want more sophisticated responses from the horse. Instead, try cantering from a shoulder-fore position, with the horse very slightly bent, looking to the inside, and with the forehand very slightly to the inside.

Once you've both learned what to expect and how it feels, it will be so easy that you'll wonder why you ever worried. Just prepare your horse well, set her up for the canter, and be patient while she learns what you want her to do. Keep smiling, keep breathing, and remember you do it for fun.

How Can I Teach My Horse to Canter?

Q How exactly does one teach a horse to canter? I have a young Warmblood who is very well balanced and coordinated for a three year old, but he is very unpredictable when we canter. Sometimes he will pick up a canter right away and canter smoothly, and other times he seems to forget how to do it and where to put his feet. When I canter him in the indoor school, he seems to get his feet mixed up whenever we get to a corner, and sometimes I even think he is trying to buck, which I guess means he is having some kind of a temper tantrum!

Also, sometimes he seems to understand my canter aids and at other times I have to pester him or trot faster and faster until I finally run him into it. I know that's not right, but I don't know what else to do. He is my first young horse project, and I want to do right by him. Should I just not try to canter him until he is older?

A At three, your youngster is still growing and developing — he's not even halfway to maturity. He is dealing with the same growth and balance problems that afflict human teenagers: His body just isn't the same from day to day, and he has better control of it on some days than on other days.

A small cross-rail on an uphill path
will encourage a horse to land cantering.

He may buck in the corners, but I don't think that "temper tantrum" applies. Bucking is a natural way for an unbalanced horse to try to reestablish his balance. Indoor arena corners are often problematic for horses that are not entirely balanced, and that category would certainly include most growthy young Warmbloods. You'll both have an easier time if you'll take him outdoors when you want to canter. In fact, for the easiest possible introduction to cantering under a rider, take him out with a couple of reliable older horses ridden by sensible people, and let him do some cantering on an uphill trail.

If trails aren't an option, let him canter around the edge of a large field. Don't worry about your canter aids at this point — your job is to remain balanced and calm, help your horse remain balanced and calm, and encourage him gently with your voice as he canters, just as you did when he was learning to canter on the longe line.

Nagging and pestering and running a horse into the canter are not good ideas, because instead of a relaxed, balanced gait, you're likely to get one that is stiff, flat, and rushed. Try this instead: Set up a tiny cross-rail in your field, preferably on or just before a slight upward slope. Then trot your horse energetically to and over the cross-rail. He will probably land cantering; you can then praise him and accept and encourage the canter. Your horse won't be resentful because you won't have pestered or run him into anything, and if he picks up his canter from a round, energetic trot, it will be a nice canter. If you don't have a cross-rail, a small log, a single cavalletti, or a highly visible ground rail will serve.

Cantering uphill, outdoors, on a straight line is really the best option of all. Your horse will find it easier to canter uphill than to trot uphill, so he's going to want to canter. He will be better balanced cantering uphill, and much less likely to canter flat and fast, and it will be easier for you to make a transition to trot when you're ready.

Canter Leads

Q I need advice about a young horse I'm training. She's almost four years old, and I backed her about a month ago. She's easy to steer and has a good walk and trot, but I've just started to canter her and I'm having a problem. I'm pretty sure she understands my canter cues, but three out of the four times we cantered, she has picked up the wrong lead. I brought her back to a trot right away because I figure she'll learn the wrong thing if I let her canter on the wrong lead. But I don't know what to do now. Is there something that I should be doing with my hands? I've been told that I should pull her head to the outside to free up her inside shoulder. I've also been told that I should pull her head to the inside!

A When horses are just learning to pick up a canter and to canter on under saddle, just getting them to step into a canter calmly is a *big* success. You've done that. The biggest mistake would be to run her into a canter from a faster and faster trot, and you haven't done that — good for you!

You shouldn't worry about her lead just yet, though. You're halfway there, because she understands that you want her to canter, and that's the most important thing right now. If you allow her to canter on for 10 or 12 strides at least, and you praise her, she won't get confirmed in the wrong lead, so don't worry about that. She *will* become absolutely certain that your canter cues mean *canter*. This is essential for all of her future work.

There are many reasons for a young horse to take the "wrong" lead when she is first learning to canter under saddle. Your mare is still trying to find her own balance during the trot-canter transition and at the canter. That's absolutely

Even if your young horse picks up the wrong lead, let her canter on.

normal, but it means that she will often give you the "wrong" lead while she is learning. Don't worry unless she is still doing this consistently after at least a month of cantering.

At this early stage, if you bring her back to a trot immediately, there *is* a danger that she will learn the wrong lesson. Since she doesn't yet understand about specific canter leads, she won't understand that you want her to canter on a different lead. Instead she may simply receive the impression that she has done the wrong thing by cantering. It's all too easy to be a little too demanding, too early, and thus convince a horse that you didn't want a canter at all.

Don't do anything with your hands. If you are traveling to the left around the arena, your mare will already be bent very slightly to the left, and you don't need to ask for any more than that. Pulling her head in *any* direction is a bad idea, and you certainly don't want to pull it to the outside. When she picks up her left lead canter, for example, her natural bend will be to the left; there's no sense in asking her to bend to the right to pick up a canter to the left. As long as your mare's nose is tipped in the direction of movement, that's all you need.

Instead of thinking about your hands and your mare's head, think about your legs and your mare's body and hind legs. This would be an excellent time to teach her to leg-yield. If you haven't already done this, take a week or so and teach her to leg-yield consistently at walk and trot. Once she understands how to move her body away from leg pressure by stepping more deeply underneath herself with her inside hind leg, she will be much easier to put into a canter on either lead.

Try this: when she can leg-yield easily at walk and trot, take her down the short side of your riding arena. *Before* you reach the long side, turn so that you are a few steps away from the walls. Leg-yield through the corner and then ask for your canter as you reach the long wall. This will put your mare in the best possible physical position to pick up the correct lead calmly and comfortably.

Keep her relaxed, and remember that the whole point of leg-yielding is that the horse learns to move away from pressure instead of leaning into it (the first and most natural response). As long as your mare is relaxed, she will remember her lesson and move away from your leg pressure. If she becomes worried and tense, she is likely to revert to what comes naturally and lean against your active leg instead of moving away from it — and *that* will put her into position to pick up the *wrong* lead.

Take your time, be patient, and make it easy for her to do what you want her to do. She'll get there in her own time, and she'll be a better horse for it.

Horse Won't Take Left Canter Lead

Q I have a 10-year-old gelding named Windy whom I use as a pleasure and light show horse. Every time I ride him I have to fight to get his left lead. He takes the right automatically, and he will eventually take the left but only after repeated tries and a lot of arguing. There is no physical reason; he has been checked thoroughly. I have tried many different ways to train him to take the lead, but nothing helps. I have had several different people ride him and he refuses to take it with them, too. I can get him to take the left lead if he is in a corner of the arena but on the straight side it takes at least three tries. This is killing us in the show ring and it is very annoying. I make it a point to train my horses so they can do everything from either side, and this is the *only* thing he is one-sided about. Do you have any training suggestions for me?

A Most horses are one-sided, some more so than others. Most horses find it easier to canter on one particular lead, and sometimes their riders don't know or care that the horse *always* canters on that lead, and they let the horse do what he prefers to do. Then it gets more and more difficult for the horse to canter on the other lead, and *then* you have a big problem. Windy may just be stiff on that side, and in the habit of cantering on the other lead.

You can use a single pole to help your horse pick up the correct lead.

If the vet has told you there's no problem (feet, joints, back, or anything else) that would cause Windy to favor one lead over the other, and if your saddle isn't digging into him when he canters on the left lead, and if there is really nothing *physical* bothering him, then you can probably retrain him. Since you don't want to have endless fights with him about this, I suggest that you treat him like a young horse just learning to canter under saddle. You don't want to punish Windy for taking the "wrong" lead, you want to arrange everything so that it will be *easy* for him to canter on the left lead, and *difficult* for him not to.

Try to work him when you're alone in the arena, and don't ask him to do it if the surface is too deep or wet. Work him at walk and trot until he is very well warmed up and is calm and listening to you. Then, at one end of the arena, trot on a big circle to the left — at least a 20-meter circle — and when you've completed your first circle, sit up straight and bring your outside leg back a little (this will put your weight onto your inside seat bone), to indicate to Windy that his outside hind leg should take that first step of canter. Squeeze him with your legs, look up, and tell him to canter. Repeat the aids every stride, keeping your head up, your seat just out of the saddle, and light contact on the reins. Remember to let your arms move with his head, because he'll need to use it for balance at the canter, just as he does at the walk.

Here's a review of the canter aids:

1. Soft contact on the inside rein, just tipping his nose slightly to the inside to ask for a little bend — the bend you already have, since you're on a circle!
2. Consistent contact on the outside rein, because it's still keeping him steady as he trots through the circle.
3. Your inside leg *at* the girth — keep your straight line through shoulder, hip, and heel — to send him forward.
4. Your outside (right) leg comes back behind the girth, to ask him to canter on the left lead.

If you don't get the canter, don't let him trot faster and faster until he falls into it, because even if he falls into a left-lead canter, it won't be a good canter. If he trots faster or takes the wrong lead, ask for a steady, balanced trot, circle again, check your own position and your aids, and ask for the canter again.

Another good exercise is to carry a whip in your outside hand. When you are ready to ask for that left-lead canter (again, trot the circle first), reach back with your right hand and smack him just behind the saddle when he's on his outside foreleg and inside hind leg in trot. His next trot step will be on the inside foreleg and outside hind leg, and the whip will encourage him to use that outside hind leg to reach up under himself and take that first step of a left-lead canter.

Another method that works well is this: If you have hexagonal or octagonal jumping poles (*not* round ones, they're too dangerous!), set up a tiny jump that will be across the path of your circle as you come through the corner of the arena. Set it very low, with the outside end of the pole on the ground and the inside end of the pole about two feet off the ground. Then trot your circle, incorporating the pole, and ask for the canter as Windy hops over the pole. Use your normal canter aids, and put a little extra weight in your inside stirrup. Windy should land cantering on his left lead. If he does, canter on and tell him what a wonderful clever horse he is. If he doesn't, bring him calmly back to trot and try again.

If he tries to canter on the right lead, it means that he'll have to take that first canter step with his left hind leg, and since the pole is higher on that side, he will probably bang the leg. If Windy is a smart horse, you won't have to do anything more than ride him through this a few times and ask for the left-lead canter as he comes over the pole. He'll figure it out, and you won't have to get out of position while he teaches himself to listen to you.

As soon as he gets it right, drop the reins, tell him he's wonderful, and put him away. *Do not* try to "confirm it" by asking him again and again — let him sleep on it. The last thing you do with a horse in any schooling session is the thing that he will remember most clearly. He'll remember that, and he'll do it better next time.

After a few weeks of this, you should be able to use just the pole on the ground, and a week later, you should be able to ask for, and get, your left-lead canter without the pole.

Horse Anxious about Cantering

Q I have a seven-year-old Arabian mare who is wonderful in every way except she doesn't want to canter unless she's on the longe (or thinks she's on the longe!). I've done all her training slowly and gently since she was a yearling, and she's wonderful to ride. She's calm and relaxed out on the trail, walks over anything I ask her to, and is very sensible about everything. She walks, trots, and canters well on the longe and walks and trots happily under saddle.

My problem started the first time I asked her to canter off the longe with me on her back. She took a couple of strides, then bucked as hard as she could.

I managed to stay on and asked her again. She went for a few more strides, and I brought her back to trot before she decided to buck. After several unsuccessful attempts to canter her, I came up with an idea. Since she cantered well on the longe line, I put a knowledgeable rider on her back and longed her at the canter that way. She went happily around. Then we changed places and I got on and she went well. Then my friend took her off the longe (but I think my mare thought she was still attached) and I asked for a canter, staying on the same circle. She went happily.

I gradually increased the size of the circle until we were going about halfway around the arena. A couple of times she started to get a bit worried and acted like she was going to buck, so I spiraled back in to our smaller circle until she relaxed again and then we worked our way out again. Eventually we were going almost all the way around the arena. She was happy and relaxed with her little ears pricked and her canter was smooth and wonderful to ride. That was in December. We were only able to do this once before the snow fell.

Cantering around the entire arena can be intimidating to a horse that is used to working on a smaller circle.

Yesterday we tried it again for the first time since last year, on the longe with me on her back. She bucked twice. I told her "quit" and we urged her on. She became relaxed and happy and went along fine after that, although we didn't take her off the longe at all. We stopped on a good note.

I'm wondering what I've done wrong in her training? Could I have gone too slowly, not introducing a rider at the canter as soon as I should have, and caused her to become set in her ways? She seems to need the security of someone on the ground as if they were longeing her. Not very practical! How can I gradually make her happy about cantering off the longe? Her teeth are fine and her saddle fits well, so I know those aren't the problem.

A Good for you, checking saddle and teeth — you must have known that those would be the first things I would suggest. Horses in pasture, running downhill, will buck to retrieve their balance when they find themselves unbalanced and moving too quickly. Some horses also buck when they're anxious; this wouldn't be the first Arabian mare I've known to have that particular reaction. Your mare's anxiety probably comes from her confusion about what you *want* her to do and her doubts about what she *can* do.

It's very revealing that she's happier with someone longeing her. It's even more revealing that she's most comfortable working on the smaller circle, even without a longe line, than going all the way around the arena. She is balanced and at ease when she's on that smaller, more familiar circle. Horses can indeed develop certain habits, and your mare may have developed the habit of feeling that small circles are "right," and larger circles and longer, straight-line canters are "wrong."

There are several things you can do about this. First, you're absolutely on the right track with your spiraling exercise. Spiral both in and out, with the goal of making your "default" circle bigger and bigger. Your goal here is to help her understand that she *can* keep herself balanced on larger circles.

But circles aren't the whole story. A circle is easy for a horse to understand, because every part of it is just like every other part of it. If a horse can canter comfortably around one part of a circle, she can get around the rest of it. Riding around the entire arena is different! Instead of a circle with every step the same, you're asking your mare to canter a long straight line, a turn, a short straight line, and another turn. And those turns, especially in arena corners, are sharper and much more unbalancing to the horse. I'm sure that your mare can get through them, but *she* may not know that yet.

Instead of riding around the arena in the "correct" way (using the corners), try riding around it in the way that drives dressage instructors crazy. Rather than going deep into your corners and riding a turn, a straight line past C, and another turn at the short end, leave the long side early (before K) and ride half of a 20-meter circle, catching up with the next long side after F. This should take care of her (and your) anxiety about balance, turns, and bucking.

Do many transitions between trot and canter. Whenever she picks up the canter, even if she's on the wrong lead, even if she's counter-bent, praise her, sit up, and loosen the reins a little. Horses *will* buck in canter if they need a longer rein or if the rider's hands and arms don't follow the horse's movement. Praise her for *any* canter, let her canter on, and then after 10 or 12 strides or whenever you feel her hesitate as though she's thinking about bucking, ask for a trot. Praise her when she trots, but make her *work* at the trot.

When you ask for trot from canter, be sure that you aren't pulling her back. Instead, think about sending her *forward* into the trot, so that you're going from a relaxed, rocking-horse canter into a big-moving, okay-*now*-we're-going-to-work trot. In other words, you're going to make a brief, relaxed, long-rein, "no demands" canter her reward for working hard at the trot. This will accomplish several things:

1. It will give both of you a different outlook on trot and canter.
2. It will help her (and you) relax, as you will be very clear with your asking aids, with your allowing aids, and with your praise, so she won't need to wonder if she's done what you wanted.
3. It will help her balance, because good, frequent transitions are brilliant balancing exercises.

If she canters and feels herself, you, or herself *and* you becoming unbalanced, and she bucks to retrieve her balance, after which you immediately drop to a trot or a walk, be sure that she *works* at that trot or walk from the very first step, otherwise the lesson she will learn is that you *like* for her to buck. (You must like it, since you reward it — that's how she knows you like something!)

If you are genuinely concerned about your mare's ability to balance at the canter, send her forward into a strong working trot, get her settled, and then ask for the canter again and allow her to relax into it. If you can help her understand that she *can* balance, that you *like* her to canter, and that you expect her to canter when she's asked and then trot when she's asked, you'll be able to make progress. She will quickly learn that the canter can be a relaxing break from that work.

One more thing: Watch yourself carefully when you ask her to canter. On the longe, we riders tend to be more careful about our position: We sit up straight and stretch our legs down and look up, because the fact that we're on a circle makes us think more about our *own* balance. Sometimes when we're off the longe and are asking for a canter down the side of an arena, we rock our bodies forward or "pump" with our seats. This is enough to unbalance a horse and make her uncomfortable, with predictable results.

And finally, I know that you've checked her saddle fit, and the fact that she canters nicely on the longe, even with a rider, tells me that the saddle isn't too far forward, but here's something to consider: Some saddles fit adequately but not wonderfully, and the horse can be comfortable when the rider is sitting up straight, but when the rider leans forward (as many riders do when anxious), the shift puts painful pressure on the horse's back under the saddle, about a hand's width behind the pommel. If this is the case with your mare, you'll notice that the bucking occurs only *when your weight shifts forward.* You may find that paying attention to your position at all times helps considerably.

Older Horse with Flat, Fast Canter

Q My horse is 14 years old. He is a Shire-Thoroughbred cross, and he was a foxhunter until about two years ago. His owner died and the horses stood around in a pasture for about a year before my trainer bought this one for me. Hero has a great walk and trot, and he is a great ride until I want to canter, and then he's fine for the first few minutes but after that everything falls apart. He has good balance and his canter usually starts off okay, but then it gets faster and faster and flatter and flatter until he is basically just running. He starts off relaxed and then becomes tense, and then I get tense because he is a big horse and when he starts running he can pull me out of the saddle if he puts his head down. I don't usually think about his size and my size (he is 17.1 and I am 5'3") except when he does this.

My trainer won't let me ride him in draw reins to slow him down and bring his head in. She says the answer is "lots of transitions" but that's what she says about practically every riding problem there is. I don't think that a million transitions are the answer to everything, and my horse doesn't think that either. I know because I have tried staying in the arena and doing trot or canter at every letter or every other letter, nothing but transitions, and it doesn't help, he just puts his head down and pulls more. What can I do to convince Hero to be more relaxed about cantering and not always want to go fast and flat?

A I'm afraid your instructor is right about the draw reins, and, sorry, she's also right about the transitions. However, the way you do those transitions matters even more than how many thousands of transitions you do. You've probably heard the expression "practice makes perfect," but in reality practice just makes permanent, so the *quality* of your practice matters very much.

The exercise you've been doing, making transitions at the letters, is something that will probably be more useful to you later on. Arena markers and letters can be good reminders if you tend to forget to do transitions, but they can also be distracting. It's easy to become so focused on the letters and so busy thinking "Here's another letter! Must change the gait!" that you forget why you are doing transitions in the first place. Your goal isn't to get a fixed number of transitions during each pass around the arena, it's to get prompt, smooth, flowing, *upward* transitions whether you're going from trot to canter or from canter to trot.

Each good transition from trot to canter builds your horse's strength and balance.

There are several ways of helping a horse relax into the canter and relax *at* the canter. Here are some that should work well for you:

Exercise 1. Work at trot, relax at canter. Since your horse has a good trot, begin by working at the trot and doing transitions within the trot itself. Start with a good working trot, ask him to stretch forward and down into a longer, more relaxed trot stride, and then slowly bring him back to his working trot. From there, ask for a little more drive from behind and a little more elevation in front, so that his working trot becomes a little more rounded, more powerful, and more elevated, and his strides become a little higher and a little shorter. Then let him stretch out to his working trot again.

Continue alternating, going from working trot to long relaxed trot to working trot to a shorter, rounder trot, and so on. Don't worry about how many strides of each trot you get, just focus on the quality of each trot and the way your horse can flow from one sort of trot to another. As you both become more proficient at this exercise, lighten your aids and continue to lighten them until you can lengthen his stride by closing your hip angle slightly and allowing an extra inch with your hands and arms, and you can get a shorter, rounder stride by sitting up straight and tall, pulsing your calf muscles against his sides briefly, and closing your fingers. When this exercise has become easy for you both, and you are barely whispering your aids and he is responding quickly and enthusiastically and energetically, *sit* that rounded, collected trot for a stride or two and ask for the canter.

Exercise 2. If the canter begins to speed up and flatten out, send him forward immediately into a strong trot, and trot circles, figure eights, and serpentines until he is once again relaxed and moving in balance, with his back lifted. At that point, ask for a canter, and allow him to canter a circle or two, or a figure eight with a simple change in the middle. If he becomes strong again, trot forward energetically and go back to *working* at the trot. When he's balanced and relaxed and listening again, ask for the canter, and so on.

Flying Changes: Timing the Aids

Q Kyra Kyrklund says to give flying change aids during the moment of suspension of the horse's legs in canter, but I wonder if the rider is really supposed to give the flying change aids just an instant before the horse's moment of suspension to give time for the horse to change during the suspension.

A I'd have to agree with both of you on this! I'm not being a weasel, really. The difference you've described may actually be partly a matter of semantics and partly a matter of perception: I think that if you watch Kyra's riding closely, you will see her shift her position and give her aids just as the moment of suspension begins. It's the best time to ask — in fact, it's the *only* time to ask. By asking earlier, you will be asking the horse for something that he isn't yet in a position to give you.

Think about the definition of a flying change: It's nothing more (or less) than a canter transition within the canter. From a collected canter, if the horse is to change cleanly, front *and* back together, to a collected canter on the new lead, it *must* make that change during the moment of suspension.

Like shoulder-in, flying changes require collection, but doing flying changes dressage-style requires a much higher degree of collection. If you want good flying changes, you need an educated, athletic horse who is truly on the aids and who is both willing *and able* to maintain his collection and impulsion and make the shift instantly. If you're going to work on single flying changes, on the rail or on the diagonal, you need a horse who is strong, supple, engaged, straight, and *listening to you*. If you are going to work on a series of changes, you need those qualities to an even greater degree.

A flying change is just a canter transition within the canter.

A horse learning his first single changes will need practice to manage a complete change during the period of suspension, and will need to be praised and encouraged to go forward and relax after each change. By the time you are asking for tempi changes, your horse should have had a lot of practice doing single changes, and have developed excellent, almost instantaneous responses.

There's another factor in play here: the rider's body language. Fit, conditioned, attentive horses can perceive incredibly subtle signals, including the involuntary muscle movements that their riders make when they *think* about (for example) a flying change. Part of keeping a horse from anticipating test movements, especially at Grand Prix where the horses do the same tests all the time, involves the rider controlling her own body. Riding at this level demands that we have the same fitness, coordination, and neuromuscular reflexes as we expect from our horses. The other aspect of keeping a horse from anticipating movements is *not* keeping him from anticipating them, it's teaching him that even when he knows what's coming next, he must wait for the rider's signal.

Let's look at a left-lead-to-right-lead flying change, step by step:

1. The horse is cantering on the left lead, with a slight flexion to the left, and the rider's position matches the horse's flexion — the rider's shoulders and hips, as always, are parallel to the horse's shoulders and hips.

2. As the horse takes his last canter stride before the change to the new lead — right hind, left hind and right foreleg, left foreleg — the rider half halts, straightens herself and her horse, and immediately asks (quietly) for a new flexion, this time to the right.

3. The rider shifts her position (again, quietly) to match the horse's new flexion: The rider's outside leg comes forward and becomes the new inside leg, and the rider's inside leg moves back and becomes the new outside leg.

4. The new outside leg asks the horse's left hind leg to initiate a right-lead canter strikeoff — and at this point, the rider's shoulders and hips are once again parallel to the horse's shoulders and hips.

Your point is well taken: The rider *does* ask for the change just at the beginning of the moment of suspension, so that the horse can change *during* that moment. Given the horse's ability to perceive subtle, involuntary movements on the part of the rider, it's reasonable to assume that the request for a change effectively begins when the rider prepares to ask. Of course, it takes much longer to read this than it does to perform it — we're talking about fractions of seconds, so in practical terms, it may be a distinction without a difference.

CHAPTER 7

Training over Fences

JUMPING IS WONDERFUL FUN, and cross-country jumping — whether at an event, out hunting, or just on a good outside course — is the most fun of all. Good, steady, predictable, accurate jumping is the result of good preparation on the flat. Many jumping problems are really just consequences of holes in the horse's training. By the time your horse is ready to begin his jumping training, he should be strong, flexible, balanced, confident, and easy to rate at all three gaits.

It's a good idea to teach your horse to jump, even if you never intend to jump him competitively and even if you're preparing him for an entirely different career.

Jumping Training and Rider Position

Q I purchased an eight-year-old bay gelding (16 hands) off the track two years ago. He was very uptight because his previous rider had slapped him in running reins and schooled him silly. I hacked him for six months and for the next year did very basic dressage and small jumps in a fun atmosphere with absolutely no pressure. Because he had a history of a bad back in racing, I have done a lot of hill work to build up his back, which seems to be fine now. I have started more serious work recently (3' to 3', 3" fences and basic dressage). He has settled down a lot but remains an excitable horse that is very sensitive to aids.

Given his previous back problems, I am wondering whether the seated jumping position is a good idea. I was taught the forward seat as a child but have trouble seeing a stride like that. My horse is *very* forward going, so the seated position seems to give me more control as I can hold him with my legs. He has an extremely bumpy canter that makes it difficult to sit very quietly all the time. Another reason I prefer the sitting position is that going into the jumps he throws his head up a lot even if my contact is soft. In the sitting position, I am not hit in the nose. Which is the better position for jumping, or is it a personal thing?

A Your six months of hacking were well spent and a very wise decision, congratulations! You may, however, need to do more. If you were working with me, I would want you to spend another three to six months doing systematic training on the flat without any jumping at all, and then continue that training while gradually adding gymnastic jumping exercises to the program for the next three months or so. Only then would I even consider taking the horse out and jumping a course. You've done an excellent job of calming this horse and helping him learn his new job, but he needs more time.

The answer to your question (full seat or half-seat?) is *both*. You should be seated unless you are galloping down a long straightaway, in which case a two-point (half-seat) is appropriate. Sit a few strides before each jump, and allow your horse to lift you into a two-point position as he takes off. The rest of the time, a full seat is correct, and much more secure.

Your saddle figures into this. If you are riding in an eventing or all-purpose saddle, sitting should not make the horse uncomfortable; in a close-contact, forward-seat, show-jumping saddle, the horse *will* become uncomfortable if you do a lot of sitting, because that saddle is designed to balance a rider in a forward seat and a half-seat, and is really not meant for much sitting or for lengthy riding sessions.

You seem to be doing a good job, and I think you will have a very nice horse if you will put more time into his basic training. I suggest that you focus on progressive flat work for several months at least, until your horse is relaxed and supple at walk, trot, and canter; is relaxed during transitions (both between and within gaits); and will stop easily and stand quietly on a long, or better yet, a loose rein.

You won't lose time by taking time. Horse training is an odd process; if you take your time and focus on building the best possible foundation, the rest of the work will come quickly and easily. If you rush through the early stages, you won't get far. It sounds paradoxical, but in training, you will arrive at your goal most quickly by going the *long* way.

Teaching a Horse to Jump

Q I need to understand the basics of educating a young horse to jump. I have a 14.3-hand, four-year-old purebred Crabbet Arabian gelding. I am definitely not jumping him for at least another year, as he is expected to reach 15 hands and is growing quite slowly. I do ride him, and I think he's going okay. Shariq is a very calm horse, contrary to the reputation of Arabs, particularly young ones, and I have never had a problem with him. He is quiet and trustworthy, and I am very confident on him.

When I bought him a year ago, I was told that he had jumping education. But when I have faced him with a trotting pole, he is very shifty going toward it. His approach is crooked, and he stops and starts and often refuses to go over the pole. He seems confused, which makes me wonder if it is possible that a year out of jumping has caused him to forget. Is it more likely that he never had any jumping training to start with?

If I do find that Shariq needs to be retrained for jumping in a few years' time, how do you think I should go about it? Being a beginner myself, and having no training experience (my jumping skills are not very advanced either), I don't think I should try to train him for fear of making a mistake and causing a bigger problem. Do you think I should have him professionally educated to jump?

A It sounds as though your young horse has no experience with either poles or jumps. If he were a trained jumper, being asked to walk over poles wouldn't confuse him. If he had been jumped badly or inappropriately, he would probably try to avoid anything resembling a jump, including poles on the ground. What you've described is a horse reacting to something that is not frightening, but that *is* unfamiliar.

Be grateful. He's a very young horse, and if he *had* been jumped as a two or three year old, I wouldn't want to bet on his long-term soundness. Jumping experience is also a variable commodity. Some horses with jumping experience "know" that their job is to run toward an obstacle, lurch over it while the rider water-skis on the reins, land painfully, and race away.

You are sensible to look for help for him and for yourself, and wise to put off his jumping education until he is older and more mature, and until both of you are more proficient at your flat work. You have plenty of time! With every new horse, regardless of his age and (alleged) experience, it's best to start him from the ground up and teach him exactly what you want him to know. He will then

Encourage your horse gently, but let him find his own way over the jump.

behave in the way that you prefer and will stay sound and happy. Here are my suggestions for teaching your horse to jump:

1. Find a good, patient instructor who can help you with flat work and eventually with jumping. The better a rider you are on the flat, the more easily you will learn to jump and the more quickly you will become proficient.

2. Spend at least a year focusing on flat work and helping your horse develop correctly. He will become stronger, more supple, and more responsive while you improve your own skills. The time and money involved will be an investment in his soundness, your security, and your mutual future.

3. When your horse is five, evaluate him carefully and decide whether to begin his jumping education or wait until he is nearer physical maturity. At the same time, ask your instructor to evaluate *you*.

4. You should learn to jump when you have good control of your own body and are reasonably proficient on the flat. Your mount for jumping lessons should be an older, more experienced horse — perhaps a horse belonging to your instructor.

5. Meanwhile, your instructor can help you introduce your young horse to the basics of jumping. The two of you can learn your jumping skills separately before you are reunited by the instructor. When you and your young horse go over your first real jump together, he should be calm and confident and you should have enough experience, and enough confidence in your ability, to help him or at least to stay out of his way.

Your instructor may be the one to ride your horse over his first single jump, his first combination, and his first line, but by that time you will already have done much of the preparatory work. There's an old saying among good jumper riders: "Jumping is just dressage over fences." It's true. Every bit of flat work you do will have an effect on his jumping, and a good jumping education begins with 100 percent flat work and ends with 90 percent flat work!

If you doubt this, consider the typical jumping round at a competition. There will be somewhere between 10 and 15 jumping efforts, and if you count the strides that the horse takes from the beginning to the end of the round, there will usually be between 100 and 150 strides depending on the arena size and the course design. One hundred strides less 10 strides over jumps, leaves 90 strides on the flat — that's 90 percent flat work.

Easy steering, quick, smooth transitions between and within gaits, straightness, bend, lateral movement, strength, responsiveness, impulsion, engagement, the ability to lengthen and shorten stride effectively and immediately — all of these are needed for jumping, and all are created through work on the flat.

There's also a psychological side to this. When you jump a course, you and your horse are exchanging the role of pilot. On the approach to the fence, *you* are in control. After landing, *you* are in control again until the next fence. From takeoff to landing, though, the *horse* is in control, and your job is to stay in balance and not interfere with him. This is more difficult than it sounds. Humans generally want to be *doing* something, but learning how and when to hand the controls over to the horse and to wait quietly is an essential part of good jumping. This, too, is something you can learn on the flat, and then practice over jumps during lessons on a more experienced horse.

Find the right instructor, invest some money in lessons, and invest your time and effort in good practice between lessons. By the time you and your horse are jumping courses happily, you will both be fit, strong, competent, and confident, and you will have an enormous amount of fun.

Jumping Terrifying Ditches

Q I am a 42-year-old rider with a nice young horse that I bought for eventing. When I was a teenager, I did mostly hunter shows and small jumper courses at schooling shows. I love the idea of learning and working on dressage and cross-country as well as stadium jumping! So far I have entered three small unrated events just to get an idea of where Alber is in his training and how much of what I remember about jumping will apply to cross-country.

There are some big differences between the kind of jumping I used to do and the things we have to do on a cross-country course! Stadium is easy for Alber, and I think he enjoys most of the cross-country course, except for one thing: terrifying ditches! At two of the events, there were ditches to jump, and Alber did everything possible to avoid them. He will leap over a large coop without a thought, but even a tiny ditch is too much for his brain. One ditch had no depth to it at all, it looked as if someone had left a telephone pole lying on the ground for a month or two and then moved it somewhere else, and the "terrifying ditch" was the mark the pole made on the ground — not exactly intimidating. But Alber freaked out when he saw it.

Fortunately it was just a little schooling event, and I was allowed to ride the rest of the course (he did fine) and come back later and try the ditch again. I still couldn't coax him over it until two little girls on ponies went over it about 10 times, and we finally followed them. I really want to continue eventing with Alber, as he is fun to ride and a good horse overall; it's just those darned ditches! How can I train him to go over a ditch instead of performing a frantic dance to get away from it?

A You've already discovered one way to move him over a ditch, which is to have another horse give you a lead. You can certainly practice that at home, but it's not very practical at competitions. At home, you can free-school or longe him over tiny ditches. Almost any mark on the ground will be noticed by a horse. Ditches can terrify horses, who can't tell how deep they are or what sort of horse-eating creatures might be lurking in them waiting to pounce.

Sometimes horses fear ditches because without realizing it, the horses' riders are sending the message that ditches are frightening. If you become tense and apprehensive on the way to a ditch, your horse has no way of understanding that you are worried because he might not jump. He feels your worry, doubles it for himself, and blames it all on the ditch itself. Riders can create problems *at* ditches by leaning forward suddenly (a sign of fear), holding their breath (a sign of fear), and looking down into the ditch (a sign that the rider is planning to stop in or just before the ditch).

If you are worried about a ditch, your horse will become convinced that it is concealing a predator.

You've heard of drawing a line in the sand? If your horse is afraid of tiny ditches you can literally draw a line in the dirt, call it a ditch, and practice riding over it. If there is any chance that you are causing some of your horse's fear, this "ditch" should improve both of your lives. Since *you* aren't afraid of a line in the dirt, you'll have the courage to approach it strongly, look up over it, and ride forward. Your job as Alber's rider and trainer will be to send him forward to (and, we hope, across) the ditch. This involves disciplining yourself first and your horse (gently) second. First, plan your approach to and your trajectory over the ditch. Ask yourself:

1. Where you will begin your approach
2. Where exactly you plan to jump it
3. What *you* will do when you get there
4. Where you will go after you've jumped the ditch

These are not trivial questions, and you *must* know their answers before you begin trotting toward the ditch. You should put a neck strap on your horse — an old stirrup leather will do nicely. You may need this later.

Plan your approach so that you will have time to establish a balanced, energetic trot, then make a straight approach to the ditch.

Aim for the center of the ditch — however long it is, aiming for the center will minimize the possibility of a run-out.

Keep your legs on your horse as you approach the ditch, ready to squeeze and encourage him forward if he hesitates. Keep your eyes *up*, looking *over* the portion of the ditch that you intend to jump, and looking *at* the place where you intend to go after the jump. If he jumps it, he'll land cantering — canter wherever you were planning to go, patting him and telling him that he's a prince.

If he doesn't jump it — if he stops — just stay where you are, directly in front of the "ditch," looking *over* (not into) it. Keep your leg on, and ask him to go forward. Don't look down; remember, this isn't really a ditch, it's a mark on the ground and you're quite safe to insist that he go forward because it truly does not matter where or how he takes off in front of the obstacle so long as he goes forward over it and gets to the other side. This is where the neck strap comes in: Hold it, or be prepared to grab it in a hurry. If your horse makes up his mind very suddenly and launches himself into the air over the "ditch," you must *not* catch him in the mouth no matter how suddenly or awkwardly he jumps.

When he lands, give him a loose rein, an encouraging leg, and praise and pats for being such a fine fellow. Take a moment to let him enjoy the praise, and then begin again. This time, he may have the confidence to trot to the "ditch," jump it, and canter away. If he does, send him forward, praise and pat him. If he stops, do exactly what you did before. He will probably take considerably less time to make up his mind this time. Continue as above until he teaches himself that jumping the ditch is a clever and profitable thing to do.

When he is trotting the ditch confidently, try approaching it from the other side, then at a slight angle, and continue to build his confidence in himself and you.

When you return to the ditch in a day or two, school it a few times at the trot. When you feel confident that he will trot to the ditch and jump it every time, begin schooling at the canter. Plan your approach: Establish a round, bouncy canter, then come in on a straight line that is long enough to allow your horse to adjust his stride if he needs to do so. As before, encourage him forward, look up and over the ditch, leave his mouth alone, and reward him when he jumps.

Is Eventing Useful Training for Show Hunters?

Q I ride hunters, which is what I've always done since I learned to ride. I love it. I am 17 years old now. My trainer is very good and has worked with my horse and me for almost three years. We do very well at the shows. Now my trainer has this idea in her head that I need to do some eventing! She took some lessons last year, went to three events, and says that she'll keep doing it once in a while because it's really good for her training and her horses.

She wants me to go to an event with my horse. I don't know if I want to do this. He is very good at being a hunter. Won't it confuse him if we do something so different? I really like my trainer a lot, and usually she is right about things but sometimes not, and I think this may be one of the "not" times.

A If you really don't want to try eventing, tell your trainer. Be honest with her. You've worked together for three years and you have a good relationship, so talk to her about your concerns. I'm hoping that you'll decide to give eventing a try. I wish that all hunter riders could try eventing. At the lowest level, it won't be frightening or terribly demanding, but you'll need skills that you may not have developed for the hunter ring. If you foxhunt in addition to showing hunters in the ring, you'll already have most of those skills. If not, keep reading.

Since you show hunters, you probably have style and poise on horseback. You've learned how to find your best spot for each jump, you're good at getting your horse to take the lead you want after each jump, and you've used those skills in both indoor and outdoor arenas. If you're very lucky, you have ridden over an outside course. If you enjoyed that experience, you will probably love eventing.

Eventing is great fun because, as they say, "Eventers do it Three Ways in Three Days!" For the price of one, you get three different competitions: dressage, cross-country, and show-jumping. At the lowest levels, all three phases may be done in one day and out of order — dressage, show-jumping, and cross-country, instead of the traditional order of dressage, cross-country, show-jumping. The order won't matter if you're giving eventing a try as a means to improve yourself and your horse. Make no mistake: eventing *can* help you improve! Preparing yourself for eventing means learning and practicing different skills, all of which will help you and your horse. You don't have to become a full-time eventer, but do take the discipline of eventing seriously enough to learn and apply the skills in a safe way.

The show-jumping phase won't be very different from the work you already do. If you want to practice making tighter turns and clearing odd-looking, colorful jumps, you can rearrange your own jumps and hang towels over them to make them look odd and unfamiliar.

On the cross-country course, your horse has no idea what is on the other side of the jump.

The dressage phase will improve your accuracy and precision. To do well in dressage, you'll have to be able to make your transitions smoothly *and* at a precise point marked by a letter. There's a big difference between "pick up a canter, then circle," and "canter at A," or even "canter between K and A." You'll be riding every stride in a way that may be unfamiliar to you, but it's a good thing to be able to do.

The cross-country phase — the heart of eventing — is probably the most difficult for riders whose only previous experience has been in the hunter ring. When you ride cross-country, you're alone and out of your trainer's sight. Unless you're an incredibly fast runner or a very clever strategist, you won't have had the chance to see horses jump all of the jumps on the cross-country course. You definitely won't have had the chance to watch one specific horse go over the course. Your horse won't have seen *any* of the jumps in advance — there's no schooling over the course or hacking on the course before the event. Every jump will be a surprise to your horse, and it will be your job to prepare him to deal with each jump by setting him up in advance.

Setting a horse up on cross-country isn't a matter of counting strides and asking for takeoff *now*. It's a matter of finding a comfortable, ground-covering pace at which your horse will rate easily, so that he can compress or stretch without great effort and without changing his rhythm or tempo. It's also a matter of riding over terrain! If you're accustomed to flat, groomed sand arenas and have limited experience jumping on grass, a cross-country course can come as a shock.

A horse that jumps 3' 6" easily in the ring may lack the strength and balance to canter up a hill, jump a 2-foot fence, and canter down the other side of the hill. A horse who sails over a solid 3-foot jump in a flat arena may not understand how to jump the same fence set up on the side of a hill. A horse that lengthens and shortens his stride easily in an arena may find it difficult to do those things while cantering uphill or downhill. Horses need to learn and practice these skills, so prepare by riding out, riding up and down hills, shortening and lengthening stride on those hills, and *jumping* before, on, and after hills. You'll need to learn a slightly different position for jumping — not as forward, but a little more upright and slightly behind your leg. And you'll need to learn to trust your horse in some new ways.

Eventing helps develop horses' minds along with their bodies. A horse who takes off perfectly when you've set him up for each jump in the hunter ring is one thing; a horse who can find his own best takeoff spot for each jump on the cross-country course is quite another. Cross-country jumping requires a different sort of partnership, one in which your horse must sometimes think for himself. You'll get into trouble if you insist on hand-riding to each jump, counting strides, and asking for a specific moment of takeoff. Instead of dictating the number of strides and the precise moment of takeoff, you ride the horse forward and round so that he can make the takeoff decision himself in the last few strides. You can't be overcontrolling, but you also can't race toward the jump and throw away the reins! When you're beginning to event, a good rule of thumb is to approach each jump at a controlled, round and bouncy canter, then add a little leg for the last three strides. This allows the horse to meet the jump on an increasing stride, in balance, with plenty of energy and power, and the ability to make any necessary adjustments while you sit quietly.

While you're learning to allow your horse to do what he needs to do, you'll still be exercising your brain, but in a different way. As you go around the course, you'll need to "think your way around" and remember everything that happens so that you can replay the course in your mind later and analyze each moment. If you have some bad jumps and can't remember which ones they were or why they were bad (approach? footing? light? distractions?), you won't be able to figure out which of your skills need more work, and you won't be able to prepare your horse better the next time. In the hunter ring, your entire ride — every approach, jump, and landing, as well as every lead change, turn, and shift in position — is visible to the judge, to the spectators, to your trainer, and perhaps also to your parents with their video camera. The second you move through that out gate, you're told exactly what you did and how it looked.

Preparing yourself for eventing means learning and practicing different skills, all of which will help you and your horse.

At an event, it's very different. You may have friends with cameras posted at a few key fences; your trainer may watch you over the first few, or the last few, jumps, but the only one who *knows* how you and your horse took every jump is *you*. If you can't remember what happened at fence five, or whether your horse actually stopped at the water or just paused to regroup before jumping, you'll only have the jump judge's record to go on, and that won't tell you *how* you jumped, just whether you did or didn't jump. Eventing makes you train yourself to notice everything, remember everything, and analyze everything. That habit will stand you in good stead the rest of your life, no matter what kind of riding you do.

What's a Five-Stride Line Exercise?

Q I went to my first jumping clinic last week and it was the most fun ever!
I'm learning to jump and I just love it, and I think my horse loves it, too.
He knows more about it than I do, but we have a lot to learn together. It was
only a one-day clinic, but I learned so much. I learned how to do a release so
I wouldn't pull my horse's mouth, and I learned to fold up my body instead of
standing up in my stirrups, which I was doing before the clinic (bad me). We did
some exercises with a cross-rail and another cross-rail and then three of them in
a bounce pattern.

When it was over the clinician said that he wished he had more time because
we would all benefit from doing a five-stride line, it's such a great training exercise, but he had to catch his plane. I've been practicing the clinic exercises at
home, and they're still a lot of fun but I really want to try the one he mentioned.
Can you tell me how to do a five-stride line exercise and why it's supposed to be
so good for training?

A Your clinician was right — that's a great exercise, very useful when you're
training a horse to jump, and very useful when a rider is learning to jump.
The five-stride line exercise will help you become more familiar with your horse's

All you need is two poles set five strides (72 feet) apart.

stride and improve your ability to adjust his stride. It will also help you learn how to approach a jump at a speed and in a balance that will let your horse meet the jump correctly without needing to chip in or jump from a long spot. Here's how I teach it:

1. Set up two jumps five strides apart (72 feet). The traditional design involves a small vertical and a small oxer, but you can set up the line as a small cross-rail to a small vertical, or even as a small cross-rail to a slightly larger cross-rail. Even ground poles will do. Keep the jumps low and inviting, because you'll do this exercise many times and you don't want to stress your horse's legs.

2. Canter to and through the first jump and continue to canter until your horse jumps the second jump. Keep cantering after that, because you want your horse to learn to stay balanced and remain in a regular, relaxed rhythm into, through, and after the line.

3. The first few times you go through the line, canter in, steer for your jumps, and see what happens. The striding may not be a perfect five, but that's okay — your horse can add a stride and pop over the second jump if he needs to. During those first few times through, feel your horse's movement and notice whether he seems to be going a little too slowly (or a little too quickly) for comfort.

4. Once you've gotten used to the jumps and the strides between them, try approaching the first one with a little more speed or, if your horse was a bit too speedy the first few times, with a little less speed.

5. Experiment! Send your horse forward and notice whether the striding becomes more even and smooth; sit up a little straighter, asking your horse to come back to you, and notice what happens to the striding between jumps. When your horse comes through the line with energy, are you feeling a landing, five even strides, and an easy jump over the second obstacle? When your horse comes through the line with less energy, do you begin to worry that you'll have to take off too far from the second jump or does your horse "chip in," putting in a sixth, short stride just before the second obstacle?

The obvious point of this exercise is to canter the line making five smooth, even strides between the two jumps. Less obviously, the exercise helps you learn how your horse's stride feels when he is speeding up a little, when he is a tiny bit slowed-down, when he is reaching, when he is holding back, and when he is "just right." This, in turn, helps you develop your ability to create a balanced five strides between jumps by riding your horse a little more (or a little less) forward through the line. When you can recognize what your horse's stride feels like when he's pushing himself a little, when he's holding back a little, and when he's cantering smoothly through the line, you'll have an important tool for your training toolbox: the ability to feel what your horse is doing, analyze how he is striding, and act to increase or decrease his stride as needed.

If you see other people zooming through the line at a mad gallop, just ignore them — they're doing it wrong. This exercise isn't about how fast you can get from the first jump to the second one. It's designed to develop three things: your feel for your horse's stride, your ability to increase and decrease your horse's stride smoothly, and last but not least, your horse's confidence *and your own*.

Need Help with Grids

Q I love to jump, and I think I finally have a sound horse with a good attitude that I can train to jump. She is willing but not rushy, so she is not a scary horse to jump, though she tends to knock down rails with her hind legs. My problem is that I need some help training her. I don't have anyone to work with or anyone to help me. My husband will set fences (sometimes), and my son will help him (rarely), but neither one of them has a clue about jumping, and I

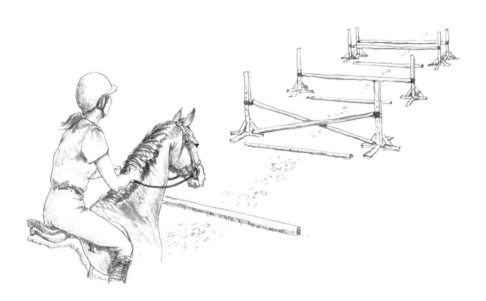

A well-designed grid is an excellent teaching tool for you and your horse.

don't know enough to tell them where to put jumps. I jumped when I rode before attending college, but that was almost 30 years ago (eek!) and now I am riding again after all those years.

I had a great instructor in those days, and I can remember him sending us through a lot of grids. He was big on grids for training horses to jump and training riders to ride horses over jumps. So now here I am with a nice horse that could probably learn to jump, and I need help with grids. My mare hangs her front legs sometimes, and I'm sure I remember my instructor setting up grids to train horses to snap their legs up better. Where should I set the fences and how high should they be? I think grids are supposed to help the rider stay out of the horse's face, but what do they do for the horse? And I can remember the same instructor working us on a bending line between jumps. Would that be a good thing to build into a grid?

A To answer your last question first, a bending line does not belong in a grid. Bending lines are all about the rider's lateral and longitudinal control of the horse between fences; grids are all about setting up a series of jumps with preset striding so that the horse can go through and figure out what to do without input from the rider (meanwhile, the rider can use the grid to practice timing and her ability to fold down close to her saddle at just the right moments). Grids should always be composed of a straight line of jumps.

Grids are wonderful training devices, but be sure that you build them accurately. Don't get creative with the distances. A well-designed grid can do a great deal to teach your horse about jumping and to increase his confidence; conversely, a badly put together grid would be traumatic and physically dangerous for a horse. If your mare is very forward, a grid with more elements can help back her off a little and encourage her to think her way through the grid and jump more carefully instead of racing through it, pulling her legs up out of the way at each jump.

Unless you have someone knowledgeable on the ground to help you, don't tweak the distances, which are based on an average 12-foot stride. If you do have someone to help, you'll be able to do some constructive grid tweaking. Horses who tend to canter on the forehand and jump flat can benefit from slightly shorter distances between the elements of the grid: This encourages them to canter with slightly more elevation and to use their hind ends more when they reach the jumps. Horses who tend to rock back and overjump as if each jump were two feet higher than it actually is can benefit from slightly longer distances between jumps; this encourages them to canter and jump with slightly longer, flatter strides.

You mentioned two specific problems that you'll need to work on when training your mare: hanging front legs and a tendency to hit rails with her hind legs.

If you can get some good help from someone who can set fences for you *and* give you the benefit of a good, accurate eye, then you might find it beneficial to shorten the distances between your jumps slightly, to help your mare shift her weight back onto her hindquarters. This will have the effect of lightening her forehand a little, making it easier for her to lift her forelegs — in other words, it should help her sit back and power over the jumps instead of cantering flat and diving over the jumps. But please don't do this unless you have help.

The other problem is one you can work on alone, and it may not even require any retraining for your mare (but perhaps just a little for yourself). If your mare is knocking rails with her hind legs, her muscles may be tight somewhere. If her muscles are *not* tight, then dropping the hind legs over jumps usually indicates that the horse is physically restricted in her neck and back. Horses who hit rails with their hind legs generally do so because they are dropping their backs instead of lifting and rounding them, and horses who drop their backs over jumps are often ridden by riders who hold their horses too tightly or try to "lift" them into the air over jumps. Setting up some grids and letting your mare go through them on a loose rein, with you holding a neck strap, should tell you fairly quickly whether her problem is caused by pilot error.

Take your time, and remember that changing the way a horse jumps means changing not only her understanding about jumping but also her muscles. Whether your mare needs to develop more muscle or develop some muscles differently, it will take time for this to happen. If you expect instant, lasting improvement, you'll be disappointed. If you understand that reshaping a body through exercise takes time, you'll appreciate the efforts she makes and allow her the time she needs.

PART II

Inside and Outside the Arena

WORKING ON TRANSITIONS

LATERAL WORK:

SUPPLING AND SOFTENING

THE GREAT OUTDOORS

CONNECTION, ROUNDNESS, AND COLLECTION

CHAPTER 8

Working on Transitions

TRANSITIONS ARE SUCH AN IMPORTANT PART of a horse's training, yet they are too often neglected by riders. If you've been training your horse and thinking only about making each transition, teach yourself to think about getting those transitions right. Make quality, not promptness, your first concern — instead of being preoccupied with moving your horse from a walk to a trot or a trot to a canter, pay attention to how your horse gets there. All transitions should be energetic, balanced, and forward. If you keep this in mind throughout your horse's training, you'll be delighted to see how quickly your horse will develop his body and mind, and how strong and responsive he will become.

Horrible Trot-Walk Transitions

Q My new horse is a good mover except when I ask him to go from trot to walk. You would never believe that he has a good walk if you only saw him when he is coming down from the trot. He has a nice trot, and I want to show him in hunter hack classes, but when my instructor says "walk," he doesn't go "trot, trot, walk," it's more like "trot, trot, stop, take a tiny little walk step, then take another little walk step, then do a real walk." It drives me crazy. I'm trying to keep him collected at the trot so he can go into a good walk with the very first step, but I'm not having any luck; my instructor just says that I need more practice. My horse is very athletic so I think I am probably doing something wrong. My instructor has me look up and pull my shoulder blades together so he'll go into walk, and that works, but he *always* stops or almost stops and then does those two little tiny steps before he gets into his real walk.

A You're right — this common problem is one of those "fix the rider and the horse will fix itself" situations. You've described a horse that is being asked to walk while still being ridden at the trot. When you trot your hunter, you ride with a shortened rein, a steady, unmoving hand, and a slightly closed hip angle. When you're ready to walk, you should sit up straight (open your hip angle), use your legs to ask your horse to step under himself a little more deeply, and close your hands for no more than a second before you relax your shoulders and elbows and allow your horse to move forward into walk, *taking your hands with him.* That's the secret and that's your problem: You're using too much hand and not enough leg.

If your legs don't act, you haven't asked your horse to engage and move energetically forward into walk; if your hands act for too long, or act and then don't give, what your horse hears is not "We're trotting, we're trotting, now we're *walking*, we're *walking*" but "We're trotting, we're trotting, we're coming down to . . . a *halt*. No, oops, I've changed my mind, we're walking, hello, that's *walking!*" In other words, your horse isn't giving you what you want, but he's giving you what you're asking for.

To change this and take the "halt" out of your trot-walk transition, you'll need to begin to ride the walk *before* your horse takes his first step into walk. Begin by asking your horse to engage a little more just before the transition, so that the sequence can be: Trot, half halt, immediately forward into walk. In terms of your position, that means going from rising to sitting trot for a step or two, using your legs to ask your horse to engage a little more, closing your fingers *briefly* when you feel him engage to make it clear that yes, you wanted him to engage, not to speed up, and then *immediately* allowing your arms to hang softly so that he can easily move his head forward and down, which is what he must do to take that first step of "real walk."

All of this happens in the amount of time it takes your instructor to say "And walk!" Try saying "Aaaaaaaaaaand" — draw it out long — to yourself and to your horse while you ask for that trot-walk transition. It means: "Get ready, I'm about to ask for something. I'm sitting straight, now engage please, pay attention, go forward, and *here*, your head and neck are free to move, now *walk* please!" The freedom *must* come first if you want a good walk. Think of it as unbuckling your seatbelt and then getting out of your car instead of trying to get out first and *then* unbuckling the seatbelt.

For a smooth transition from trot to walk, let your arms move forward and lengthen . . .

. . . and let your horse reach forward and down with his head and neck as he takes that first step into walk.

Horse Falls from Canter into Trot

Q When my horse is cantering and I ask him to trot, he seems to fall forward into the trot. I experience a clear sensation of falling forward and slowing down, and then he does a couple of steps of something, I suppose it must be a trot but it just feels like hiccups, and then he'll trot forward again. My husband says that Dorje needs transmission work because he has a really rough downshift. I think he is right because that is exactly what it feels like.

I've tried using my shoulders instead of my hands to bring Dorje down to the trot, and I don't think I am pulling the reins, but there's something I'm just not doing right. I end up scrunching my shoulders together or bringing them up to my ears, and he still falls into the trot and I tip forward and bounce around. Is there another way I can ride those downward transitions so that he moves into the trot more smoothly?

A That's an excellent description of a canter-trot transition where a horse loses energy and power, falls into a short, shuffling trot, and then, at the rider's urging, shifts into a more active trot. This is hard on the rider, because it does feel as if you're falling forward, then lurching (or hiccuping — another good description!) before finally shifting into the new gear and trotting on.

I suspect that like many riders, you've developed a habit of thinking of the canter-trot transition as a "downward" transition, and you've developed the physical habit of asking for the transition with your reins and then just waiting for the horse to "come down to trot." That typically produces exactly what you've described: a jerky shift from canter to a short, uneven, bumpy trot, sometimes preceded by an almost-walk or an almost-halt that throws the rider forward and then backward in the saddle with a whiplash effect.

You're absolutely right that fixing the rider is the first and most effective step toward fixing the horse. This is how you can train your horse to go from canter into a balanced, energetic, forward trot:

▸ First, choose the right spot for your horse to make the transition from canter to trot — set him up for success. In an arena, choose a spot on the long side, shortly after the turn to give him a long, clear straightaway. In the field or on the trail, choose a spot where the footing is good and the path is level or, better yet, sloped slightly uphill.

- One or two strides before you reach that spot, check your position. You should be balanced, looking up and out, with your weight shared by your seat bones and thighs, and your lower legs resting against the horse, knees relaxed, heels down.
- When you are ready to trot, bring your shoulders back slightly (back and down, not up around your ears), close your hands firmly on the reins, and when your elbows come back to your sides, *keep them there.*
- At the same time, let your heels drop a little lower and use your calf muscles to give your horse a brief squeeze that says "Forward!" Then give your horse a clear, two-word command: "Trot forward!"
- When he trots forward with energy — which he will — praise him without looking down or dropping the reins. Your message should be, "I like what you're doing, keep doing it," rather than, "Okay, you can stop now."

If you visualize a steep hill at the point of transition and think of sending your horse energetically up that hill at a trot, you'll be thinking "forward and up" and that's where your horse will go. In *that* sense, every transition should be "upward." It will help your horse's training and your own position and balance if you can teach yourself to think in that way.

A horse that feels himself being wrenched down the trot may give a little buck in protest.

Difficult Transition: Trot to Canter

Q I ride a 17-hand, five-year-old Hanoverian mare that belongs to my instructor. The mare is still green because she is so large that my instructor has been taking her training very slowly. She is also a little bit lazy, especially in trot-canter transitions. Right now I am the only one riding her because Sally (my instructor) is pregnant. When Sally watches me school her mare, she doesn't like our slow, sloppy trot-canter transitions. I am doing my best to get the mare jumping forward into canter, but she really is a slug and it takes all the leg muscles I have to keep squeezing her forward at all. There is just no way I can squeeze any harder.

Sally has told me to kick her a couple of times, but this goes against everything I believe, and it also goes against everything Sally has taught me about horse training. I want to be able to train horses for myself and other people, so I don't want to get into any bad habits now. I think Sally is just frustrated because she would rather be riding her own horse. Is there a training technique I could use to install a better "trot-to-canter" button on this mare?

A If this mare is 17 hands at age five, she is going to be enormous. She will probably continue to grow and develop for at least another three years, so it's not quite fair to look at her today and evaluate her for all time.

I'm not in favor of kicking horses, but I have some sympathy for Sally. At five, this mare is certainly ready to do some under-saddle work, and you're dealing with one of the most important aspects of a horse's early training under saddle: the development of a quick, eager, strong reaction to the rider's aids for "Go forward now!" You're going to need to go back several steps and teach the mare that a light pulse of the calf muscle — *not* a constant squeeze or a hard kick — means "Forward!"

If you are squeezing the mare with your legs all the time, then something is wrong. I suspect that because she is large and not so easy to maneuver, you've been using more and more leg to keep her moving forward, and you've inadvertently trained her to accept your leg aids without reacting to them. In other words, instead of training her to respond to the leg, you've been *desensitizing* her to the leg. Eliciting forward movement from a horse should be like turning a tap and allowing the water to come rushing through; the energy and potential forward movement should always be present.

Go back to transitions in and between the walk and the trot, and work to develop a big "Yes, here I go!" reaction to your leg aids. Most of the time, your legs should lie softly against the horse's sides, acting lightly and briefly and only when necessary. Instead of kicking the mare, use the whip to reinforce your light leg aid. Teach her that leg means "Forward!" Make yourself absolutely clear to the horse in a way that will help her training and not block it. Before you resume schooling trot-canter transitions, the mare should have developed a strong interest in going forward and a quick response to the aids. Once those essential elements are in place, you'll be able to school trot-canter transitions correctly.

Those trot-canter transitions may seem like baby beginner work, but they are important, because when the mare is finally able to do them energetically and well, they will be the foundation for much of the advanced canter work, including flying changes, that will come later in her training. What matters now is that this mare learns how to balance and how to work from behind. Your correct work now will make more complex work easy; incorrect work now will make that later work impossible.

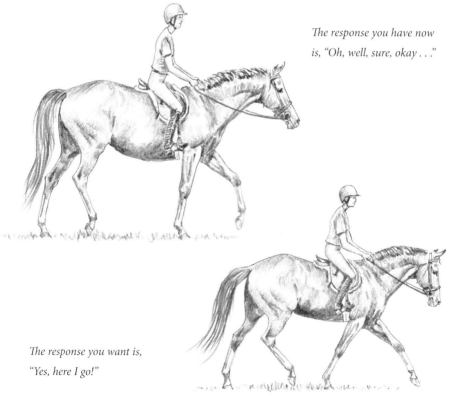

The response you have now is, "Oh, well, sure, okay . . ."

The response you want is, "Yes, here I go!"

The job you are doing is important — the greatest effort in training goes into the first work, the basics, all the things that too many riders dismiss as "easy exercises" or "beginner work." Those riders think they have discovered shortcuts, but they haven't. They will never train a horse to the upper levels, because the holes in their training — the things they rushed through or skipped entirely, thinking, "Oh, we can come back to that later" — will come back to haunt them. Holes in training always do.

Mare Puts Her Head Up

Q Sometimes on transitions up, but almost always on transitions down from canter to trot, my mare puts her head up in the air. I don't know how to make her stop doing this. I have tried holding the reins more tightly during the transition, but she does it anyway. I read your article about giving with the inside rein when asking for a trot-canter transition, and thought it might also work for a canter-trot one, but it didn't, it only made her head toss bigger. Now I am back where I was, holding the reins evenly and hoping that she won't toss her head. Is there anything else that I can do to fix this problem?

A Giving with the inside rein is very effective for the trot-canter transition because the horse needs freedom to step through with the inside hind leg. It won't help a canter-trot transition because what's needed for *that* transition is a steady contact that your mare can move into as she starts the trot and steadies her gait. Make your hands absolutely steady — hook your little fingers through your grab strap if necessary — as you ask for the transition to trot.

You should also check the quality of your seat — do you sit down hard just before asking for the trot? Stay *light* in the saddle. As an experiment, try cantering in a half-seat and remaining in your half-seat during the transition to trot. A sudden bump on the back will cause most horses to drop their backs and toss their heads.

If your mare still tosses her head during the transition, here's one more exercise you can use. The previous ones were for you, to keep your hands steady and your seat soft and allowing. This one is for her, to keep her stepping through and using her topline. It's much easier for a horse to toss her head when her back is stiff and dropped — a horse with a stretched, lifted back is more likely to drop her nose than to flip her head in the air.

If your mare continues to toss her head, instead of wondering how you can stop her with your hands and reins, try this: Just before you ask for the trot, use your inside leg and allow your outside leg to slide back slightly. This is similar to what you would do if you were asking for a canter depart, but in this case your *outside* leg is passive, and your *inside* leg is the active leg. At the same time, bring your inside shoulder back very slightly, and barely increase the tension on your inside rein — you should see her inside eyelashes, not her whole eye. You are creating and maintaining a very slight inside bend that will help keep your mare focused and on your aids.

If your horse's back is relaxed and lifted, he won't flip his head in this manner.

Picking Up the Canter from the Walk

Q How can I teach my horse to canter from the walk? She is four years old and can canter pretty well from the trot although she gets a little strung out, and she canters great when she's turned out. I know she can take the canter from the walk because I've seen her do it from practically a standstill when she's turned out. How can I train her to do it when I'm riding her?

A Your mare is young and the two of you don't have much experience cantering together, so before I explain how to canter from the walk, I'd like to give you some thoughts to consider. Your mare's canter is fine when she isn't carrying a rider; it's not as good when she's under saddle. That could be perfectly normal for a young, green mare; alternatively, it could indicate a problem with the fit of the saddle or the balance of the rider. Ask your instructor, or someone else with a good eye for a horse's movement, to watch your mare cantering on her own and then watch you ride her at a canter. Ask them to consider the following questions:

- Does she pick up the canter easily, moving as if she were going uphill?
- Or does she tip downward, trotting with faster, longer strides until she more or less falls into the canter?
- What is your position (a) before the canter, (b) during the transition, and (c) at the canter?
- Does she canter easily, comfortably, and in good balance?
- Can you sit easily, comfortably, and in good balance; or are you leaning forward, pushing the saddle with your seat, or using your legs constantly to keep her cantering?
- How does she look when she is cantering? Is her mouth relaxed or tight, are her nostrils relaxed or flared, how does she carry her head, and what are her ears doing? What is her tail doing?
- As she canters, does she become more relaxed or more tense?
- What is the sound of her canter — noisy footfalls or nearly silent?

Your watcher doesn't have to be an expert — anyone with a good eye (and ear) and a list can look for and identify signs of tension and signs of relaxation at the canter. If your mare starts out with a slightly tense face and a noisy canter, but then her face relaxes, her tail lifts a little, and her canter becomes quieter as she continues to canter, that would indicate that she is a little unsure at first but quickly relaxes into the rhythm of the gait and canters more easily as she goes along.

On the other hand, if she starts out reasonably well but canters more loudly as she goes along, and shows more tension in her mouth and neck and tail, that would indicate that she becomes less comfortable and finds cantering more difficult as she goes along.

Until your mare can canter cheerfully and easily, she probably lacks the strength to go from walk to canter. You can help strengthen her by working her in ways that use the muscles she will need to take and maintain the canter. Do many transitions between walk and trot, many transitions within the walk and the trot, and many smooth, accurate transitions (quality counts!) between trot and canter. When she does those things easily, begin spiraling her in and out at the canter. When you can canter her on a 20-meter circle, spiral out and expand it to a 25-meter circle, spiral in again to a 20-meter circle, briefly spiral in a little more so that you're cantering on an 18-meter circle, then spiral back out again onto a circle that's 20 meters or larger. When she does this with a relaxed mouth, a lifted back, a swinging tail, and good balance, you will know that she is now strong enough to canter from the walk.

If she is a very large mare, you may want to expand this whole exercise, beginning with a 25-meter canter circle and spiraling outward to 30 meters and then back to 25 meters, inward to a 20-meter circle and then back to your 25-meter circle. Some big, lanky young Warmbloods cannot canter easily on a 20-meter circle, especially in an indoor arena. Depending on the construction of the arena, a 20-meter circle might bring the horse into contact with the wall; also, some arenas are not really 20 meters wide. For reasons of economy, many arenas are only 60 feet wide instead of the 66 feet that are the equivalent to 20 meters.

That difference can be very significant to a large horse. If the horse works on a track that is a foot or two to the inside of the rail, the effective width of the arena may be nearer 56 feet than 66 feet. For a large youngster, that can make all the difference between a comfortable, balanced canter circle and a desperate scrambling to remain upright through a series of too-tight turns. Measure the width of your arena, just so you'll know and can take that into consideration. You might prefer to do most of your cantering outside in a larger arena or in a field, wherever there is reasonable footing and plenty of space. Your mare must be relaxed to learn,

The rider's position should not change when going from walk to canter.

and she can't possibly relax at the canter if she feels that the canter threatens her balance and her ability to remain on her feet.

The final factor is *you*. Before you ask your young mare to canter from a walk, you'll want to be sure that you are a relaxed, balanced, non-interfering rider. While your watcher is analyzing your mare's canter, ask if she or he will take a few moments to watch *you* and be analytical and truthful about your seat, position, balance, and use of the aids.

If your mare is happy and balanced and strong, and canters eagerly and easily from the trot, cantering from the walk shouldn't be difficult for either of you. Don't try too hard — with an eager, energetic horse, you'll need to do nothing more than indicate "Yes, now you may canter." From walk to canter, your own position should barely change — you'll still be sitting straight, with your arms following your horse's head and neck. If your mare is on the aids at the walk, check your position to see whether it's right for canter; if it is, put a little more weight on your inside stirrup, bring your outside leg back slightly, and *allow* your mare to canter. Getting a canter should be a matter of putting your mare into a position from which stepping into canter will be easy and comfortable, and then allowing her to do it. *Set it up and let it happen.*

Want a Smooth Canter Depart

Q Sometimes I have trouble getting a smooth transition from a trot into a canter. I'll think I have my horse collected but when I use my outside leg and rein, he'll go into a faster trot instead. Also, if I try to use my seat to get the canter, he usually inverts and either trots really fast and short, or sometimes goes into a really horrible, rough canter (he stays inverted). I seem to have better luck going into a canter from a walk! What could I be doing wrong?

A There are several possibilities here. First, when you say that he feels collected but your canter aids just make him trot faster — guess what? If he's truly on the aids, your canter aids should produce a canter. If he doesn't understand that you want a canter, he should offer you a longer stride at the trot, not a faster trot. You already know that going from trot to canter doesn't necessarily mean going faster, it just means changing to a different sequence of footfalls. Now you need to make that clear to your horse.

If your horse is trotting in balance and with impulsion (controlled forward movement, not speed), he'll be able to pick up the canter easily. If he is on the forehand and inverted (topline dropped instead of lifted and hind legs not coming up underneath his body), he will find it almost impossible to reach far enough under himself to canter. He will trot faster and faster and eventually fall into a canter, but it won't be the canter you wanted. He can't go into a nice balanced canter from an unbalanced trot, but as you've noticed, he *can* go into a flat, strung out, four-beat canter.

When your horse speeds up, he's saying (a) that he isn't balanced enough to canter, either because he isn't sufficiently balanced at the trot or because he loses his balance during the transition, or (b) that he honestly thought you *wanted* him to speed up.

Improve his balance at the trot by doing many transitions from walk to trot and back again, and transitions within the gait: from shortened trot to normal trot to lengthened trot and back again. This will keep him attentive and balanced and will make the trot-canter transition easier for him.

If the transition itself is the problem, it's interesting that he takes the canter more easily from the walk. That generally indicates that the rider is in a better position to ask for and get the transition when the horse is walking. This makes sense, because your walk and canter position are similar: upright, seat in the saddle, hips moving with the horse, arms following the movement of the horse's head and neck. It's easy for a horse at the walk to move into the canter — the rider doesn't have to change anything about her position.

At the trot, on the other hand, the rider is holding her back and arms steady and either shifting forward out of the saddle (rising) or dropping down (sitting) at each stride. Either way, it takes a little more to inform the horse that you want a canter. If you are doing rising trot, first balance your trot and then sit the last few strides before you ask for the canter. That's because before you *ask* for the canter, you have to begin to *ride* it: sit deep into your saddle, straighten your shoulders, lengthen your legs, and half halt. Squeeze briefly with both legs, and when you feel the horse step more deeply under himself, squeeze your fingers and then relax your fingers again. Repeat as necessary — never *prolong* a half halt, just repeat it.

When you feel that your horse has become more balanced (that's the purpose of a half halt, after all), half halt again, but this time, after you squeeze your fingers, relax only your inside hand. At the same time, bring back your outside leg to ask your horse to strike off into canter. Keep your eyes *up*, looking ahead of you, and as your horse begins to canter, *give* softly with your inside rein. You aren't going to drop him on his nose or throw the rein away, but horses use their heads and necks to balance themselves at the canter, just as they do at the walk, and you must be ready to follow his movements. Don't lean forward — that simply asks the horse to keep trotting and go more onto his forehand, and negates the effect of your half halts. Stay perpendicular to the horse and ask him to jump into canter from your aids.

If he doesn't canter immediately, repeat your half halts, regroup, and ask again. And when he does it, even if he doesn't take the correct lead, praise him and let him canter at least 10 strides or so. It's important for him to understand that you *did* want a canter and he *was* a good horse — if you bring him back to a trot too soon, he may think that he was mistaken and you didn't want a canter after all.

To avoid an inverted canter depart, relax and lighten your seat; keep most of your weight in your thighs and lower legs instead of trying to "use your seat." Deepening your seat means allowing your seat to become deeper, it doesn't mean pushing or grinding your seat against the saddle. If you push the saddle into your horse when asking for canter, your horse's natural reaction will be to drop his back away from the unpleasant sensation. He'll invert, and you'll have exactly the problem you've described.

Instead, use your legs (inside leg at the girth, outside leg just behind it) to ask him to canter, and let your seat be soft to encourage him to jump up into canter with a rounded back. If you need to reinforce your leg aid, use your whip — don't

A balanced trot can ensure a balanced transition to canter.

push with your seat. If you have problems lightening your seat, try going from a full seat to a light, almost imperceptible half-seat (two-point position) when you ask for the canter — make it clear to your horse that his back is being invited *up*.

Once he is cantering, run down your position checklist:

- ▸ Inside leg at the girth?
- ▸ Outside leg just behind the girth to keep the hindquarters from falling to the outside?
- ▸ Body quiet — hips and seat following but *not* pumping?
- ▸ Inside hand asking for just enough bend for you to see his inside eyelashes, *not* his whole eye or his forehead?
- ▸ Inside shoulder back so that his shoulders and yours are parallel?
- ▸ Contact, rhythm, and balance appropriate to the gait and to the figure (straight line? circle?) you're working him on?

Most under-saddle training is rider-dependent — the rider can help the horse, confuse the horse, or interfere with the horse, even to the point at which the horse cannot do what the rider wants him to do. When you're training, always check your own position and balance and mentally review your aids before you ask the horse to do something. This will make you a better trainer, save your horse much confusion, and save yourself a lot of time.

Transitions within the Gaits

Q I rode in a clinic recently and had the best time ever, learned tons, and wrote down a lot of exercises to do with my horse at home. I don't have a riding instructor so I wanted to get enough stuff to work on for the next four months until there's another clinic nearby. This clinician said I should do a lot with transitions within the gaits. I'm not sure I know how to do those. The transitions I know how to do are walk-trot-walk, trot-canter-trot, and I'm starting to do walk-canter and halt-trot transitions. How can I do transitions within gaits?

A Transitions within the gaits are wonderful exercises to increase your horse's balance and strength and responsiveness. They're also exercises that most horses enjoy doing because they incorporate stretching and moments of relaxation. A transition within a gait means accomplishing a transition by changing from one sort of trot to another sort of trot instead of going from the trot to the canter or the walk.

Do these exercises outdoors if possible, on the trail or in a field. The best way to do transitions within the gaits is to focus on the quality of the gait itself, and not on whether you get the transition at *this* letter instead of *that* letter, or at *this* tree instead of *that* rock. At this point, promptness is far less important than quality.

Begin at the trot, because it's easy for the horse and he'll be able to remain calm and balanced while doing some fairly hard work. Ask your horse for his normal trot, the one that's most balanced and comfortable for both of you, and then ask him to lengthen his strides slightly. Feel what is happening underneath you, ask for only as many strides as your horse can comfortably offer you, then ask him to return to his easy trot for a few minutes before you ask him to lengthen his stride again. Alternate his ordinary trot with the longer, stronger trot, always focusing on the quality of the gait.

Over time, you can ask for more strides and more frequent transitions, but over the weeks and months that you practice transitions within the trot (and later, the canter and the walk), never ask for more than your horse can do *well*. These transitions will simultaneously build up your horse and improve your own timing and feel, so both of you will enjoy the exercises and their benefits.

If your horse's trot is slow and short-strided, transitions within the gait . . .

. . . can help him develop a more forward, purposeful trot.

CHAPTER 9

Lateral Work:
Suppling and Softening

LATERAL WORK IS AN ESSENTIAL PART of your horse's training, and it doesn't have to be difficult or complicated. You can use lateral work to teach your horse to multitask (moving sideways and forward at the same time), to teach him to use his muscles and joints in new ways, and to teach him to become more attentive to you and your aids. Focus on quality, and keep reminding yourself that less is often more. That is, one or two correctly executed steps will be of more value to your horse's education and physical development than twenty or thirty badly executed steps.

Softening a Stiff Horse

Q My barn owner has a new horse that she has asked me to work with. He is gentle and cooperative on the ground, and obedient under saddle at walk and trot. I haven't cantered him yet, partly because he is the stiffest horse I have ever sat on in my life! He doesn't bend at all, in either direction. I need to train him to bend and not be so stiff, but I don't know what to do. I don't believe that the problem is the saddle or the rider, because this horse is also stiff when he is turned out in the arena.

A If the veterinarian hasn't found anything overtly wrong with him, then training is definitely this horse's best chance of becoming flexible. The traditional, classical way to help a horse become more supple is to encourage him to reach and stretch in all directions. All of your training exercises should have the goal of suppling your horse both longitudinally and laterally.

In this case, since the stiffness is so pronounced, I think it would be an excellent idea for the horse's owner to invest in some massage and perhaps in some chiropractic treatments as well. This horse needs all the help he can get, and you should probably plan to incorporate some light massage and passive stretching into every ride.

It would be a good idea to go back to basics and put the horse on the longe line, not necessarily every day but at least every other day at first. Elementary longeing will help you understand more about his stiffness and movement, and will enable you to *see* his problems and progress. It will also enable the horse to relax and stretch on the largest possible circle (no smaller than 20 meters, or 66 feet). Encouraging the horse to reach with its neck while moving forward from behind will eventually help the horse achieve a stretch through its topline. Transitions in a "long and low" position will also help the horse learn to stretch a little more.

Horses' spines don't flex much from side to side, but their muscles will generally stretch enough to allow them to flex in both directions. When a horse is noticeably stiff and inflexible, the problem is likely to lie with the muscles, not with the spine itself. Muscles can be stretched over time, although some naturally supple individuals will always be more flexible than others even if they are given exactly the same training. If you have ever taken a class in gymnastics or skating or dance, you already know this. Some individuals are "naturals" and some are not, but the exercises can benefit them all.

Lateral work is wonderful for all horses, and you can certainly incorporate leg-yields into this horse's sessions under saddle. Since the objective at this point is to lengthen and stretch the horse's muscles, don't attempt shoulder-in, half-pass, or even shoulder-fore until the horse has become comfortable with basic leg-yields at walk and trot. You can do leg-yields on a straight line or incorporate them into ring figures; both will be useful. If the horse doesn't know how to leg-yield, teach him to turn on the forehand first, then teach leg-yielding. If he already understands leg-yielding, you can begin the exercises immediately.

One of the most valuable exercises you can do is spiraling in and out on the circle. For this horse, I'll suggest a very low-key, modified version of the exercise, in which you begin, probably outdoors where there will be plenty of room, by walking the horse on a 20-meter circle. Be attentive to the horse's responses, and remember that for a truly stiff horse, even a tiny amount of bend will be difficult and stressful. Once you are satisfied with the degree and consistency of the horse's bend on the circle, use your inside leg actively and shift your weight slightly to the outside to encourage him to shift slightly sideways, away from your inside leg, expanding the circle. Relax on the larger circle for a moment, allowing the horse to continue circling with his new, easier bend. Now use your *outside* leg to encourage him to return to the original circle. Enlarging the circle and returning to the smaller circle may involve only a few steps of leg-yield in either direction;

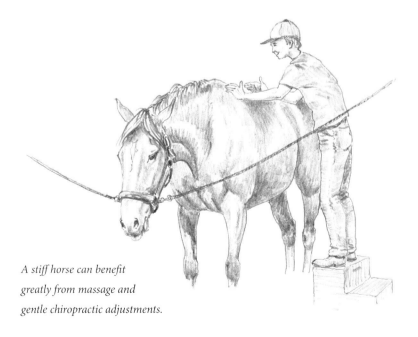

A stiff horse can benefit greatly from massage and gentle chiropractic adjustments.

that's fine. Keep the horse relaxed and praise him — he's working hard and will only be able to perform the exercise and benefit from it if he is relaxed and able to reach with his hind legs and allow his muscles to stretch.

You can do this spiral-in, spiral-out exercise every day, indoors and outdoors. Make the horse's relaxation and forward movement your top priorities; sideways movement is a secondary priority. It matters, but relaxation and forward movement must come first. Whenever you shift the horse inward, reducing the size of the circle, remember that this is difficult for him, and be sensitive to his reactions. If he slows, stiffens, and lifts his head, he is telling you that the exercise is becoming physically uncomfortable and perhaps a bit frightening as well. When that happens, instantly encourage him to shift outward, maintaining his bend but increasing the size of his circle while reaching forward and down. As he spirals out onto a larger circle, he should straighten a little; as he spirals in onto a smaller circle, his nose-to-tail bend should increase. He should always be bent along the curve of the circle itself.

If you ride outdoors on a trail, you can practice asking the horse to leg-yield from one side of the trail to the other. If you ride indoors, don't attempt to leg-yield the horse from the centerline to the rail. Instead, beginning on the centerline, walk energetically straight forward, then ask for two or three steps of leg-yield. *Before* the horse slows, send him energetically forward again in a straight line for several steps before asking for another two or three steps of leg-yield.

Focus on relaxation, forward movement, and cooperation rather than on the ultimate size of the horse's circles or the number of steps he takes in leg-yield. Relaxation is key because a tense horse cannot stretch — you can test this yourself with your own muscles, by tensing your legs or arms and then trying to reach and stretch them without relaxing them first. You can't do it, because it can't be done.

Over time, as the horse becomes more relaxed, you will be able to ask for longer steps and greater energy. Eventually you will also be able to spiral in to a much smaller circle before spiraling out again; at some point, you'll find that your horse can now spiral from a 20-meter circle down to a 15-meter circle and back again. It may take many months to reach that point, especially if the horse is as stiff as you've described him. Be patient, do the exercises slowly and energetically in both directions, and be generous with your praise. When you make comparisons, be sure that you're comparing *this* horse as he is now with the way he was a month, two months, or six months ago. It does no good to compare him with some other horse. Your goal is not to make him more supple than the slinky horse in the next stall; it's to help him achieve a "personal best" in flexibility.

Horse Doesn't Yield to Leg Pressure

Q My horse Duke and I have shown successfully in Arabian and open shows since he was a foal. When I started college, I just wanted to ride him for fun and not have to work so hard at achieving the right look. He is six years old now, and we are learning dressage at a local training center. I ride him two or three times a week and have a lesson on him every other week. Dressage is helping us really enjoy each other more and makes what I had been doing seem ridiculous. I now realize I was taught all wrong from the beginning. Among other things, I learned to pull and maneuver the horse's head into position and ride with mainly my knees and thighs in contact with the horse.

My problem is teaching him to yield to leg pressure. Until we started the lessons, he was not used to having my lower legs touch him unless I was asking him to move in a specific direction. Now when I try to use my upper calf as my instructor tells me to do, he becomes less soft; I apply more pressure, he becomes less soft — you get the idea. This goes on until we drift completely off course or until I give up because my leg is so tired. All the bend in his body disappears. I know this is totally wrong. My instructor says she has five-year-old kids who can do it, so not that much pressure is needed. She suggested nudging him rather than using steady pressure until he gets the right idea. This works better, but not well enough. I'm so tired of arguing with Duke over this, and I know he is trying to tell me something by stiffening up, but I just can't get it. How can I train him to yield to leg pressure?

Work with your horse on the ground to help him become more responsive to leg pressure.

A   Begin by taking a moment to think about Duke's history. What you are asking him to do is new to him and not entirely comfortable. He has to learn to understand new aids, and he has to develop a different way of going; this is mentally and physically *very* demanding. In his previous life, he was not asked to bend or stretch or yield to your leg. In fact, your lower leg rarely touched him at all. You're asking him to make a big effort and to stretch his muscles in new ways. Eventually he will develop new muscling from doing correct work; right now, he's confused and sore.

Ask your instructor to give you some lessons on a horse that's been correctly schooled, so that you will *know* when you are applying your aids correctly. It's not easy to learn a new way of riding and teach a horse at the same time; *one* of you should know what is going on. Riding a horse that responds well to the leg will teach you to recognize the response that you'll be able to expect from Duke once he knows what response you want and can give it to you.

Horses become stiff when they hurt — and when they don't understand. You can make the whole retraining process easier for Duke by taking some precautions:

1. Before you ride, give him an extra-thorough grooming. This is good for his circulation and muscle tone, and will help you notice any tight or sore muscles that could interfere with your ride.

2. Give him a long, slow warm-up (10 or 15 minutes of walk, 15 or 20 minutes of trotting and ring figures, a little canter *if* you are working him at canter) before you begin real work. Don't ask him to stretch any muscle that hasn't been warmed up first, or what he will get is not a stretched muscle, but a torn one.

3. Give him many breaks so that he isn't asked to hold a particular position for more than a couple of minutes at one time. Muscles being used in unaccustomed ways will quickly become tired and stiff, and they will cramp unless you let them relax and stretch frequently.

4. After the ride, cool him down slowly and then groom him again.

When you put your leg on a horse and he moves away from it, there are many factors involved. The horse has to feel your leg, interpret the leg pressure accurately, and respond correctly. You have to ask, then relax and give the horse a chance to do what you asked him to do. If he makes a mistake, be sure that he knows what you are asking, that he is in a position to do what you want, and that he *can* do what you want — then ask again, quietly.

Duke *feels* your leg, but he doesn't know what it *means*. Kicking won't help, and neither will grinding your leg into his side. Show him what you expect while you're on the ground. When he is standing still, put a hand on his hip and say "over." Most horses have been taught to move away from this combination of pressure and voice. If he does it, praise him, give him a moment to think, then ask him again. Repeat the request, each time moving your hand a little closer to the place where your leg will lie against his side when you are in the saddle. Do this from both sides. That's his first lesson: Pressure and "over" mean that he should step over even when the pressure is on his rib cage instead of his hip.

Repeat the lesson for a few days, then do it while you are grooming. Repeat it in the arena at a walk; again, do it from both sides of his body. When he understands and responds well, mount and do the same thing at a standstill, and this time use your leg and your voice. If you sit very straight and still, so that only one leg is active, and that leg gives a quick squeeze while you say "over," he will eventually make the connection. When he does, praise him and let him walk on. After a few minutes, stop and do it again. Ask him to do it once more at the very end of your riding session, then *dismount at once*. He'll remember it very well the next day.

When you take lessons on a school horse, ask your instructor to teach you the most basic form of leg-yielding: the turn on the forehand. When you understand how this works, you'll be able to help Duke learn it, and once he has the idea, it will become steadily easier.

When you are in the correct position to ask your horse to move away from your leg, use your leg *briefly,* then relax it again and give Duke a chance to respond. All of your aids should be quiet and brief — if you need to ask him again, *repeat* the aids. A leg-squeeze lasts only a second — don't prolong it.

When you prolong an aid instead of repeating it, your aid becomes hard and your body becomes rigid; your horse doesn't get a chance to respond; he becomes stiff because *you* are stiff; and eventually, you become tired and stop pushing. When this happens, the horse learns nothing useful: As far as he knows, you stiffened and clamped down on him for no apparent reason, and then stopped, also for no apparent reason. Try this instead:

- Sit straight, tall, and relaxed.
- Tip your horse's nose very slightly away from the direction in which you want him to yield — left for a leg-yield right, right for a leg-yield left.

- Bring your inside leg back slightly, from the hip, and give him a soft squeeze. Stay tall, keep your weight balanced, and keep your other leg relaxed, barely touching his side, creating an "open door."
- Tell him "over" and squeeze, then relax, wait for his response, and try again.

A stiff *rider* will create a stiff horse. Your horse is your mirror; if your body becomes stiff, your horse's will, too. Remain relaxed while you give your aids. Practicing deep, steady breathing is a good way to remind yourself to loosen up.

Too Much Turn on the Forehand?

Q My Icelandic mare is the most intelligent horse I have ever met, but I am having a strange problem with her. Three weeks ago I began to teach her some lateral work, beginning with turn on the forehand because that is how my instructor taught me to teach horses many years ago. She was devoted to dressage and said that it was easiest to teach horses to step away from the rider's leg at a standstill, so it is best to teach turn on the forehand first, before even leg-yielding. I would like to know if you agree with this.

My mare learned to move away from my leg in just a few sessions, but now whenever I ask her to stop, she begins to turn one way or the other and I can't make her stand still. When I ask her to stand still she becomes anxious and agitated, and I can tell she is unhappy. She never had a problem standing still before. I am afraid that I have trained her badly, and I don't know what to do or where to go from here.

A Your teacher made a good point: It is often easier to teach a horse the basics of moving away from the rider's leg by beginning with turn on the forehand. If your mare is a typical Icelandic, she is very clever and sensible and won't have any difficulty learning this lesson. And you are right — her intelligence may be part of the problem you're experiencing.

Your mare learned very quickly that she was meant to move away from your leg. That's good. The problem is that she moves too much and goes too far, and that is something you can change. Clever, enthusiastic horses that like to please will often offer us more than we want of whatever we are teaching them at the time. That's not a problem, really — it's just the horse being overly generous. In your case, you probably focused on teaching your mare to move away from your leg, and were happy when she offered a step, then two, then three steps, and now she is showing you how well she understands and how well she has learned her lesson. Your mare has done nothing wrong; in fact, she is trying to give you what she thinks you want. Now it's time for you to refine the lesson and teach her to move away from your leg *one step at a time.*

Throughout the lesson, keep everything positive and friendly. Her response is already excellent — you just need to make your aids softer and more specific and precise. First, practice walking, halting, and then walking on immediately; this will remind your mare that not every halt involves stepping sideways. When she goes easily from walk to halt to walk again with no sideways movement, you can return to the turn on the forehand.

This time, do it a little differently: Begin by creating the best, most balanced and square halt you can manage. At the halt, sit tall, relax, and breathe deeply for a moment — this will help relax you and your mare. Then, very gently squeeze your ring finger on one hand (say the left) to help tip your mare's nose ever so slightly to the left. At the same time, bring your left leg back from the hip, just a few inches, and apply a brief inward squeeze with it while relaxing your right leg so that it puts *no* pressure on your mare's side. The pressure from your ring finger and your leg should last for only a heartbeat. When your mare takes a step to the right with her hind legs — in fact, as soon as you feel her inside hind *begin* to step over — relax your ring finger and your left leg to remove the pressure. As she completes that step, take a deep breath, sit tall, center yourself, relax, and exhale. Your mare will stop and stand.

You can repeat this as often as you like, whether you want one step or ten. Ask for a single step, release, relax, reward, *pause*, and then ask for the

Ask for and reward one step at a time.

next step. The first lesson you taught your mare was: "When I ask you to move away from my leg, move away and keep moving." Now the lesson is: "When I ask you to move away from my leg, take a step — I'll stop asking as soon as I feel you begin to move. Thank you! That was a lovely step. Now I would like you to take another step sideways, so I am asking again in the same way: ask, release, relax, reward. Thank you."

This is a wonderful exercise for your mare, as it will help her become even more alert and attentive, paying close attention to your wishes and signals. It is also a wonderful exercise for *you*, as it will make you very aware of your balance, your position, your breathing, and your aids. Above all, it will make the two of you pay attention to and respond to each other.

This is a good general lesson for you as a horse trainer, because it shows you how one skill can appear to deteriorate while you are building up another one. This is not cause for concern; the problem is temporary and self-correcting, and the deterioration is usually deceptive. It is important, though, that you be aware of the fact that training involves swings and roundabouts, peaks and valleys — progress is *not* achieved in one steady ascending line.

When to Teach Leg-Yield

Q I have a young Appendix-Quarter Horse gelding, and I have been working him on the longe line for about four months now. At this point I am doing very short longe line periods and spending more time riding. He is a well-built horse and very strong, so although he is only three, I don't think I am hurting him by riding him for 20 minutes every afternoon (I put him on the longe line for 10 minutes first). I know that my longe line work is correct because I learned it from someone who learned it from you. When would I be able to begin to teach him to leg-yield?

A You seem to be doing everything right with your horse — good for you. Leg-yield is something you can teach your horse quite early in his mounted work. It's useful for him to learn to move away from your leg when you ask. This lesson will do him good now and help in his later training, so don't hesitate to start teaching leg-yield just as soon as your horse achieves balance at walk and trot.

Here's how to teach him leg-yielding to the right:

- ▸ Start walking around the arena, on the rail, tracking left.
- ▸ Walk up one long side and across the short side, but then, instead of going all the way to the next long side, turn early and come down the quarter line.
- ▸ Walk your horse straight down that quarter line, parallel to the wall.
- ▸ Then just tip his nose very slightly to the inside, leaving his body straight, and use your left leg just barely behind the girth to ask him to move over to the right.

Be sure that your arms are still following his head movements at the walk, stay tall in the saddle, and give him time to respond to each leg squeeze. Arena walls have a magnetic effect on horses, so if you ride on a straight line parallel to and near a wall, it's easy to teach the horse to leg-yield toward the wall. To leg-yield to the left, reverse the procedure: Track right around the arena and use your right leg to shift your horse toward the wall on his left side.

If your horse has any difficulty understanding this, check your own position. If a rider uses her left leg too energetically to ask a horse to move to the right, she may inadvertently collapse over her left hip, pushing her body weight to the left. When that happens, her leg is telling the horse to move to the right but her *body* is telling the horse to move to the *left*.

The result is a confused horse that can't guess what the rider wants, and a frustrated rider who thinks that her horse isn't responding to the leg aid. If you are sitting tall and straight, and your horse needs a little extra help, you can shift your weight slightly toward the wall. After a time, you will no longer need to do this as he will no longer need the extra help.

By taking your outside leg slightly away from his body, you offer your horse a space to step into.

Whys and Hows of Shoulder-In

Q Can you explain the whys and hows of the shoulder-in? I'm not a dressage rider, but a hunter/English pleasure rider who likes to do lots of flat work with my horse, who has been very patient and cooperative with me as we have been learning to do lateral work together.

A Shoulder-in is a very valuable exercise for your horse, and you don't have to be a dressage rider to take advantage of it. Hunter and jumper riders use it, too — I've never attended a George Morris clinic where shoulder-fore and shoulder-in weren't part of the flat work. Shoulder-in is a *wonderful* exercise to increase the engagement of the inside hind leg, improve flexion and balance, and teach the horse to move from the rider's inside leg into the outside rein.

Having said that, I should point out that shoulder-in is *not* for the horse that is still learning early basic flat work. It requires that the horse be reasonably balanced and supple and muscled. Before you begin shoulder-in, your horse should understand and accept the aids, have a relaxed, supple back, and move eagerly forward from the leg. There is no point in asking a horse for shoulder-in if that horse doesn't accept and move happily into contact. The shoulder-in will be faulty, and instead of being a good exercise, it will be useless or even harmful. The purpose of the shoulder-in is to increase the engagement of the inside hind leg, and this will happen only if the horse is already comfortable with and proficient in the basics.

Shoulder-in requires a tiny bit of collection, and that is *too much to ask of a green horse*. If your horse has had a solid year or so of good basics, and is forward and balanced and rhythmic, then you're ready to begin. In shoulder-in, the horse is bent, very slightly and evenly, around the rider's inside leg. The horse is looking away from the direction of movement — in shoulder-in to the right, the horse will be bent around your left leg and looking to the left.

In shoulder-in, you bring your horse's forehand *off* the track and toward the inside so that his inside foreleg crosses over his outside foreleg and his inside hind leg crosses over his outside hind leg. Shoulder-in involves asking the horse to move very definitely *forward* and sideways.

As with any other movement, you'll teach this gradually, asking for one step, then two, then three, and so on, sending the horse energetically forward after each request, so that the impulse to go forward remains strong in both horse and rider. You'll do this at the walk at first, and later, at the trot.

I've found that the easiest way to teach shoulder-in is to ask the horse to circle at a balanced, rhythmic walk. If you have access to a dressage arena, it's easier; at one of the short sides, walk a circle that brings you back to the rail on the long side.

Keep your horse bent on the circle as you reach the rail — this is the bend that you want for the shoulder-in. Then take a step off the rail, just with the forehand, as though you were still circling. Ask the horse to take the next step *down* the rail, still bent as for the circle, but with the idea that you don't want to circle in precisely the same place. If you do this correctly, your horse should take a step down the rail, away from the circle, still bent around your inside leg, but reaching across and underneath his body with his inside hind leg. You will feel his inside hip drop as the leg stretches underneath him.

Maintaining the bend, send the horse actively forward on the circle. Your new circle will be located just one or two steps down from your previous circle. When you return to the rail, take the first step into the new circle and then ask for a step or two steps down the rail, maintaining the bend, then send the horse actively forward on the new circle. You can work your way down the long side alternating circles with a few steps of shoulder-in.

Alternating circles with steps of shoulder-in helps keep the horse relaxed, rhythmic, and forward.

Your position for shoulder-in is similar to your position during a turn: shoulders and hips parallel to your horse's shoulders and hips, inside leg at the girth to maintain the forward movement, and outside leg behind the girth to support the horse through the turn and "catch" his hindquarters if he falls out of the turn.

Stay tall and balanced. If you collapse over your inside hip while you try to *push* your horse over with your inside leg, you will only confuse him and stop both the forward and the sideways movement of his legs. If you collapse or lean over your outside hip, and try to pull your horse sideways under your body weight, he won't be able to move correctly. Your weight should be very slightly on your inside seat bone, but this will happen automatically if you are tall and balanced, and bring your outside leg back.

Keep your reins and hands where they belong: one on each side of the horse's neck. You *cannot* help your horse learn shoulder-in by crossing your inside hand over his neck and pulling.

Your inside rein should be soft and elastic, just indicating to the horse that you would like his nose to point *this* way, please. The outside rein will have a stronger but still elastic contact, as this rein affects the degree of the horse's bend and the position of the horse's outside shoulder. The ideal shoulder-in angle for dressage tests is 30 degrees, and you should always be aware of your angle because the greater the angle, the greater the engagement and the greater the effort. Don't ask for too much too soon.

Ask for a little at a time, be clear, be soft, reward the effort, and go forward onto your circle immediately afterward. If anything goes wrong, go forward onto your circle, regroup, then try again. Circles and corners are the best places to begin asking for shoulder-in, because the bend is already established, and if you are riding your circle or corner correctly, you are already asking the horse for a little inside flexion and a little more engagement of the inside hind.

If you try to teach shoulder-in by taking the horse straight down the rail, bending him, taking his forehand off the rail, asking for shoulder-in, and then straightening him again to go down the rail, you will confuse him and lose impulsion. Working from circle to shoulder-in to another circle lets you send him forward while maintaining his bend and your position, and it's less confusing for both of you.

Take your time, be patient, and focus on correct position and good contact. Remember to *breathe* and to let the horse stretch down on the circles between efforts — and have fun!

Introducing Lateral Work to Young School Horse

Q I live in the UK and have returned to riding over the past five years, having ridden a little during my twenties. I am now a rather overweight male in my late fifties so am more into pleasure riding than anything else. Last December I bought my own horse, Domino, a five-year-old, 16.1-hand cob mare that is used as a lesson horse where I board. She is willing and pleasant and popular with many of the other adult riders for lessons and hacking out. I ride her once a week in a one-hour group lesson and once a week on a hack of one to one-and-a-half hours. The lessons tend to have different attendees and dwell on transitions more than anything else. I suspect I might benefit from something else.

Domino knows the basics and has reasonable transitions between halt, walk, trot, and canter, although they aren't always reliably performed! She can make a good attempt at 20-meter (and perhaps 15-meter) circles and serpentines at trot though they are harder and more misshapen at canter, and she sometimes will do a flying change when asked on a cantered serpentine. She has really no "lateral" knowledge as yet, and I have found it almost impossible to get her to leg-yield, though I've tried only once or twice because I am a little afraid that she may be too young for that. She tries when asked for shoulder-in. (Or is it shoulder-fore when the angle is very small?)

She is obviously still young and green, and I am old and green! I am keen to find ways in which I might prevent her from being too affected by the routine of school use and wonder if trying to get her to do some lateral work on the trail might help. It certainly would help in getting gates closed and in avoiding tree trunks and gateposts! However, I don't know what age is appropriate for introducing her to such things, whether I can expect such a drafty type of horse to do this sort of thing well anyway, or even whether that is the best thing to bring on a horse who is used so much in the school.

A Lucky you, Domino sounds like a lovely mare. She's certainly not too young, and you're certainly not too old, to begin some lateral work. I agree with you that it would make riding out more enjoyable, especially through woodland. Horses tend not to take *our* knees into account when they estimate how near they can get to a tree.

It's easy for a horse to become confused when she is ridden by many different riders. The problem is not that Domino is inattentive or unwilling, it's simply that she has become an excellent school horse. Some of the qualities that make a lovely school horse are *not* qualities that make a sensitive, one-owner riding horse.

A sensitive, well-trained, one-owner horse — a competitive dressage horse, for example — is a horse that has thoroughly learned the language of the aids. To such a horse, each shift of the rider's weight, each vibration of the boot and movement of the leg, each closing of the rider's fingers and tilt of the rider's shoulders means something specific. The rider also learns to interpret the horse clearly and accurately. This kind of understanding is accomplished through many hours of work during which the horse and rider develop and refine their communication.

A good school horse, on the other hand, cannot afford this degree of sensitivity — it would go mad. A horse that is ridden by many people with varying backgrounds, physiques, and riding experience and ability, including some who are absolute beginners, *must* be calm, kind, and able to generalize and extrapolate. If she has even four different riders — one tall and thin, one short and stout, one weak and slow, and one athletic and coordinated — she will receive different signals from each rider even if they all ride at the same level and are taught by the same instructor. If a horse is regularly ridden by 10 riders of widely varying abilities, she may become entirely unresponsive (a *bad* school horse) or she may learn to generalize (a *good* school horse) and interpret, for instance, any sort of push, squeeze, kick, or shove from shoulder to flank as a signal to move off or keep moving.

Domino can certainly learn to leg-yield and to do a turn on the forehand. These are the beginnings of lateral work, taught first because they are simple and easy for both rider and horse. Shoulder-fore and shoulder-in come later, once the horse has understood the basic concept of moving sideways away from the rider's leg.

A good beginning: The mare steps under herself in response to the rider's leg.

You don't need an arena with markers to do this work, but it does help at first. However, if you'd rather work on these exercises while you are riding outside, that is fine. I generally prefer to do them outside, especially leg-yielding, which must be done at a good forward walk (and eventually at the trot). Indoors, where the horse has no particular incentive to move forward, it's easy to fall into the trap of thinking "sideways" instead of "*forward* and sideways." Outdoors, everything changes and "forward" tends to take care of itself to a great extent. The horse that can't understand the point of moving away from the leg on the centerline of an indoor school is usually quite happy to leg-yield from one side of the path to the other, over and over again. Many things come more easily to a horse that is moving forward energetically in the fresh air, with something to look at.

The key is to sit straight and use your leg (take it back behind the girth but move it back from the hip, not the knee!) to press the horse's hindquarters *over*, away from the leg pressure. If the horse has difficulty understanding what you want at first, put a little more weight into your seat bone and stirrup on the outside, in the direction that you want the horse's hindquarters to go, and keep your outside leg light so that the horse can move into it comfortably. Other key points are:

▸ The horse must be straight, and so must you. Leg-yielding and turns on the forehand require a minimal bend — you should just be able to see your horse's inside eyelashes and nostril.

▸ Signal, don't force. You aren't *moving* the horse's hindquarters with your leg, but using your leg to tell the horse that you want her to move her hindquarters. There's a big difference!

▸ Leg pressure is a signal — pressure is a brief "squeeze and release," not constant pushing.

▸ Sit straight. If you focus on trying to shove the horse sideways with your leg, you are likely to collapse over your hip on that side, and that will confuse your horse! Your leg aid will be telling her "Move over *that* way" while your weight will be telling her "No, no, come back *this* way."

▸ Ask for *one* step at a time. After each step, relax your legs, praise the horse, equalize your leg pressure and your weight and balance on the horse, then prepare the horse, and ask for the next step. This is essential if you want the horse to move step by step — otherwise, she will quickly learn to whirl in response to constant pushing by your leg.

The good news is that if you can ride Domino regularly, even for shorter periods, and if you can be consistent and clear with her, she *will* become more responsive to you and to your aids. At first, you may need to spend the first 20 minutes of every ride refreshing her memory and reminding her that *your* signals, unlike the ones she receives from all and sundry at the riding school, mean something specific. But as the two of you develop your partnership, it will take less and less time to refresh her memory.

Turn on the Haunches

Q   Now that my horse and I have mastered turn on the forehand and leg-yield, I think we're ready to learn turn on the haunches, but I have no clue how to begin, and for some reason it seems much harder and more complex than turn on the forehand. How the heck do you get a horse to move its front end and leave its rear end in place, or is it okay to move the rear end some? I think the horse would have to move one hind foot even if it's supposed to pivot on the other one, wouldn't it? I would really like to understand this and also to have a mental picture of what we should be doing.

A   You ask good questions. Indeed the horse's rear end *does* move, and so do his hind feet. Turn on the haunches is more difficult than turn on the forehand for two reasons: First, the horse is being asked to carry more weight behind and "lighten the load" on his front end; second, the horse is being asked to move into the direction of his bend. That's more physically demanding than moving *away from* the direction of the bend, as he does during a turn on the forehand (leg-yield doesn't involve a bend, just a slight positioning of the horse's head away from the direction of travel).

For your mental picture, visualize your horse walking down the long side of the arena. When you reach E or B (the halfway point), think of your horse almost, but not quite, coming to a halt. He is already positioned with his head very slightly to the inside, but his body is straight — you now need to create a bend, just as you would if you were turning (which is, in fact, what you're going to do). Keep thinking "walk" at all times, because turn on the haunches does *not* mean pivoting around a fixed hind foot — your horse should maintain the rhythm of the walk throughout the turn. This is very important: Maintain the walk rhythm even if your first few turns on the haunches take up much more space than you think they should. The degree of bend and the tightness of the turn can be increased later, but whether a turn on the haunches is wide and sloppy or tight and precise, it will be correct *only* if the horse maintains his bend and his walking rhythm throughout.

Turn your own body from the hips, just as you would during an ordinary turn at the walk.

If you could see yourself from above, you would notice that the horse's hind end is turning by taking small, short steps, while his front end is turning by taking much larger, longer steps. If you could freeze this aerial view, you could draw a small circle where your horse's hind legs are walking through the turn, and a larger circle around it to show the larger movement of your horse's front legs and shoulders. Eventually, as your horse becomes good at this exercise, the smaller circle will be very small indeed; your horse's inside hind leg will almost be taking steps in place.

In a full turn on the haunches, you would soon find yourself parallel to the rail again, facing in the opposite direction. This is fun and a good exercise, but as usual you should begin by performing an easier, less demanding version of the movement.

Let's go back to the moment when you are riding down the rail and bring your horse to an almost-halt before bending him and asking him to take his shoulders off the rail. Ask him for a step, then another step, then — if he did the first two steps easily — one more step, and then send him forward at

After one or two steps of turn on the haunches, send your horse actively forward.

an energetic walk, straightening him as he takes that first step across the arena. Don't worry if you have performed only one-eighth or one-sixteenth of a turn on the haunches — this is exactly what you *want* to do. You're going to use the same technique you used when you taught your horse to leg-yield, making it easy for him to understand and to perform, asking for a little at a time, and making "forward" a priority.

"Forward" in this case means maintaining the horse's energy and walk rhythm. Check his walk, his energy level, his bend, and your own position, then ask for a few steps of turn on the haunches. Monitor his energy and rhythm. As soon as he begins to lose either (and this might be after three steps, or two, or just one), send him forward, straighten him, and continue until he is moving energetically forward in a steady walk rhythm. Then come down the rail and begin the exercise over again.

Asking for too many steps at once is likely to cause your horse to slow down, stiffen, and pivot on his inside hind foot, and you don't want any of those things. It's also likely to make him uncomfortable in the short run and sore in the long run, so be patient and follow the old training principle: Ask often, ask for very little, and reward generously.

Done correctly, this exercise will help strengthen and supple your horse's hindquarters and increase his degree of engagement.

I Can't Get My Horse to Do Haunches-In

Q I'm finally where I could start doing haunches-in with my horse, and it turns out that he absolutely positively doesn't want to have anything to do with it. He's nine years old, and I have owned him since he was five. When he was six I taught him leg-yield and turn on the forehand. When he was seven he learned turn on the haunches, and this year we've been working on shoulder-fore and shoulder-in and improving his turn on the haunches. So he ought to be ready for haunches-in, right? But he just won't do it, and when I try to make him, he does a little crow-hop or else he kicks out with a hind leg. Last night I tried again when my teacher was there, and he jumped around and then finally he reared. That scared both of us. He is usually nice about learning things, so what could be his problem?

A   It sounds as though there are three possibilities here: Your horse may
    be physically uncomfortable, he may be confused, or he may be both
uncomfortable *and* confused. Let's consider physical discomfort first.

Haunches-in makes new demands on your horse, and puts stress in new areas
of his body; specifically, his stifles and his outside hind leg. If you have access to
trails with hills, you can help strengthen your horse's stifles by doing more work
up and down hills; if you don't have hills, work over cavalletti can accomplish the
same thing.

Confusion is common when horses first begin to learn haunches-in, and
horses don't deal well with confusion. Some become anxious and hypersensi-
tive, some become annoyed and resentful, and some kick or rear to express their
frustration and emotional discomfort.

A rider who is teaching haunches-in for the first time and isn't sure what to
expect from the horse may be confused and frustrated as well, and she may hold
the reins too tightly, hold her breath, and stiffen her own body. If her intentions

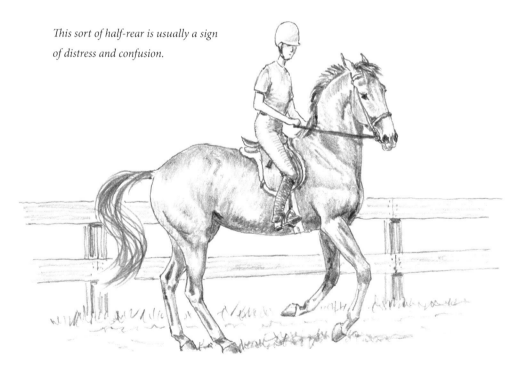

*This sort of half-rear is usually a sign
of distress and confusion.*

are to send the horse forward in haunches-in, but her body language is telling the horse "halt" or even "halt and back," mutual confusion and frustration are likely. As with all other exercises, less is more — asking often, asking for little, and rewarding generously is beneficial to the horse *and* the rider. Both need small, manageable challenges quickly followed by breaks during which they can regroup and retrieve their "forward" and their positive attitude.

The combination of discomfort and confusion should be avoided at all costs, because it can lead to the horse feeling trapped by the exercise and losing confidence in himself and his rider. Haunches-in is a "trappy" exercise anyway, because until now, the horse's lateral work has required listening to *one* of the rider's legs; the other leg has merely been resting against the horse's side in case the rider needs to prevent the hindquarters moving in that direction. Now, for the first time, your horse is being asked to pay close attention to both your inside and outside legs, and obeying your signals will feel strange and possibly uncomfortable at first. If he's uncomfortable, doesn't understand what you want him to do, and feels pressured and trapped by your legs and reins, almost anything can happen.

Never be afraid to back off and ask for less bend or fewer steps, and always praise your horse.

Use your feel — your equestrian tact — to sense when your horse is being pushed too far, and immediately send him forward on a longer rein, then relax, regroup, and try again. Never be afraid to back off and ask for less bend or fewer steps, and always praise your horse. This is a progressive exercise — you must ask for very little at first, so that your horse can understand what you want and feel confident in his ability to do it. Asking for too much or asking for too long will make the exercise physically difficult and mentally overwhelming for him.

There are several ways you can begin working on haunches-in. One of the easiest is to walk a straight line, on the inside track, the quarter line, or even the centerline, sending your horse forward and making him straight, then displacing his haunches to the right for two or three steps, then sending him forward and straightening him again, then displacing his haunches to the left for two or three steps, and so on. This exercise is simple but very effective. It helps you avoid asking for too much at once and gives you many opportunities to praise, reward, and try again.

Some horses relax more on a circle than on a straight line; for such horses, a better exercise would be to ride a 10- or 12-meter circle at the walk at one end of the arena, then continue down the rail using haunches-in to maintain the bend for a few steps, then moving forward on another 10- or 12-meter circle. During the steps of haunches-in, the horse must reach forward and across with his outside hind leg; on the circle, you can maintain the bend and energy while giving the horse's outside hind leg — and his brain — a break.

If you are like many riders, your outside leg may have been primarily passive until now, whether it was at or behind the girth, and your horse may not be entirely sure that he needs to follow directions from that leg. If this is the case, and you perceive him to be resisting or evading your leg, be compassionate when you interpret his reaction. Instead of "I hear you and I don't want to do it," his reaction may mean "What? You're already using your inside leg, and now your *outside* leg is giving orders too? Are you serious?" If that's the case, this would be a good time to regroup, relax, and then practice the straight-line exercise described above, asking for only a single step of haunches-in at a time.

Lateral Work: What's the Training Order?

Q I've had several instructors give me different advice over quite a few years. Now I'm starting over, on my own, with a new six-year-old horse and absolutely nobody in the area to help me. I think I probably own the only horse in this area that isn't a ranch horse. I'd like to do some more lateral work with Sandor because he's learned turn on the forehand very well. I think the next things to teach him would be leg-yield, half-pass, and shoulder-in. My question is, in what order should I teach them?

A Since your horse has already learned turn on the forehand, you can safely stop doing that exercise or, at least, save it for times when it may be useful (opening gates, for instance). The main point of turn on the forehand is that it's an easy way to teach the horse to move away from the rider's leg. After that, the "order of go" for your lateral work should be leg-yield, shoulder-in, haunches-in, and *then* half-pass. This makes sense when you consider what you're trying to achieve through your work: the progressive training and development of your horse.

Leg-yield is more difficult and complicated than turn on the forehand because the horse has to learn to move sideways, step by step, away from your leg while remaining straight and *continuing to move forward*. Once your horse has learned to leg-yield smoothly and evenly in both directions, you can take him to the next level of difficulty and introduce the shoulder-in.

Shoulder-in is more complex and demanding than leg-yielding because it requires that the horse move sideways while maintaining a bend. Shoulder-in asks the horse to step forward and sideways with his inside hind leg, reaching underneath himself, and *that* requires a tiny bit of *collection*. The horse will work on three tracks, with his inside hind tracking his outside fore (in shoulder-fore, the inside hind tracks between the two forefeet). Next, you'll tackle the other important three-track movement: haunches-in.

Haunches-in helps your horse become supple, strong, and energetically responsive to your outside leg in the same way that shoulder-in helps him stretch and strengthen his hips, become energetically responsive to your inside leg, and develop some carrying power behind.

Half-pass is quite demanding and should be tackled only after your horse can move easily and comfortably in and out of shoulder-in and haunches-in. Half-pass may not look much more difficult than shoulder-in or haunches-in, but in half-pass, the horse is moving in the direction of his bend and making full use of his hips and his shoulders.

The training order matters because each successive exercise makes greater demands on the horse's mind and body.

CHAPTER 10

The Great Outdoors

TRAILS ARE WONDERFUL FOR HORSE TRAINING. You can train your horse *for* trails — that is, prepare him for the new challenges posed by footing, terrain, trees, hills, and so on. You can also train your horse *on* trails, using the outdoor environment to help your horse develop his strength and flexibility, his endurance, his attention span, his ability to work in company, his ability to work alone, and his trust in you.

Trails can be just as wonderful for the trainer. Training on trails can increase your confidence and help you develop your ability to focus on your horse and ride him correctly regardless of the terrain, the weather, and many types of distractions.

Training for Trails

Q My husband and I have just moved to a beautiful part of the country where we hope to enjoy a lot of trail riding. Our little "farmette" is on just three acres, but across the road from our property is a forest preserve with plenty of trails to explore. We brought our horses with us, and I am now faced with the need to train my mare to go on trails. Lucy-fur is definitely a city girl, or at least a subdivision girl. All the trail riding I've done has been on vacations, on other horses. Lucy-fur is intelligent and balanced, so I don't expect that she'll get into trouble on the trail, but I would like to know if there is some "trail prep" training that I could do with her at home before we venture out into the big exciting forest world.

A A good trail horse is intelligent and balanced, attentive without being nervous or spooky, and careful without being worried. It sounds like your mare has those traits, but I like your idea of preparing her for trails; most people wouldn't even think of that.

I suggest that you first go out on the trails riding someone else's experienced trail horse, and take notes about the various places you go and the obstacles you encounter. Whenever the horses have to do something out of the ordinary — if they need to bound up a steep hill, for example, step over a series of logs, or pick their way through a path consisting primarily of mud and rocks — take notes about those things, too. Go through your notebook at home and ask yourself how your mare would deal with each of those situations. Wherever you answer "I don't know" or "I'm not sure," mark that item with a star, because those items are the ones you'll want to try to approximate at home before you take your mare on the trails.

While you're on the trail, notice what other horses do and how you react to their actions. Some horses pick their way carefully around rocks while others just sort of stumble along, hooves clanking against stone. Some horses look carefully at the ground on the other side of a log before picking up their feet and stepping across, while others seem unaware of the log until they trip over it, suddenly snatch their feet out of the way, or make a sudden, violent leap to one side when they finally notice that there's a big long thing near their feet that might be a log or possibly a crocodile!

Spending hours each day in a hilly pasture helps prepare a horse for trails.

If you have been riding your mare in rough, rocky fields and uneven pastures, then the footing probably won't come as a complete surprise to her. If you've ridden her only in boarding stable arenas, where the footing is, if not manicured, at least consistent and predictable and flat, then she may need to learn how to carry herself *and you* on the trails. If you have friends whose horse pastures include hills or rolling terrain, try to arrange a week or two of "hills holiday" for your mare. Learning how to balance traveling up, down, and across hills will be excellent preparation for trail rides.

Introduce her to anything you think she might meet on the trail, from dogs to llamas to ATVs. Anything that can be made familiar *now* will help her make an easy transition from arena horse to trail horse.

Spooky Horse Has Short Attention Span

Q My horse becomes inattentive and spooky on trail rides. He is slow to respond to my leg and rein signals, he rubbernecks right and left like a tourist, and he's a pain to ride. He can focus on me for about two seconds at the beginning of the ride, but then he's distracted by this, he's distracted by that, and he's not really "with me" any more. He sees scary things everywhere — last time we were out, I had to get off and lead him past a tree that had him all worried and upset. I have to do this several times every time we go on the trails, and it's getting to be a pain. If the tree didn't eat him when I led him past it, why was it so scary when I tried to ride him past it? Was it just the same attention span problem and he didn't remember it was the same tree? I was hoping that we could explore area trails together and have a great time, but now I'm not sure that can happen. It surely won't happen as long as he has a two-second attention span! How can I train him to keep listening to me even if I'm not bugging him all the time? Sometimes I think he pays better attention when I'm leading him than when I'm riding him.

A   You may have answered your own question. You don't need to be "bugging him all the time," but you do need to be in constant communication with your horse. A big component of training consists of teaching a horse to listen to the rider and let the rider's directives overcome his own instincts; another important part of training involves teaching the horse to trust the rider so much that he'll do this *happily*. If your horse takes his attention away from *you* for more than a second or two at a time, perhaps *you* took your attention off him! Consistent riding means constant communication — not the kind where you issue orders and your horse says, "Yes, right away, sir!" but the kind of communication that lets you carry on a steady, gentle, low-key conversation.

Being spooky and timid and looking around at all the things that might be dangerous doesn't indicate a short attention span, that's just a horse being a horse. Horses are naturally timid; they worry about strange objects that might attack or harm them. That's normal. Through the training process, we teach horses that they can rely on us to take them safely past strange objects.

The last line of your question is on target: Your horse trusts you more when you're leading him than when you're in the saddle. This is very common. Horses are accustomed to focusing on a visible handler, and they trust and feel safe with you when you're working them from the ground. They won't necessarily trust you as much or feel as safe with you when you're in the saddle. When you're in the saddle, your horse can't see you, he can only feel your aids and hear your voice. When you're standing on the ground communicating with your horse, you are secure, balanced, and focused on him. In the saddle, you may be less secure, balanced, and focused. This makes you less of a leader, and your horse may be less relaxed and trusting. Once you dismount and take charge again, he can relax, trust you, and believe you when you tell him that it's safe to walk past that horse-eating tree.

When you're out on the trail, practice reassuring your horse and encouraging him to pass things (scary rocks, horse-eating trees, and so on). Stay safe! If he is so frightened that he won't approach or pass it unless you dismount and lead him, dismount and lead him. This isn't reinforcing bad behavior, it's showing him that when he is afraid, you will help him out. Walk past the item without making a fuss about it, paying close attention to your horse and to yourself. How are you standing and walking and moving, how are you handling your horse, what reassuring signals are you sending with your posture and movement and voice? You'll need to give him — or find equivalents for — those signals when you're back in the saddle.

Horses aren't good at entertaining two thoughts at once. If you focus on your horse and keep your horse's primary focus on you, then other objects of interest (e.g. scary rocks and trees) will capture his attention only briefly, because you'll recapture his attention almost instantly. Your horse needs to know that you are listening to him. If his attention wanders briefly, or if he warns you about a

Your horse may need you to lead him past a dangerous-looking bush.

potentially horse-threatening rock, tree, or abandoned hubcap, don't ignore his warning. If you pretend not to notice, he will warn you again — by spooking again, perhaps more strongly this time. If you comfort him, he will know that he was right to be afraid — since you acknowledge that the object is frightening, spooking is obviously the correct reaction! Avoid both of those scenarios. Instead, give your horse the clear message that you've seen the object, evaluated it, analyzed its potential for harming horses, and are unimpressed. Your horse won't understand your words, but he will understand the body language and breathing that goes with them.

Say, "Yes, a rock/tree/hubcap, I see it, thanks, no worries, everything's under control, now I believe we were about to leg-yield for seven steps, yes?" Let him know that you've registered his warning, that you don't perceive a threat, and that there is something you want him to do *now*. Practice attentive riding. This resembles defensive driving, in which you are looking at the road, aware of what's in front of you, in back of you, and along the sides of the road. You're aware of oncoming traffic, your own speed, and the indicators on your car's various dashboard gauges. You're aware of what you're doing and of things that might happen. Defensive driving makes you a more effective driver with better judgment and more reliable reflexes.

Contrast this with the inattentive driving you may do when you are preoccupied with other thoughts and drive almost on "autopilot," barely noticing anything at all. You may find yourself in the parking lot at your office, then suddenly realize that it's Saturday and you had intended to drive to the bank and the grocery store. We do that kind of driving when our minds are elsewhere. Passive, unfocused, "passenger" riding is the kind of *riding* we do when our minds are elsewhere. For better driving and better riding, think Zen: Exist in the moment; *be here now.*

Training a Horse to Neck-Rein

Q I'm training my new horse on trails, and he's taking to everything like a duck to water. I've always wanted a real ranch horse, and this one was brought up on a ranch where all the horses live out and run around together, run up and down hills, and do natural horse things until they are five years old. He was eight when I bought him last year, and he's just wonderful. There's one thing he doesn't know how to do that I don't know how to train him to do and that is neck-rein. He's 15.2, very sure-footed and balanced, and very intelligent and willing. (Can you tell I think he's just about perfect?) How can I teach him to neck-rein?

A Neck-reining has little to do with necks and reins, and everything to do with the rider's seat and balance and leg and weight aids. Neck-reining doesn't mean turning the horse with the reins, it means riding with loose reins and turning your horse with your legs, weight, and balance *as usual.* Your rein hand's default, neutral position will be

Laying the outside rein against your horse's neck can become a cue that means "turn."

about four inches above the withers and will move only a few inches right and left ("turn"), up and forward ("go forward"), up and back ("step backward").

The prerequisite for teaching your horse to neck-rein is that your horse must be responsive to your weight shifts and to the movements of your legs and seat, and be happy and confident when ridden on very light contact. Begin by using your reins as "leading reins." Hold one rein in each hand, keeping your hands wider apart than usual so that the reins don't touch the horse's neck. Ride with as little rein contact as you possibly can, preferably with no contact at all.

When you want to turn left, ask for the turn with your body, look left, and move both hands a few inches to the left so that the right rein lies loosely against his neck. Do not pull the rein, shorten it, or do anything that would move the bit or create or increase contact with your horse's mouth. The only signal at this point should be the *loose* right rein lying against his neck. If your horse is very responsive to your leg and weight aids you'll probably get that left turn.

If you don't get it, exaggerate your weight and leg aids and repeat the cue. At the same time, take your left hand farther away from your body and use the left rein to lead him into the turn. As soon as he turns, take the right rein away from his neck. Once he understands this, you'll be able to hold your loose reins in one hand and cue your horse by moving that hand a few inches in the direction you wish to turn.

He already knows how to turn; he's just learning a new signal that means "turn *this* way, *now.*" The neck rein cue is the rein itself lying against his neck, but you'll also be using your legs, seat, and weight to indicate the turn, and you'll reinforce the cue by using the inside rein as a leading rein. Hold the "cue" rein against the side of his neck only during turns, so that he can learn to associate it *only* with turning.

Be patient. If you are impatient, you'll think too much about the reins, and you'll try to pull your horse into a left turn by moving your rein hand strongly to the left — *don't do it.* This doesn't help the horse understand and it won't help him turn; it will just put pressure on the *outside* rein, tilting his head and twisting his neck to the outside and throwing his weight onto his inside shoulder. This confuses and upsets the horse and convinces riders that neck-reining is mysterious and difficult. It isn't, and you'll find it easy if you remember that, as always, the reins should be the *last* and *least* of your aids.

Repeat the routine often, first at a walk, and then at a jog. Many repetitions at each gait will help confirm your horse's response to his new cues.

Can Trail Riding Improve Gaits?

Q Everyone I know says that it's important to take horses out on trails, even show horses. I agree that it's good for them to have a break from the arena and that it's fun to go somewhere else. What I don't understand is why people say that trail riding can make your horse's gaits better. Most of the exercises we do in the arena are supposed to make the horse's gaits better, and you couldn't do most of those on the trail where you wouldn't be sure about the footing and there would be people passing you and doing crazy things. My horse is in training and I don't want to risk his legs by racing him around on rocky or muddy paths. I go on a trail ride once a month for fun. If it's going to make my horse's gaits better, I will do it every week, but I want to understand exactly why and how his gaits are supposed to improve.

A Going out on the trail a couple of times a week would be a great thing for your horse and for you, too. Trail riding has a lot to offer horses, and some of the things that you like about the arena — the walls and roof that protect you from the weather, and the predictable, even footing — are not necessarily the best thing for your horse's all-around development.

Horses can become bored and unenthusiastic if all they see is their stall, the barn aisle, and the walls of the indoor arena. Even an outdoor arena is a welcome change, and they're even happier if they can go out on the trails. Your horse's training won't be harmed if you take him out on the trails once or twice a week; his body and mind will both be improved even if all you do is walk him down the trail on a loose rein.

You're right about the footing — most trails have rocks here and there, and many have seasonal mud. They may also have fallen leaves in autumn and snow and ice in winter. But those variations in footing, with the exception of ice, don't have to threaten your horse's soundness; in fact, they can add to it. Just as walking on a treadmill doesn't develop your muscles in the same way as walking outdoors on grass or on a hilly path, working in the arena doesn't develop your horse's muscles in the same way as working outdoors on grass or on trails.

Horses were born to be outdoors, walking and running over varied terrain, breathing fresh air, and looking into the distance. A trail ride can give them a welcome break from their regular work, but that doesn't mean they'll be lazing along

Trail riders can combine a refreshing break with good exercise.

making no effort. In fact, they may be working harder on the trail, but they'll do it because they want to. Walking up and down hills is low-impact work, but it's demanding in its own way and it's *fun*.

The ground poles and cavalletti that many of us use in arenas can help us develop our horses' coordination, musculature, and stride, but you can also develop those things outdoors, on the trail. Trail riding allows you to ask your horse to take longer strides when there's a reason for it that he understands: to move up, move out, catch up, or keep up with other horses, or just to see what's around the next bend or over the next hill. Horses worked exclusively in an arena have to be taught to step over poles and cavalletti. Horses on trails understand exactly why they need to step over a branch or log or fallen tree — it's between where they are now and where they want to go.

To encourage your horse to stride out at the walk, you can increase the distance between walk poles in the arena, or you can walk your horse down the trail next to another horse with a more ground-covering walk. If you and the person on the other horse are carrying on a conversation, you'll find yourself automatically using alternating leg aids to keep your horse level with the other horse.

If your horse needs encouragement to go forward in the arena, he may take much more interest in going forward on the trail. You can sit quietly and enjoy the "forward" that you don't have to ask for, or you can make use of the "forward" and ask your horse for longer strides at walk and trot — something that may be difficult in the arena where there is no horizon in sight, no hills to look up at or valleys to look down into, and always a wall coming up in a few more strides.

If you have become frustrated with basic lateral work in the arena and you are finding it increasingly difficult to move your horse laterally from the quarter line to the rail, use the trail to refresh your lateral work. The problem with horse and rider boredom, indoor arenas, and lateral work is that often the "forward" component is lost or minimized because both rider and horse are focused on the "sideways" component. Of the two, "forward" is more important, because without that, the value of "sideways" is lost. Most horses show much more reach and enthusiasm in leg-yield, for example, when they are being asked to shift sideways while going energetically forward on the trail.

If your trails offer opportunities to canter and gallop, take advantage of the chance to let your horse extend at the canter and to practice letting your horse go and bringing him back to you. "Rating" a horse is more easily learned as an outdoor exercise on a trail.

Trails provide reasons for horses to do things — that's a nice change from doing things just because we ask them to. Trails are fun for horses and should be fun for riders as well. Horses tend to come back from enjoyable trail rides showing renewed interest in their work. They've had a chance to stretch and look around and breathe, and they're refreshed and ready to do whatever you ask them to do. Go out on the trail with your horse, have fun, and watch his enthusiasm return, his musculature develop, and his gaits improve.

Working on a Slope

Q How exactly should I ride my horse on hills? I ride on trails four or five times a week, and we're always either going up a hill or coming down one. I want to train my horse to take those hills smoothly in his stride (literally), but I don't know how to do this because I don't know how to ride up and down hills. I've ridden all my life in Kansas where everything is flat, but now I'm in Washington State, where *nothing* is flat.

Your position and balance should be the same whether you are riding uphill or downhill.

If I want to teach Commander to power up a hill at the walk (but not trot!), what should I do? And how can I teach him to go down the hill (at a walk) in balance? Mostly, I'd like to train him to stride the same (rhythm and time-wise) going up and down hills. Right now he seems to rush up the hill and do stop-start, stop-start down the hill. I try to help by leaning back a little and taking my legs off him on the way down so that he won't think he has to rush, but it isn't helping.

A Your horse needs to develop more balance and more muscle to help him move comfortably on hills. If he's from Kansas too, he's probably just as confused as you are. You can help him learn to stride evenly by staying in balance in a light half-seat and using your legs to push him on a little when he climbs the hill — don't worry, you aren't asking him for speed, you're just using your legs to ask *his* legs to reach forward a little more.

When you're coming down the hill, it's okay to sit straight (don't lean back), but your best position would be that same light half-seat. Use your legs — remember, they aren't asking for speed, they're reminding your horse to use his hindquarters more energetically and reach forward with his hind legs. He'll find it much easier to balance if you remind him to keep his "engine" underneath him and not out behind him.

Help for Young Horse Bucking

Q My young horse (he's four) has started bucking on trails. Not every time we go out, but often enough that I'm nervous about riding him if I don't have someone else with me. He seems to start off okay, then get more and more annoyed with everything, and by the end of the ride I know some little thing is going to set him off, and he'll throw a buck. I need to train this out of him — I'm too old (43) to be riding a horse that bucks. I would hate to give up my trail riding because that was one of the main reasons I have a horse. For me, arena work is prep for trails.

The back story on him: His breeding is mostly Morgan and Saddlebred with some Connemara and a little bit of Standardbred (I think) thrown in. He is a little bit long in the back, and he has a big trot. About a month back, I started working him on the level trails just around our property. They are not all that interesting to me, but they are obviously full of things that frighten Finnegan.

The odd thing is that he never used to buck until I started taking him on the trails, but nothing is different. I use the same saddle and bridle on the trail that I do in the arena, and we're doing the same work for the same amount of time. Walk, trot, and canter, nothing fancy, for one hour (sometimes a little bit longer on the trail). We do exactly what we would do in the indoor but without the dust! We work in the ring on Monday, Wednesday, and Friday, and on trails on Tuesday, Thursday, Saturday, and sometimes on Sunday, but not for very long. By Sunday he is usually mellow enough that things don't bother him and he doesn't spook or buck, so our Sunday rides are nice even though they're usually short because I have to be home in time for church.

A You're very observant and you've answered most of the questions I would have asked. I think your work on trails may be causing your horse to become tired and possibly slightly sore as well. Four years old is still young and immature. Doing the same work on trails that he does in the arena means using his muscles differently, because he has to adjust to the footing and the slope. Trails don't have to be steep to cause physical stress. If you've ever used a treadmill and adjusted the angle so that you were walking slightly uphill, you know that even a tiny change of angle makes you work much harder. If your trails have

Don't ask yourself, "Why is he being bad?"
Ask yourself, "What is he trying to tell me?"

any slope to them at all, your horse is working considerably harder on the trail than in the arena. That isn't a bad thing in itself; trail footing is good for developing a horse's balance, and walking up and down hills, even gradual, tiny hills, is good for developing a horse's muscles and power. A horse has to work his hindquarters and belly muscles to tackle hills in stride — that's where all that lovely muscling comes from.

I don't think it's the nature of your training that's at fault here — I think the problem is the *duration*. You're asking him to work on the trail, which is fine, but you're asking him to work for the same amount of time that he works in the arena. That's not so fine, and it's probably making him tired and sore, which provokes that end-of-the-ride buck. The fact that he doesn't do this on your Sunday trail rides tells me that something is different and there's only one difference you've identified: On Sunday mornings, your trail rides are shorter.

An hour or two of gentle walking on trails on a long rein may not be any more demanding than a full hour of walk, trot, and canter in an arena, but when you move your training program, unchanged, from the arena to the trail, you're asking too much of your horse. When you go from *working* indoors to *working* on the trail, you need to cut the time back. If you spend two hours on the trail, spend most of that time walking. You can encourage longer strides at the walk and do some light lateral work. But when you're *working* on the trail, try to do no more than half an hour of your arena routine. When you figure in the footing and the slope, that half hour is probably giving your horse as much of a workout as a full hour in the arena.

In time, you'll be able to do more work on the trails, but for now, go easy and focus on enjoying the outdoors with your horse. Remember, he is still growing and developing. You want him to grow up to love trails as much as you do, so drop back to a level of work that will allow him to relax and enjoy the outing.

Horse Wants to Graze under Saddle

Q Our horse Snortin' Norton is a cremello Missouri Foxtrotter that was a pasture ornament for 12 years. When we bought him, we took him to a professional trainer who did a wonderful job in 60 days refreshing his training. Snort is quite a character; he has a very playful attitude and doesn't mind anything you do to him. We've ridden him on trail rides, and he's just a doll.

Sweet, compliant, happy-go-lucky, isn't afraid of "monsters" in the woods, and is respectful of the other horses on the trail. There's just one thing.

When we took Snort to the trainer's, he was taken off of pasture and put on hay and a *slight* amount of grain for weight reduction. Since we've had him we have changed from a boarding situation where he had *no* pasture to owning eight acres of our own. We used to have battles when we would ride or lead him past grass. He was quite obsessive about grazing.

We figured that once he was on pasture at the new place, he would drop this compulsion about grazing. As long as we keep him moving on rides now, he's fine. But if we are in a meadow and stop to rest or get our bearings, the battle starts. He gives us several clues, mainly pawing and dropping his head. Our solution to this is to keep him moving. We give him the go forward cue, or ask for side movement, backing him up, whatever we can to keep his feet going our direction and keep his attention on us. He seems to challenge me more than my husband, but I ride him less often. My level of awareness with him isn't what my husband's is.

My question is this: Is this just something that we might need to cut him some slack on? He was on pasture 24/7 for 12 years; can he learn that under saddle means *no* grazing? And in general is it such a bad idea to let them graze under saddle provided you tell them when they can and when they can't? What about a grazing cue? I've heard that some endurance riders do this.

On one hand, I think that for the few hours he's being ridden a week, he can stand not to graze. On the other hand, I'm trying to consider the horse and be compassionate. He feels about grazing the way I do about chocolate! Is it a possibility that once spring gets here, and he's getting his fill of pasture at home, he won't be so determined when we pause on a trail ride?

A It's natural for horses to graze all the time, but you're right, it's not practical to allow the horse to graze at will while he is being led or ridden or handled in any way. You're also quite right about him being able to survive *not* grazing during those few hours a week when he's being ridden.

Your horse is nice and predictable, since he paws and drops his head when he wants to graze — that's a convenient built-in warning system.

There are several ways to deal with the grazing-under-saddle problem. But first, be reasonable. While you're teaching your horse a new skill — the skill of standing *without* trying to graze — don't make his life more difficult by tempting him. If you're going to stop and stand on a trail ride, and you know perfectly well there is no good grass in his pasture, but lovely grass in the meadow, don't ask him to stand in the meadow — find somewhere else to stop and stand. Make it easy for him to do what you want him to do; stop and stand when you're on a less appealing, less edible surface. Later, when he understands his lesson and has had a lot of practice, you can begin testing him in more tempting spots.

I would cut him some slack, yes, especially when you are riding him through grass that he doesn't have at home. Horses adore new grass, and it *is* their natural food — you don't have to allow Snort to stop and graze every two minutes, but do understand why it's natural for him to *want* to do it. And don't expect him to learn that "saddle = no grazing" — if "saddle" comes to mean "no grazing *plus* lots of pulling and whacking," Snort will soon learn to run the other way when he sees the saddle, and that's not what you want.

Especially on long trail rides, it's reasonable to let horses have an occasional pause that includes, when possible, a chance to graze and a chance to drink. You most certainly *can* teach a horse a grazing cue; it's actually quite easy to teach a signal that means "Okay, *now* you can graze." This will make it easier for you to say "no" at other times, because he'll understand that he needs to wait for the signal. The signal itself can be anything you like, just so long as it's a clear signal that you

A horse that plunges his head down to graze can pull you out of the saddle.

don't use for anything else. With one horse rather like your Snort, I used a combination of three taps on the right side of the crest and a verbal "eat grass now" signal; after just a few days, the taps were the only signal the horse needed. People-oriented horses really enjoy showing how well they understand signals, and Snort is obviously people-oriented. If you have any difficulty with this, try using a little bit of clicker training — it's a very good way for you to make it clear to your horse that you are rewarding the moments when he stands quietly on a loose rein with his head *up* instead of the ones when he is diving toward the ground.

And speaking of reins, one problem that afflicts the riders of horses that dive for grass is the tendency to hold the reins too tightly and keep them far too short. If those lengths of leather and your biceps are all that's keeping your horse's head off the ground, you will quickly develop sore hands, arms, and back, and your horse will quickly develop a sore mouth and neck (and, eventually, back). You need to be able to ride — even on the trail, even through the meadow — on a long rein with soft contact (if you are riding English) or on a loose rein with no contact (if you are riding Western). It's difficult to make yourself give the horse a long rein if you are worried about him snatching *all* the rein and eating grass, but here's what happens when a rider shortens the reins and keeps them short:

1. The horse's mouth hurts, and he tries to push away from the pain by extending his neck forward and/or dropping his head.
2. This *seems* like an attempt to graze, so the rider typically jerks the reins and tries to pull the horse's head up.
3. The horse becomes steadily more uncomfortable in his mouth and neck. His muscles begin to cramp, at which point the horse, desperate to avoid *that* pain, will typically plunge his head and neck forward and down.
4. At this point, the rider either drops the reins or falls forward, allowing the horse to stretch its neck (and grab some grass while his nose is close to the ground).

What the rider "learns" from this is *wrong*, but understandable: that she or he is helpless and needs to keep the horse's head as high and pulled-in as possible at all times. What the horse learns from this is inconvenient, but accurate: that the misery of a sore mouth and cramped neck are caused by the rider constantly pulling the horse's head up and in; that the way to escape from the situation is by deliberately plunging his head and neck toward the ground; and that this action is rewarded with a longer rein, a chance to stretch the neck, and even, on particularly happy occasions, some grass.

Endurance riders do let their horses graze and drink during long rides, but their horses aren't allowed to graze as they please throughout the day. The riders use signals for "Okay, eat *now*" or "Go ahead, have a drink *now*." This saves time, effort, and argument along the way. Endurance riders also, as a group, are always preoccupied with their horses' soundness and comfort and enjoyment of the sport — it's one reason that I love working with them! Because of this, they make every effort to teach their horses to work efficiently, at a steady gait, with head and neck extended and lowered (or raised) to whatever degree the terrain requires. They are always listening to their horses; their horses, in return, listen to them.

That's what you should do with your horse: Ride him forward, easily and steadily, letting him find a comfortable working position for his head and neck. Since he doesn't get ridden very often, you must make a special effort to ensure that he is allowed to relax and stretch his neck regularly. Carrying tack and a rider is work for a horse, and moving in a way that it would not normally move is also work. During work, it's nice to have good working conditions, and it's nice to have breaks. That's where the "Okay, graze *now*" signal will come in handy, because you don't want to have your horse's nose plunge toward the ground automatically, as soon as you stop.

You are clever to have figured out that your level of awareness matters. When you shape the behavior of any animal — equine or human — you need to be very aware of everything that is going on, so that you can make any needed changes or corrections *before* the problem develops. If your horse has been moving forward steadily but begins to slow down and move with a less regular gait when there is grass underfoot, he is saying, "I'm beginning to think more about this grass than about *you*." At that moment, you have choices. You can wait to see if your horse stops and grabs the grass, or you can keep him moving and praise him for paying attention. When you stop and stand, on the other hand, you probably won't want to have to put him back into motion just to keep him from diving for grass. That's where a grazing cue can help, because if your horse is waiting for the cue, he won't dive until the cue is given.

You may also want to teach a cue for "We're going somewhere, head *up now*, please" to get your horse's head off the ground when he's been grazing (with permission) — this will make your life easier and your horse's life more pleasant. No rider should need strong biceps to ride a horse; no horse should have a rider hanging on his mouth.

While you're at it, work on your leading! Practice leading Snort all around your property, make your wishes *very* clear, and reward him for staying in position and paying attention to you. When you stop, he should stop, too. When you give the grazing cue — and not before — he should be allowed to graze. When you move on, he should be in position at your shoulder, paying attention to you, and not varying his pace or making attempts to go anywhere you haven't asked him to go or do anything you haven't asked him to do. The work you do when you are leading him will be reflected in his increased attention and improved cooperation under saddle, and your work under saddle will improve his behavior on the lead rope.

Snort sounds like a lovely horse, and I'd say he's lucky to have you for an owner. I hope you enjoy many pleasant years together, with regular grazing on cue and no heavy pressure on either end of the reins.

Horse Misbehaves on Trails

Q My beautiful trail horse is very well trained in the arena, where instructors are usually found. She is fine on the trail with one or two other horses she knows. Unfortunately, I get only two riding days per week, and the weekend day often turns into a seven-horse trail ride, wherein Dolly sometimes acts like an idiot in one of the following ways:

- ▸ If she wants to lead the herd and I do not let her do so, she will act pushy; she will canter sideways, with her hind feet off the trail.
- ▸ If I bring her to the back of the herd, she will jig nonstop unless I let her run up to the head of the herd. This is often combined with the sideways and backwards problems.
- ▸ She attempts to back off the trail.
- ▸ She refuses to stand still while tied for lunch stops or even gate openings and closings.
- ▸ She cuts in front of all the other horses in an S-pattern while moving fast.

There seems to be no rhyme or reason to which days she is good and which days she is naughty. I have learned the theory behind the "fix" for each of these problems but am probably not applying the solutions fast enough as I am still figuring out what to do. It would be ideal to ride with an instructor, but I do not know any such person who would go out for an all-day trail ride with the crazy New Yorkers in my circle. And that is the only time she does the bad stuff described! By the way, Dolly is a big red Appaloosa, age 13, who lives in a stall but gets turnout every day plus 25 miles per week of trail riding. I have been riding for almost three years, and have owned her for nearly two.

A It's not at all unusual for horses to behave beautifully in an enclosed arena and much less beautifully outdoors, on the road, or on the trail. Dolly needs to be trained to understand (a) exactly what your expectations are *on the trail*, and (b) that she must meet those expectations.

You need friends; the all-day trail ride group would be perfect. They undoubtedly find your mare's behavior irritating or at least distracting and probably won't mind helping you out. If they mind, try bribing them with food — brownies make excellent bribes for humans.

You're going to play a game for as long as the trail ride lasts, and you'll need several riders (at least) and a wide trail or field. The game is called "leapfrog" or "lead, follow, stop, and go," and its purpose is to teach your horse that when she is out with other horses, she must go quietly whether she is first, last, or in the middle, and whether she is going ahead alone or stopping and standing while the others go out of her sight. She must do all of this calmly and at all gaits, *and* she must be calm about passing other horses and being passed by them.

A horse that misbehaves on trails is annoying to everyone, not just her own rider.

The game is played like this: The riders take their horses out in single file at a walk. The last horse in line trots past all the others and takes his place at the head of the line, and then the new "last horse" does the same thing, until every horse has been first, last, and in the middle many times. The next phase involves the last horse stopping, standing, and *then* trotting past the others, passing them and continuing until he is out of sight of the others, at which point he is made to stop and stand until the others arrive. Then it all begins again at a trot, with the last horse cantering past the others, and when everyone is comfortable with that, the last horse is asked to stop, stand, and *then* canter past — *well* past — the others, and stop and stand again and wait for the others to come up.

It sounds simple and easy — it isn't. But it's worth doing, and I think you'll enjoy the process. Consistent practice is the key to this situation, and that requires other horses and their riders, so start bribing your friends! It will be good for the other horses, too, and their owners can either enjoy showing off how perfect they are, or work quietly on any problems that *their* horses may have.

Bringing a Horse's Back Up

Q My husband's gelding, Beau, was trained English (worked on the bit) and used to do a lovely trot in the ring with his neck arched and presumably his back rounded. Since we don't do any ring work any more, but only trail ride, Beau is starting to carry himself more like a Western pleasure horse, with a flat neck, and I am worried that he is hollowing out his back. My husband thinks he should always ride with a loose rein, Western style, rather than keeping the contact with the snaffle bit that would be needed to round him up. Is it possible to trail ride Western style but have the horse's back rounded up?

A If Beau is hollowing his back and flattening his neck, it is *not* because he needs more contact in front. The pretty arched neck that you like to see can be very pretty indeed if it comes from a happy, strong horse that is carrying himself and his rider well, and carrying his neck and head. A horse that moves well will develop his back muscles and the muscles of his neck. You'll feel those muscles under you when you ride, and you'll see them in front of you, from the withers to just behind the horse's ears.

A good topline comes from correct movement over time.

You can't bring a horse's back up by doing *anything* with the bit or the reins. If you use a curb, or try to make Beau's neck arch through constant contact with a snaffle, you may succeed in teaching him to overbend his neck and tuck his jaw back toward his chest, but this will not lift his back, or help it in any way — in fact, it will hurt his back.

If Beau is working well, using his hind end, and is relaxed through his back and neck, he will develop his "carrying" muscles over time. If his back (or his neck, or his mouth) hurts for any reason, he is going to drop and stiffen his back. Beau is still young, and he may be telling you that his back is sore. Check his back and the fit of his saddle, and remember that if you want Beau to carry someone up and down hills and for long periods, you need to develop the muscles that will allow him to do it in comfort.

If he isn't sore, than perhaps he just isn't being asked to step up under himself from behind. Your husband needs to use his legs to ask Beau to move forward energetically — not *faster* but with larger steps and more engagement. If Beau shuffles along with a sagging back, he won't develop the muscles that will allow him to move differently — if he moves forward in an active walk under a balanced rider, he *will* develop his topline.

The important thing to remember is that all of this takes time; muscles don't develop overnight. Be patient, and build each good ride on the foundation laid by the previous good ride.

CHAPTER 11

Connection, Roundness, and Collection

ON A HORSE THAT IS STRUNG OUT AND DISCONNECTED, riding can make you feel as if you are pushing a very heavy wheelbarrow. If you're in this position and you're worried that your horse seems unable to collect, shift your focus. Collection isn't a starting point, it's a goal. Collection can come only from strength and balance and roundness, and those qualities can come only from *connection*, so that's your starting point: Begin by getting your horse connected! Whether he is strung out, inverted, overbent, or just heavy on the forehand, your initial training goal will be the same: to teach him to use his hindquarters more actively and energetically.

Making a Horse Round

Q I have been studying dressage from books for several years but only started taking dressage lessons a few months ago. I love it, but I find that I am often confused by the differences between what I thought I understood from reading all those books and what I am told by my dressage teacher.

My latest confusion has to do with roundness. My teacher says that I should always ride my best and ask my horse to move the best he can. It is okay for me to warm up at a trot on a loose rein (hanging rein with no contact), but as soon as I shorten the rein, I am supposed to put my horse on the bit and make him round. I am a fairly physically fit person, but this activity takes a lot of strength! It's not that I get tired, but I wonder how people manage who are less fit. Can you not ride dressage unless you are very fit and strong? So, I am confused about rider fitness. On the same subject, I am also concerned about the horse, because all of my dressage books and my teacher agree that a horse has to be strong and fit to go on the bit and stay round. Is it right for me to expect the horse to be round immediately after his warm-up, and stay round until the end of the lesson, or is that too much to ask? And how much strength should the rider need to accomplish this?

A Those are good questions — all that reading has been good for you! I hope that once you've read this answer, you will take it to your instructor and sit down and discuss it together. It's not my purpose to undermine your instructor in any way. I suspect that the divide between you is largely to do with semantics and perceptions, so the two of you will need to talk. First, you'll need to define your terms.

Riders do need to be fit and strong. This is *not* so they will have more power over their horses and be able to kick and hit them harder but so they will have excellent control over their *own* bodies. Without good control of your own body, you can't give light aids or develop an educated, "listening" seat. A rider with good tone and good control, both of which require good core strength, can be quiet and gentle and use the aids precisely and clearly. The fit, trained rider can ask the horse for, say, a balanced, cadenced trot in a rounded outline; the fit, trained horse can offer exactly what the rider wants, then continue to offer it until the rider asks for something else. The horse maintains his outline and cadence not because the rider is holding the one and forcing the other, but because he *understands* what the rider wants and is *able* to stay round.

The warm-up on a loose rein is only the beginning of a proper warm-up. Those first 10 or 15 minutes will stimulate the horse's circulation, begin to lubricate his joints, and help his muscles to become physically warm and capable of being stretched safely. Once this is accomplished and you pick up your reins to begin the rest of the warm-up, your horse should be working on the aids, but he should not necessarily be *round* at this point. The remainder of the warm-up should consist of everything the horse can do, from the simplest exercises to the most complex and sophisticated ones. By the time you reach the end of the warm-up, the horse will have done everything he knows how to do.

Throughout the warm-up, as the exercises become more difficult and demanding, the horse should become more round, so that by the end of the warm-up, he *is* round. That is when the *new* lesson may begin. So, yes, if your horse is working at a sufficiently high level that he is able to become round and sustain his roundness, then he *should* become round toward the end of the warm-up, and through the lesson that is to follow (with regular breaks for stretching, of course). But that's not quite the same thing as trotting on a loose rein for 5 or 10 minutes and

Rider strength is useful for many activities other than riding.

then saying to the horse, "Okay, become round right now, and stay round for the next 40 or 50 minutes!"

I vastly prefer the expression "on the aids"; "on the bit" seems to put the emphasis on the bit, the reins, and the rider's hands, which encourages front-to-back riding. "On the aids" begins with seat and legs and ends with the rider receiving the horse's energy into her hands and either holding that energy as roundness or allowing it to flow forward as an increase in stride length, speed, or both. The bit and reins and hands become, as they should be, the last and least of all the aids.

"Put," like "strength," is a word that can create misunderstanding and miscommunication between teachers and students and between riders and horses. You need to know exactly what your instructor means when she tells you to "put" the horse in position. There is nothing wrong with showing the horse what you want him to do and encouraging him to continue doing it as long as he can do so comfortably. There is a great deal wrong with forcing the horse into a position and attempting to hold him in that position by force.

Imagine yourself as a young ballet student. The instructor asks you to perform a plié in fourth position (the equivalent of you using light aids to ask the horse to round up for a moment). You don't quite understand the idea, so the instructor puts one hand on your lower leg and one hand on your back to show you — lightly and gently — precisely what your posture should be. Then she lets you carry on, and steps in to adjust your position only when you lose your alignment. *That* is good teaching.

Now imagine the same situation, but this time, the instructor pulls your leg hard and shoves your back into a position that you cannot achieve without pain, or that you can barely achieve and cannot maintain for more than a moment before beginning to cramp. That would be bad, and it would be even worse if she then tried to use force to *keep* your body in that position. At worst, you could be physically injured. At best, you would feel hurt and frightened, you would not trust the instructor, and the next time she approached you to adjust your position, you would become tense and rigid, which would ruin the whole point of the lesson and of ballet lessons in general.

Your horse is in the same situation as that young ballet student. You can ask him to move in ways that will help create the posture you want, but you must do so very gently, and not ask him to sustain the effort to the point of cramping or pain. If you allow the horse to develop his strength and balance gradually, and if you ask for no more than he can give at any time, you will find that the horse will steadily become stronger and better able to do what you want him to do, more easily and for a longer time.

Back muscles have to stretch, neck muscles have to stretch, belly muscles have to contract . . . this is *work*! Ask for a few strides, then allow the horse to relax and stretch forward and down, then put the horse back together (gently!) and ask again. Being a soft rider doesn't mean being an incapable or indecisive rider. It means being a strong and definite rider who can offer clear requests to the horse. If you have to hold more, it should be a clarification ("Please put this *here*"), not an amplification.

Think in terms of asking and allowing the horse to do what you want him to do, not about "making" him do this or "putting" him in a particular position. Good riding means riding in a way that enables the horse to become steadily stronger, more flexible, and more clear in his understanding, so that when you ask him something, he is able and eager to offer you a quick, enthusiastic response. If you do that, and your training follows a logical progression that develops your horse's body and mind over time, then your horse will always be ready for each new demand and will find it easy to do whatever you ask him to do — including "getting round."

Helping My Mare Use Her Hindquarters Better

Q I have had my first horse for about a year and a half now, and I have reached the point where I realize I must do something about her not using her hindquarters properly. I am wondering what type of ground work and saddle work would be best to help her use her hindquarters instead of pulling with her front legs. She is not using her muscles to hold up her back either. We do trail and arena riding but are not working in any particular discipline.

I am thinking of doing hill work with her but want to know what I should be looking for to make sure she is doing it in a way that is suitable. Also, I have heard cavalletti work would be useful but wonder if just by doing the cavalletti work, would she automatically use her muscles and hind legs correctly, or do I have to watch for something there, too? Is lateral work, such as turns on the hindquarter or leg-yields, useful? How much of this type of work should I do per day for it to be useful?

A   If you know how to longe a horse, some longe work on the largest possible circle will get you off to a reasonable start. Use just a longeing cavesson and longe line — no bridle or saddle or side reins. Longeing will accomplish several things:

- ▸ It will let you watch your horse and see exactly how she moves *without* a rider.
- ▸ It will let you improve the way she uses herself — still without a rider.
- ▸ It will let you build up her strength and musculature so that she will be able to move more correctly and efficiently *with* a rider.
- ▸ Even when you're doing good under-saddle work, you'll be able to go back to longeing periodically, both to warm up your mare and to give you a visual impression of her improvement.

On the longe, ask her to walk, trot, and halt but not to canter. Cantering won't be helpful until she is strong enough to move comfortably and correctly at walk and trot, making smooth transitions and halting easily. By that time, you'll be working under saddle. Before you ride, check your tack very carefully; a saddle that doesn't fit comfortably can make a horse understandably unwilling to use her hindquarters and back. A horse that engages her hindquarters

Longeing lets you exercise your horse and watch him at the same time.

will *lift* her back. If the saddle causes pain, the horse will learn to keep her back flat or drop it away from the saddle, and her hindquarters will never become engaged.

Check your mare's mouth and her bit. A horse with a painful mouth may be reluctant to move forward, especially at walk because of the need to make balancing gestures with her head and neck. If your veterinarian confirms that the mare has no mouth problems, then you can begin mounted work.

Once your mare has gained strength and coordination on the longe line, and when she's comfortable carrying a saddle and rider, take her out for long walks, incorporating any slopes you can find. Low hills would be ideal, but you can build her up on the flat ensuring that she walks actively — a *marching* walk. Riding her consistently from back to front will strengthen her hindquarters and back and enable her to use them better.

Focus on sending her forward energetically. Don't worry about trying to "set" a particular head or neck position — that's how horses learn to move incorrectly in the first place. Ride with the longest rein that allows you to keep a soft, stretched contact with her mouth. The more she uses her hindquarters and back, and the more actively she walks, the more head and neck motion there will be, so remind yourself to let your shoulders and arms, not just your fingers and wrists, follow the movements of her head and neck.

Whether you are longeing or riding your mare, always look for three things:

1. You want her to step up underneath herself with her hind legs.
2. You want that energy and movement to go through her back, which will lift and swing.
3. You want that energy and movement to continue through her neck, so that she will reach forward and down at each stride.

Riding over cavalletti requires a horse to lift and reach with her legs, but I suggest that for now, you make use of hills instead. Cavalletti need to be adjusted (height and spacing), and the work can be very stressful for a horse that isn't fit or well muscled. Before using cavalletti, give your mare a month of longeing and several months of active walking up and down hills. She'll be happier and healthier if you're riding outdoors, and you'll both have a better time. It helps psychologically, as well — it's easier to get a horse thinking "forward" if you are actually going somewhere. The active, purposeful, reaching walk your mare will offer on the way home, especially at feeding time, is the walk you want her to do *all* the way through your ride.

Lateral work can also help, and you're right to want to do it. But wait until your mare is stronger and moving better, and until you have a good instructor to help you. Like cavalletti work, lateral work can be of great value if done correctly, but can be counterproductive and damaging if done incorrectly. Do a little leg-yielding on the trail when your mare is moving forward nicely, but wait until she is stronger before asking for more than that.

One more thought: You will be asking your mare to use muscles she isn't in the habit of using, and you will be asking her to use other muscles differently. Don't push her too hard or too long, and don't ask for a lot every day. Spend a lot of time grooming — the massage will help improve your mare's circulation. Turn her out as much as possible (ideally, *all* the time) so that she will get constant, gentle exercise. Always warm her up before working her, either by walking and trotting her on the longe or by walking for 15 or 20 minutes under saddle.

Finally, remember that a big effort on Monday will make her stiff, sore, and perhaps a little unwilling on Tuesday or Wednesday, so don't be surprised or annoyed if her progress isn't perfectly linear. Any exercise progress chart for human or horse should show an overall upward trend but if you look at it closely, you'll see peaks and valleys, not a straight line slanting upward. If you're working correctly, the overall trend will be *up*, but it won't be "better every single day"; it will be "two steps forward, one step back." Help your mare by alternating her days of hill work with days of long relaxing rides on the flat.

Heavy on the Forehand

Q My gelding (14-year-old Thoroughbred cross) is heavy on the forehand. When I bought him, I was younger and just wanted to race around on my horse, jumping all sorts of things and having fun. We did that and he was wonderful. He stayed in a pasture while I went to college, then I took him to graduate school with me. Now that I'm older and know a little more about horses and riding, I realize that I let him slop along on the forehand and that this isn't good for either one of us. I have a nice place where I can keep him turned out all the time in the gelding pasture and can ride him in the arena when my schedule allows it (four days a week). What can I do to change things and get him off his forehand? Or is it too late for him to change?

Even a horse in his late teens can be reshaped through systematic, correct training.

A It's certainly not too late for either of you to change, and your situation sounds ideal. To make the best use of your time and the arena, spend the next year riding thousands of transitions between and within gaits. Ride them carefully, and feel what your horse is doing underneath you. Just making the change from walk to trot or short trot to longer trot can't be enough — you need to feel that the transitions are being initiated by your legs and that he is responding with his hind legs.

In addition to transitions, plan to do thousands of half halts. Ride all of your transitions and all of your half halts *forward* and *from the leg* so that your seat allows the horse to do what your legs are asking him to do, and your hands do as little as possible. Retraining and redeveloping your horse will require riding him strongly forward into a soft, light hand. You'll do no pulling, and you'll give him nothing to lean on — just soft, light communication.

When you become bored with this, add some lateral work. This too must come from your active legs and allowing seat, *not* from your hands. Periodically, to test yourself and your horse (do this when you're alone in the arena), hold the front of your saddle, a lock of mane, or a neck strap so that you aren't using your reins *at all*, and work on transitions. If you're interested in finding out just how much hand and rein you habitually use, this is an amazingly revealing exercise.

Lightening Eventer's Forehand

Q I have a Training-level event horse, a Thoroughbred, who I would like to move to Prelim soon. He consistently scores in the top three in the dressage phase. His personality is very cooperative, and he is uncomplicated to ride. He easily comes on the bit and gives a nice elastic feel on both sides. He is fairly responsive to my leg, but as he is pretty laid-back, sometimes I have to insist on sharp responses.

My problem is that although he feels light and responsive, he likes to carry his neck fairly low. The slightest squeeze on the reins causes him to drop his head. (I really believe that he thinks that is what I'm asking for.) It is much more pronounced at the canter. This was not really a problem doing Training-level tests, but it will not be acceptable as we move along. Trying to get him more "in front of my leg" by strong leg and frequent half halts is only somewhat helpful. We have tried lots of exercises, the most beneficial being zigzag leg-yields, spirals, and downward transitions in shoulder-fore, as recommended by my instructor, whom, unfortunately, I can't afford to go to as often as I'd like.

Needless to say, this is preventing us from accomplishing lengthenings and 10-meter circles, and we have problems rocking back for jumps after a galloping stretch. (Note: He is very responsive to a medium snaffle and does not try to run through my hands.)

A Your horse may not yet be quite strong enough to lift the base of his neck and carry it. You've done well to allow him to move naturally — many riders would have tried to "lift" his head and neck with pulling hands or a more severe bit. You've identified the problem: Your horse isn't really balanced in front of your leg and still has some work to do before he can become lighter in front. If you've done the basic straightening work with him, and he's low in front and heavy on your hands, ask yourself whether he is *evenly* low and heavy or *unevenly* low and heavy (leaning more on one hand than on the other).

If he is even, then his problem is simple: He isn't using his hindquarters enough, and he isn't using them correctly. Until he uses them, he won't really stretch his back muscles. Those muscles must stretch to develop his topline and enable him to lift and carry his neck and head at all three gaits. A horse's head and neck can weigh 150 pounds, and it's all unsupported structure. It takes muscle to lift and carry that structure, and the muscle you need him to add in the neck will develop only *after* he learns to step more deeply under himself, stretch his topline from hind heels to ears, and shift his balance back.

The exercise you mention is a good one: Use half halts to ask him to rebalance and stronger legs to ask him to step up with his hindquarters, moving them closer to his forehand. If he continues to move down rather than ahead, he will be pushing his weight (and yours) instead of *carrying* it, and he will continue to be long, strung out, and unable to do the things you want him to do: lengthenings, smaller circles, and "coming back" after a jump.

Stand back and take a good look at your horse. Are his muscles and joints stiff, weak, or both? For the sake of his soundness and future as an athlete, exercise him in ways that will strengthen his joints and build his muscles while stretching and suppling both. Here are some exercises that will probably help:

Trot-canter-trot transitions. Count your strides, beginning with, say, 30 strides of trot, then 15 strides of canter, then another 30 of trot, 15 of canter, and so on. As he adjusts to the exercise, becomes more supple in the back, and learns to shift his weight back, off his forehand, gradually reduce the number of strides until you are doing 20 and 10, then 16 and 8, 10 and 5. You'll know when to reduce the number of strides, because your horse's balance will tell you what he's ready to do.

The markers along your galloping track can help with stride-counting exercises.

More shoulder-fore. Incorporate this into your trot-canter transitions. A horse that is bent laterally cannot become stiff and resistant as he moves forward — take advantage of that! Position him, bend him, put him into shoulder-fore, and *then* ask for the canter transition.

Half halts. Transitions are the key to everything, and half halts are the key to transitions. Always remember that the test of a half halt is whether you *feel* the horse rebalance underneath you — lacking that, it was not a half halt.

Remember that your leg, whether it's a soft leg or a strong leg, is used *briefly* and then relaxed. Don't try to sustain a constant squeeze; that will only make you tired and teach the horse to ignore your leg aids. Ask, wait, and if you don't get the response, ask again using the same soft, brief pressure. If you don't get the response this time, instantly use your dressage whip to reinforce your leg. *Reinforce when necessary, but never escalate!* Always maintain your own balanced position; this will help your horse balance.

These are simple exercises, but over time they will build up your horse's topline. The way to lift your horse's neck is to build up the muscles that lift the base of his neck, and the way to do *that* is to get the hind end engaged so that his back will stretch and build up. Once that's happening, his neck will build up as well, and eventually you will look down and see that his neck has lifted, widened, and filled up the space between your hands. Before you compete at Prelim, you need to be able to play your horse like an accordion at all three gaits, but especially at the canter. Your work over jumps is going to be exactly as successful as your work on the flat. When you can play your horse like an accordion, those 10-meter circles are easy and the jumping is *much* safer.

You probably have a galloping track marked out so that you can check your speed against markers. Take your horse outside and use this track for trot-canter transition exercises. Go back to doing 30-15-30 strides to start with, and work your way down to 10-5-10. When you're comfortable with the process, do the same transitions exercise at the canter, but this time all transitions will be *within* the gait. Begin with as short and round a canter as your horse can do, then come forward in your two-point and ask him to roll on and lengthen his stride at the canter. Settle back into your saddle, sit up, *keep your legs on*, and ask him to maintain rhythm and tempo while returning to his shorter, bouncier stride. After enough repetitions, you'll feel that he is cantering uphill even when he isn't, and that he could stop, change gaits, turn, shorten stride, lengthen stride, or jump as soon as you asked him to.

Fortunately, most horses love canter-canter transitions once they understand them, and they enjoy alternating between the short, round canter and the longer, flatter one. This exercise will give you the habit of doing effective half halts, and you'll be glad of it later. All of your work will pay off when you're approaching a jump on the cross-country course knowing that you can add a stride by sitting up and closing your fingers and legs, or leave out a stride by folding forward and kicking on!

Frustration: Horse Won't Collect

Q I'm having a lot of trouble with my horse. He is three years old but he is a Quarter Horse, so even though I know you advise not to ride horses until they are four or five, I have been riding him for one year because he was completely mature. Quarter Horses are mature when they are two years old, so it is fine to ride them then.

My problem is that after more than a year of riding I still cannot get him to cooperate with collection. He should be flexing at his poll and engaging his hindquarters. I am a dressage rider, so I have to ride in a snaffle for shows, but I have been riding him in a curb at home to get a better headset. He will arch his neck with the curb, so I know he is just being stubborn when he refuses to do it with a snaffle. He does better when I use spurs but he still refuses to collect. He will collect some at the walk but is too stubborn to collect at the trot and canter. Also, unless I wear long-shanked spurs with rowels, he will not move away from my leg for leg-yields.

I rode in a clinic last week and the clinician was not very nice about my horse and me. He said that if I used a crank cavesson then my horse would not find it so easy to avoid collecting. My arms are sore from trying to get him to collect, and I am tired of kicking with the spurs, but this clinician didn't offer any useful suggestions at all, just the new cavesson idea. One more problem is that my horse has an upside down frame. (The clinician said he was inverted, which is the same thing, isn't it?) How do I make him go in the right frame and collect up at trot and canter? I am so frustrated. The clinician said we were not ready for a show, but that's why I am training my horse in dressage — to go to shows. If we can't show unless he will collect all the time, and he refuses to collect, what can I do? Maybe you can tell me what bit and noseband would help me.

A I hope you're sitting down, because I have some news that may surprise you. Your Quarter Horse was not fully mature at age two and now, at three, he's still not even halfway to maturity. The person who told you that Quarter Horses were mature at two was very badly mistaken. Two-year-old Quarter Horses can be their own worst enemies — you aren't the only rider who has been fooled by that lovely combination of muscular appearance and sweet, laid-back, "just tell me and I'll try" attitude. But their bodies simply can't handle a hurry-up, get-an-instant-frame-with-hardware style of "training." They may look strong, but they're babies, and if they're too cooperative for their own good, it's up to us, their riders, to look after them and avoid asking for too much, too soon.

Once you understand this, I hope you'll begin to understand why your youngster can't collect. Please note that I say "can't," not "won't." "Won't" implies that your horse *could* collect if he wanted to but is refusing to do so. That's simply not the case. Your horse isn't being stubborn or uncooperative. He simply can't do what you want him to do. Collection isn't something that we ask of very young horses or horses early in their training. Your horse's first year or two of training should involve slow, carefully planned work designed to help him stay sound while he learns to balance himself under a rider and then perform all three gaits at the rider's request. Collection doesn't come into it.

Asking too much too soon will cause a young horse to invert, shuffle, and develop incorrectly.

Here are my thoughts on your situation:

First, don't look for equipment to solve a problem, unless it's an equipment problem. If your stirrups are bent or your bit is broken, replace them. If your horse's saddle doesn't fit, adjust it or replace it. *Those* are equipment problems. What you have is something else entirely, and you need to reorient your thinking a little. Spurs won't give your horse the ability to collect. A more severe bit may cause him to arch his neck, but he's not becoming collected, he's just trying to get away from the pain in his mouth. That's not training. Trapping your horse between spurs and a severe bit is not training, either, and although it may shorten his frame and his steps, the result will *not* be collection, but inversion and shuffling.

Second, force isn't training, nor is it conditioning. Your arms shouldn't be sore, because you shouldn't be pulling your horse's mouth. He feels it much more than you do — if *you* are sore, how do you think *he* feels? Your horse is inverted (has an "upside down" frame) because he is very young and not very well muscled. To work comfortably in a more rounded frame, he'll need muscles in the right places, and you can't add those by pulling his head in or using a stronger bit or kicking him harder with — or without — spurs. Stop pulling and kicking your horse, and put away the severe bits and spurs. He doesn't need them and neither do you. What you both need is time and a sensible training plan.

The way to build a physique that will let your horse become collected when he's *ready* for collection is simple. It's also virtually equipment-free. Spend a lot of time walking and trotting him on a loose rein, sitting lightly or in a half-seat so that his back can move freely. Teach him to work calmly, stretching his back and reaching forward with his head and neck. He needs to move with long strides and without interference from the rider. This will let him begin to develop the strong muscles (and tendons and ligaments and bones) that will make it *possible* for him to collect much later in his training.

Focus on walk and trot, and don't worry about cantering. Until your youngster has developed good balance and muscles, he won't be able to do any work at canter in any case. If he offers a few steps of canter now and then, accept them, but then ask him to trot. Don't try to work him at canter for another year or so.

Before you can begin to think about collection, you need to think about *connection* — all the parts of the horse's body working together. When your horse moves forward calmly and energetically, stretching over the topline and reaching forward and down with his head and neck, continue to work him in this long frame while you focus on helping him develop his balance and his ability to maintain a steady, clear rhythm at walk and trot. When he's doing that easily, steadily, and consistently, and has the muscles to prove it (and this could take six months or more, even with a good rider and trainer, so don't be impatient), you can begin to ask him, very gently, to start to shorten his frame. If he's very talented and you're a wonderful rider, the process of shortening the frame might take one or two years. If he's anything less than naturally brilliant, or if you're not an extremely well-balanced, knowledgeable, and technically proficient rider, it might take much longer.

I don't know who that clinician was, but I hope you can find a better one. Instead of suggesting that you add a crank cavesson to your arsenal, he should have suggested that you get rid of the curb and spurs. I strongly suggest that you look for a good, sensible instructor who understands equine physiology and conditioning. You need help. Stop worrying about where your horse's head is and how arched his neck is. Go back to the snaffle, ride your horse forward, and help him develop his body, balance, and rhythm. As he learns and matures, he'll be able to begin to shorten his frame, flex more, and generally do the things you'd like him to do. But don't expect or ask those things of him right now — he can't do them.

There is no place in horsemanship for the kind of "training" that makes a horse unhappy in the short run and unsound in the long run. If you want to become a good rider, try bringing your horse along at a speed and in a way that will be pleasant for him and help him last for many years. It's the right thing to do, and it will save you money, too. You won't have as many vet bills and you won't waste money on bad clinicians or crank cavessons or stronger bits or longer spurs.

"Holding a Horse Collected"

Q I recently audited a clinic with a big-name dressage person who told us that side reins were an abomination and should never be used. I was shocked because I use them all the time, and my instructor even has me ride in them to help with my horse's collection. I'm not strong enough to hold my horse collected all the time. My instructor (she didn't go to the clinic) said that she has heard of this man and that he prefers to work with top-level horses and riders, and that he has probably forgotten how to train horses to go collected because most of the ones he sees at clinics are already collected.

But I am not so sure about this, because another thing he said at the clinic was that side reins could ruin a horse's walk, and my horse's walk has gotten a lot worse than it was when we first started out. This is after two years of training, so you would think it would improve, not get worse. He said that side reins would ruin the walk and put the horse forever "behind the bridle," and his description of a ruined walk sounded a lot like my horse's walk the way it is now. So, now I am wondering if my instructor's methods are wrong.

A A horse that is still in the early years of his training is not ready to perform collected gaits. It's not possible to force true collection, but it is, alas, quite easy to try to pull a horse into an outline and make the horse so uncomfortable that he tightens his back and neck and walks with shuffling steps and a short stride. You've probably heard that a horse should be ridden from back to front, not from front to back. This applies to *all* of a horse's training — on the ground, on the longe, on long lines, and under saddle. I am sorry to say that your instructor seems to be teaching you to ride and train your horse from front to back. Wrong early training will invariably come back to haunt the horse, if not the trainer, later in life. If you want your horse to achieve collection some day, you are on the wrong path and need to get on the correct path and start your journey again.

Collection is the product of years of systematic, correct training aimed at developing the horse's body and mind, and it cannot be achieved from front to back. Trying to "collect" a horse from in front will have many consequences, none of them good. A horse ridden from front to back can't use his back muscles correctly. The discomfort caused by the restricted position of his head and neck and by the impossibility of using them to stretch and balance causes the horse to tighten his back muscles in self-defense and become stiff. When the horse's back is locked and dropped, there is no possibility of him engaging his hindquarters. A horse like this has been thoroughly disconnected. *Connection* is missing, and

Side reins should not be part of your regular riding tack.

collection is absolutely out of the question. Unless a good trainer steps in, discards the extra equipment, and restarts the horse's training, that horse will never feel comfortable under saddle — he will always be tight in the back and neck, and he will always be behind the bit.

The only way to reclaim a horse like this is to get him moving forward freely; often the only way of doing this is to put the horse on the longe line where he can trot without having to deal with the extra burden of a rider. If the trainer is adept at longeing, and the horse does enough trotting and trot transitions, eventually also over poles and cavalletti, the horse *may* be able to learn to relax his back. If he can do this, he'll then be able to use his hindquarters more effectively, engage his belly muscles, and finally stretch over his back and through his neck to his jaw.

Once he is confirmed and comfortable using his back muscles again at trot, which may take months, it will be possible to begin retraining him at walk and canter. This will require a sensitive trainer who uses minimal equipment: longe whip, longe line, and longeing cavesson *only!* All of the gadgets, gizmos, and "auxiliary" reins will have to remain in the tack box if this horse is to be successfully rehabilitated, developed, and educated.

Side reins are not intrinsically evil, but they can cause great damage when they are misused, and sadly, they are misused often, by many people including your instructor. I don't fault that clinician for warning riders against using side reins, although it might have been more useful for him to show how and when side reins could be used correctly.

One more thing: The degree of collection is determined by the horse's training and development, not by the size of the rider's biceps. Collection isn't something that you create by main force, and it shouldn't take force to maintain it. A horse's head and neck represent quite a lot of unsupported structure — at least 150 pounds. The gripping and pulling necessary for a rider to force a horse's head and neck into a particular position and hold them there with the reins would be similar to the effort you would make if a 150-pound person had fallen off a cliff and you were trying to hold her up by her bra straps.

No instructor should ever tell a student that she should be riding with tight side reins, much less that she should do this because she lacks the physical strength to "hold the horse collected." That concept shouldn't even exist. You should carry your hands; your horse should carry himself; and the reins should provide a steady, soft, elastic connection between his mouth and your hands.

How Can I Collect and Extend My Horse's Stride?

Q How can I extend and collect my 11-year-old Thoroughbred gelding? I am 14 years old and have been riding for about three years. I took lessons each week for about a year and a half, then leased my gelding for six months. Now that I own him, I've taken a total of three lessons in seven months. So, I am pretty much left on my own for riding and never was taught how to collect and extend my gelding's stride.

A Don't be in a hurry to attempt collection and extension. You've done very well for someone with so few lessons. I know that you want to train your horse correctly, but training correctly takes time, because it depends on the rider's ability to develop a horse's body and mind slowly and carefully.

There's a whole sequence of achievements in training, and collection is at the top of that list. However, we start at the *bottom* of the list and work our way up. Here's the list (and remember, read it from the bottom up, not from the top down):

- ▸ Collection
- ▸ Straightness
- ▸ Impulsion
- ▸ Contact
- ▸ Suppleness
- ▸ Rhythm

Riding to music is an enjoyable way to maintain a steady rhythm.

As you see, it all begins with rhythm. The four-beat walk, the two-beat trot, and the three-beats-and-pause canter are three different rhythms, and it takes time to train the horse to walk in a clear, even, four-beat rhythm; to trot in a clean two-beat rhythm; and to canter so that you can feel, hear, and see the one-two-three beats of the hooves and the pause (the silent fourth beat) when all four feet are in the air.

Once your horse is moving reliably in an accurate, steady rhythm in each gait, you can begin to do exercises with him to increase his suppleness. Large ring figures, turns, circles, figure eights, changes of direction, transitions between gaits — All of these, if your horse is going steadily forward in the correct rhythm for his gait, will make him more balanced and flexible.

When your horse finds it easy to maintain a steady, pure gait and can bend and flex and turn without speeding up, slowing down, or losing the beat, you can begin to focus on sending him forward into a soft, steady contact. And when you're at this point, it's useful to start "gymnasticizing" your horse by asking for longer and shorter steps, though this is *not* extension or collection, it's *lengthening* and *shortening*. Lengthening means asking your horse to keep his steady rhythm while reaching a little more forward with each leg and taking a slightly longer step with each leg. This is harder work than you might think, for your horse and for you.

When your horse can maintain his gaits, bend, flex, turn, and go forward into a steady contact, and when he can lengthen and shorten his stride without losing the rhythm or the contact, you can begin to work on straightening him. This means keeping his body straight on straight lines, so that his hind feet track his front feet on the same side. It also means keeping his body bent like a banana from nose to tail on turns, curves, and circles, so that, again, his hind feet will track his front feet. This is so that he will be evenly balanced over his four legs; the only way he can do this on a turn is to bend his body and track his forelegs with his hind legs.

A horse with a stiff, straight body goes around a turn like a motorbike, by leaning into the turn and putting more weight on his inside legs. This isn't stable and it shows a lack of suppleness. It also shows a lack of straightness, because "straight" means "tracking straight," which means that the horse's body and balance match what he's being asked to do — straight body on a straight line, curved body to match the curve of the turn or circle.

By the time a horse reaches this point, he's usually in his second year of good dressage training. As he begins the third year, he takes all of these qualities and learns to put them together and offer more energy to the rider — but *controlled* energy, not running-away energy. The rider continues to develop the horse's body and mind, strength and balance, and ability and understanding, so that the horse has a great deal of energy that he makes available to the rider. "Impulsion" is *not* the same thing as "speed." "Impulsion" means that *if* the rider adds a little pressure with her legs, the horse will step more actively and more forward with his hind legs; *if* the rider then opens her fingers or moves her hands forward an inch, the horse will surge forward to fill up that space.

This is the very beginning of collection: the rider's ability to ask the horse to step up underneath himself, rounding his back and carrying more weight on his hindquarters. But the rider is only able to get this from the horse because of all the previous work. That's what has made the horse *able* to do what the rider is asking. It takes strength, balance, suppleness, and impulsion for a horse to be able to collect when the rider asks. And when the horse is able to collect — when the horse finally has that power — *that's* when the rider can begin to ask for extensions.

Extension comes from collection. The horse that is round, powerful, carrying himself with a lifted back and paying great attention to the rider's aids, the horse that steps more deeply under himself and becomes even more round when the rider asks for collection, is a horse that has developed the ability to reach forward with longer strides when the rider allows him to move out. A horse going from collection to extension isn't being *pushed* to extend — he's being *permitted* to extend.

So, don't be in a hurry. Start at the bottom of the list and work your way up. There are so many different factors involved that there cannot be a fixed schedule for any of this. What matters most is the progression: Your horse must be rhythmic, forward, on the aids, straight, reliable in his bending and balance, and performing good, clean, balanced transitions between and within gaits before he will have the power to collect and extend when you ask. Master one step at a time, and let your horse tell you when he's ready to move on to the next one. You'll need help along the way, so try to find a good instructor or even a good clinician who can work with you every few months. That's not as helpful as good weekly instruction, but it's much better than trying to work completely on your own. We all need someone to check on us periodically to keep our riding and training "on track."

PART III

Training Tools

THE RIDER'S BRAIN

TRAINING EQUIPMENT:
METAL AND LEATHER

LONGEING, WARM-UPS,
AND YOUR HORSE'S TRAINING

CHAPTER 12

The Rider's Brain

ADVERTISING IS CONSTANTLY EXHORTING us to buy this or that piece of equipment that will magically do the work of training for us. In fact, most training happens between the rider's brain and the horse's body, by way of the rider's body and the horse's brain. The trainer thinks of what the horse should do, puts the horse into position to do it, asks the horse to do it, allows the horse to do it, and lets the horse know when he has done it correctly. That's it — that's training. The only absolutely necessary piece of equipment is your very own functional, alert, aware, intelligent, compassionate (and protected) brain.

"Instant Training" at Clinics — Can It Work?

Q I audited a clinic where people brought green horses, wild horses, basically all untrained horses to be trained in an hour. It was pretty impressive to see horses that didn't know anything learn to be handled and tacked up and ridden. The clinician picked four different horses and spent an hour with each one. He spent the first half hour teaching the horse respect, and then taught the horse all the other stuff in the second half hour, and at the end he got on the horse and it would just stand there without any bucking or anything. I was amazed because I've always been told that training can't be instant, but I saw this happen and these horses were fine with a saddle and bridle and rider.

Now that I've seen this kind of instant training and I know it works, I don't know what to think. I didn't really watch closely for the first half hour because all the running made me dizzy, but I listened to what the clinician was saying about how important it is to be firm but kind, and why you have to get the horse's respect, and how you do this by making the horse move his feet until he will stop and drop his head (respectful) instead of looking at you (challenging). I started watching again when he let the horses stop, and I could see that the horses were very respectful, they dropped their heads and didn't try to make eye contact, and you could see them licking their lips (that means they respect you a lot).

I watched the second half hour carefully and it was amazing. When my old horse was being trained, the trainer worked with him for weeks before he even put a saddle or bridle on him, and then he did more stuff another week before he ever got on, and then even after he rode Dynamo for a week he still couldn't get him to stand totally still like the horses at the clinic. I wasn't allowed to ride my horse until the trainer had worked with him for three months! I'm going to be buying a new horse this spring, a two year old that isn't trained to ride yet, and I think I could save a lot of time and probably money (having a trainer work with your horse for three months is expensive) if I could take him to a clinic like this and bring him home trained to ride. What do you think?

A I think there are too many so-called trainers on the road today, doing demonstrations in which a confused young horse is put through far too much in a very short time. There is no advantage to pushing and rushing a horse. There are many disadvantages to the horse (to his comfort and understanding and soundness), to his owner, and to anyone who will ever ride that horse, throughout his life.

Three months is not very much time. A "90-day horse" is not considered to be a fully trained horse. In that time it is possible to start a horse; introduce him to a saddle, bridle, rider; and teach him basic stop, start, and steering. After 90 days, that horse is at the beginning of his training, not at the end of it. Ask yourself how proficient you would be after three months of piano or ballet lessons, and you'll have a better understanding of how much a horse can achieve in

Good training is quiet, gentle, and slow — not an exciting spectacle.

that amount of time. As for "instant training," many people would love to believe in that! It would be convenient to take a horse to a clinic and bring him home trained. But the truth is that *training takes time.*

Training develops the horse's body and mind, improving his understanding, increasing his confidence, and helping him become more fit, strong, and balanced. Training is meant to prepare a horse for the work he will be expected to do, whether that means racing, driving, jumping, dressage, reining, working cattle, or carrying a rider on trails. There is *no* work for which a horse can become physically and mentally prepared (trained) in one hour or one day.

With that answered, let me answer the question you would probably like to ask now: What exactly *was* happening at this clinic? It sounds to me as though what you saw was a series of horses being run off their feet until they were physically and emotionally exhausted and drained. If you were dizzy from watching them race in circles, how do you think they felt?

When a horse is frightened, tired, and probably more than a little bit sore, which would be the normal result of running in a small circle for half an hour, it's easy to get him to stand still, and even to accept a saddle, a bridle, and a human on his back. But the horse has not been *trained* to stand still or to be tacked up or ridden; he's simply been *pushed* to the point where he's completely out of his comfort zone, and the combination of fear, confusion, discomfort, and exhaustion has caused him to shut down or tune out. You're probably familiar with the concept of "flight or fight," but that's not the whole story. A horse's first reaction to something frightening and unpleasant is to flee, but if running away is impossible (he can't get out of the pen) and fighting is impossible (there's a predator chasing him — that's why he's running), and he reaches the point where he can't really run any more, his body and mind will automatically embrace the third option: freeze.

A horse's comfort zone is a delicate thing — I like to describe it as a big soap bubble around the horse. Blowing gently and carefully into the bubble makes it larger and larger, but blowing too hard or too suddenly will destroy the bubble. If you think of your horse's comfort zone in that way, it may help you learn to push very, very gently so that you are always expanding the zone a little but without ever breaking that bubble of security and trust. A good trainer will always progress by systematic "advance and retreat" — always pushing the horse a little, so that he makes progress, then letting him relax, then pushing again, but never pushing so much or so constantly that the horse becomes uncomfortable or fearful. It's advance *and retreat,* not "advance, advance, and advance some more." Training is not a strictly linear progression; it's a dance. Over time, as the horse's training continues and his education progresses, his comfort zone "bubble" can be expanded to a very large size indeed.

What you saw, alas, was a classic "freeze" or broken-bubble scenario, in which a horse was chased and chivvied into exhaustion, at which point he shut down and appeared calm and submissive. The trainer then proceeded to saddle, bridle, and sit on the horse, which stood quietly, apparently accepting whatever the trainer did. But the truth is the horse was not accepting anything — he had mentally and emotionally gone somewhere else. A good trainer would have recognized this, apologized to the horse, and left him alone, or explained to the audience how this had come about and why it was bad for the horse and should not be attempted by anyone, at home or elsewhere — but then a good trainer would not have pushed a horse to that point in the first place.

When a horse is in that state, it's possible to do just about anything to him, but it's like doing something to a person who is in shock, unconscious, or anesthetized — as far as that person knows, the experience itself doesn't exist. The person won't have a clear memory of the event and certainly won't understand it and learn from it. Experiences like that contribute nothing to a person's education or to a horse's training, although their consequences (e.g., soreness and possibly lameness for the horse) may matter very much indeed.

This is one of the greatest dangers of misusing the round pen. Used sensibly, it can be a very useful tool, but what you saw was an example of round-pen work being done wrong.

It's not wrong to keep a horse's feet moving; horses do think better if they're allowed to move. Sometimes it's very wise to allow a horse to move around a little instead of demanding that he stand like a statue — for example, when you're introducing a rub rag or a spray bottle to a horse that has never seen such things before. But making a horse run — not just move his feet, but run — for half an hour will not help the horse think better or learn anything at all. At first, the speed will agitate the horse; then it will tire the horse. A tired horse will slow down; an exhausted horse will stop, a horse in "freeze" mode will stand still. At that point, he can be bridled, saddled, and mounted, and may even walk around the ring if the rider insists, but it isn't truly *accepting* any of this, and he certainly isn't being *trained*.

In this case, you couldn't even learn anything useful from the clinician's half-hour monologue, as he was giving wrong information about equine body language. After half an hour of running in a circle, a horse that stands still and drops his head is expressing fatigue, not respect; when he licks his lips he is expressing thirst, not respect. Lifting his head and looking at the trainer does not imply a challenge, it only indicates that the horse's brain is still switched on. *Not* looking at the trainer is partly a function of the dropped head, and partly an announcement that the horse has given up. An exhausted horse that hangs his head and doesn't look at the trainer is not saying "You have my respect, please train me now"; he's saying "I can't get away and I can't deal with this, so good-bye, I'll just be in the quiet place in my head."

Once that has happened, the clinician may sit on the horse, or even stand on the horse and dance a jig, fire a rifle, or play the violin. He may also get the horse to travel around the pen at various gaits, but the *thinking* part of the horse's brain is not involved. When the horse's owner takes him home and tries to tack him up and ride him a day or two later when he's had a chance to recover somewhat, that horse is unlikely to remain passive, submissive, and quiet. In fact, he's quite likely to explode — running and bucking are common reactions — when his brain "reconnects" and suddenly realizes that he's wearing tack and carrying someone on his back. There's been no gradual, gentle, confidence-building desensitization — no *training* — so the horse's reaction is likely to consist of surprise, shock, fear, and a great desire to escape. At this point, both the rider and the horse are in immediate physical danger.

For your horse's sake, continue to audit clinics before signing up as a participant. For a low cost, auditing lets you discover what a clinician is like without hurting your horse, your bank account, or your principles.

"Name" Training Programs

Q I'm no great expert on horsemanship, but I'm going crazy from the pressure at my boarding stable. A few months ago, the barn owner went to one of the clinics held by a national "guru" (I don't want to use his name), and she and all of her friends are now totally devoted to this guy and his methods. She bought books, videos, and equipment, and she's already planning to go and participate in a program that will get her a fancy (and very expensive) certificate with his name on it. I can deal with all of that, but she and the others are hassling me and the other two people at the barn who aren't following this guy's "magic" methods and buying all his equipment and such. The real kicker is that the three

You'll do best to follow one method or system that makes sense.
Training is about the horse, not the trainer's name.

of us "holdouts" are now the only people at the barn who actually ride our horses! Everybody else is playing games on the ground and having mental communication with the horses. In your opinion, is there any reason for anybody to ever follow any of those training programs?

A I'm just as glad not to know whose training program this is. I suppose my answer to your question would have to be "yes and no."

I'll give you the "no" part first. No program is magic, no equipment is magic, there's no such thing as instant horsemanship or instant training, and anyone who believes that horsemanship can be packaged, purchased, and paid for by check or credit card is more than slightly delusional.

But there *are* sensible, rational training systems in the world. If someone chooses one of those, follows it carefully, thoughtfully, and thoroughly every day, year in and year out, and makes adjustments according to the specific conditions and needs of his or her specific horse, then yes, that person will do a much better job of training a horse than someone who takes a scattergun approach to finding a mentor or a system. The person who wants it all, and who tries a little bit of this one's approach, a little bit of that one's approach, and a few bits and pieces from other systems, is likely to end up confused, and confusing his or her horse.

If you want to learn to train horses and train them well, you will do best to embark on a nonstop study of The Horse, coupled, if possible, with an apprenticeship with *one* person whose training methods you admire in principle *and in practice.*

Does My Horse Know When I'm Training Him?

Q Last year I got my own horse, and I want to understand exactly what is the right way to train him and how to handle problems that come up. I don't have a specific problem to ask you about, but I would like to know whether my horse knows when I'm training him and when we're just goofing around. I think I act differently when I'm training him, but does he understand the difference? What's the most important thing I should think about when I'm training my horse? I like to spend time with him, but I only do real "training" with him four days a week so that he won't get bored.

A You may act differently when you're consciously training your horse, but you are actually training your horse whenever you interact with him, even if you think you're "just goofing around." There's nothing wrong with having fun and enjoying your horse, but whether you're grim and serious and pushing hard, or laughing and giggling and playing games, it *all* counts as training time.

You didn't mention your horse's age, but that doesn't matter — he's learning from every interaction with you whether he is six months old or twenty-six years old. Your mission as his trainer should be to make all of your interactions positive and pleasant and to teach him only those things that you want him to learn.

Your horse doesn't distinguish between training and nontraining interactions with you. When you longe him or do trot-canter transitions or teach him to step over logs on the trail, that's training. When you put feed in his feeder and clean out his stall and cross-tie him and groom him, that's training. When you give him treats, that's training, and if you hit him, that's training too, because he learns from it all. He doesn't go into "learning mode" when you use a particular sort of equipment or have specific intentions to teach him something — he just learns.

It's very easy to teach a horse something that you don't want him to learn. Think of all the horses that trot and canter nicely but fall apart at the walk because their riders have always worked at trot and canter and used the walk as a way to have a moment's rest between efforts. They've effectively taught their horses

Your horse learns from you all the time. To him, it's all training.

that they don't care what the horses do when they walk under saddle. Think of all the horses that are trotting or walking on the longe line or in the round pen, hear "Good boy!" and then cruise to a stop, because they've learned that "Good boy!" means "You don't have to do anything now." No rider ever initiated a formal training session with the plan of teaching a horse to wander aimlessly at the walk or to stop working when he hears "Good boy," but those lessons were taught and learned just the same.

Since your horse doesn't know which interactions with you count as training and which ones don't, it's your obligation to make *all* of your interactions with your horse as pleasant and enjoyable as possible, whether you're training him or "just goofing around."

Telepathic Training?

Q This probably sounds insane, but I'm convinced that my horse has some kind of telepathy with me. She seems to know what I am feeling and thinking even when I haven't said anything about it to my trainer or to myself. Here's an example: I was riding down to a rolltop, which is a jump that I find more intimidating than practically any other design of jump. My horse started to slow down and then stopped right in front of it. My instructor yelled, "Hit her with the whip, she needs to go right now!" but I didn't do it because I know that she read my mind and didn't jump because I was afraid. I said, "Wait, we'll go again," and the second time I cheated and closed my eyes about four strides out and just thought, "Okay, mare, let's go, take care of me," and she jumped it like a dream. I told my instructor what I did and she said, "You're crazy, but it worked." I don't know why but I can never see my stride to a rolltop, and that just makes me more nervous. Can all horses do this with all riders,

Your horse will always sense your confidence or your fear.

or is this just something weird with my mare and me, that we are doing some kind of telepathic training?

A   If I were you, I wouldn't make a habit of closing my eyes four strides out, because something might happen and your mare might need some help or direction from you. I've been to events where spectators have been yelled at to remove their purses or backpacks or even their dogs from jumps when there were horses just a few strides out, preparing to jump those jumps. But you did it and it worked, and there *is* a lesson to be learned from that.

All training from the saddle is "telepathic" in the sense that you are sending your mare all kinds of messages that you may not realize you're sending. Whatever you're feeling and thinking will be expressed by your body language. If you're approaching a jump convinced that you're going to fly over it, land, and turn right for a good approach to the next jump, that's likely to be what happens. If you're approaching a jump convinced that it's going to be a disaster and you can't imagine where your horse will take off and you would really prefer not to go over that jump at all, then you probably *won't* go over that jump — not because you gave your mare a telepathic order, but because the changes in your posture, breathing, and leg aids told her "Danger, I don't want to do this, let's not do this!"

Congratulations — you've had a *very* useful insight into riding and training, which is that your thoughts and feelings affect what your body does on every level, and thus affect your horse. This is what the instructors of a previous generation used to mean when they said, "Throw your heart over the fence and your horse will follow it."

"Hi, I'm Star and I'm Going to Be Your Horse Today"

Q   A few years back you gave a clinic at our barn and something you said really cracked me up. You were talking about checking your horse every day and how a horse can be very different one day than he was the day before, and you said we needed to let our horses introduce themselves to us every day, like the waiters who say, "I'm Brad and I'm going to be your waiter tonight." Would you mind explaining more about how this helps a training program? Four of us have bought new (young) horses since you were out here, and we sure could use the help as we're still getting to know them.

A I'm glad you liked that. The point I wanted to make was that you should let your horse introduce himself to you on a daily basis. At the beginning of every ride, find out just who your horse is today, because he may not be exactly the same horse you rode yesterday. You're probably familiar with the difference in responsiveness and overall attitude that you see in your horse when he is either happy and full of enthusiasm from the previous day's work, or tired and sore from the previous day's work. That's only part of what you need to observe each day.

If your horse is young, then he may quite literally be different from day to day because he is growing, his balance is changing, his cartilage is turning to bone; the horse you ride on Monday may be very different on Tuesday and even more different on Friday. If you will let your horse introduce himself to you each day, you will know him better, understand him better, and train him better. If you don't like his Wednesday or Friday self as well as you liked his Monday self, you can think back to the work you did on those days, and you may find some legitimate reasons — not excuses — for him.

When a small child fails to behave perfectly in public, we are quick to offer a litany of excuses, reasons, and explanations. The child is tired, uncomfortable, anxious, coming down with a cold, cutting a tooth, overstimulated, stressed, exhausted, or experiencing growing pains. All of these words and concepts could equally be applied to horses, but we're often much less sympathetic with our horses. When a young horse fails to behave perfectly in public, riders often have a different reaction: "He's resistant, he can't get away with that, I'll show him who's the boss!"

A little compassion and sympathy will go just as far with a horse as they will with a child. Use them! A little empathy can work wonders, too. If your horse has a wonderful training session on Monday and then a less than wonderful, or even a barely adequate training session on Tuesday, ask any bodybuilder what happens when you push hard one day and then come back and stress the same muscle groups on the following day. (Hint: "Improved

Always ask how your horse is feeling today.

performance" is not the answer.) Humans and horses build muscle by tearing it down and building up the same muscle in a larger size. Periods of rest are necessary to build up the muscle — if you push hard day after day after day, all you will accomplish with those muscles is to tear them down.

Begin every ride and every training session by finding out what kind of horse you have to work with today. While you're grooming your horse and checking him for cuts, lumps, and loose shoes, check his TPR (temperature, pulse, and respiration) as well. Just as you check the gauges in your car before you start out on a trip — even a short trip to the market — you should check your horse's "gauges" before you embark on a training session or schooling session. Asking "Does my horse have a fever, is his heart rate normal, is he breathing fast?" will help you notice a minor problem before it becomes a major one. You know how even a slight fever makes you feel achy, painful, tired, and out of sorts. If someone demands a brilliant performance or even extra effort from you when you're feeling that way, you won't be able to manage it. Your horse can't manage it, either.

When you're mounted and warming up, you don't have to run through the dressage training tree — a quick checklist will suffice. Ask yourself these questions, and test as you go:

- Do I have "go"? Do I have "whoa"?
- Do I have position? turn? bend?
- Do I have forward response to my legs?
- Do I have lateral movement response to my legs?
- Do I have these things at the walk? at the trot? at the canter?
- How balanced is my horse?
- How even are his gaits?
- How flexible is he?
- Is he hurried or relaxed?
- Is he distracted or attentive?
- Is he eager, or just accepting?

You can customize the checklist to make it your own, and then make it your habit to check it at the beginning of each ride, at some point during the ride, and whenever anything feels wrong. By getting to know your horse every time you ride him, you will be a more effective trainer, and if there is ever any real problem developing, you will notice it before it has a chance to become serious.

Patience, Persistence, and Praise

Q I attended a colt-starting clinic of yours last fall and have been trying to use all of the information to help me with my 10-month-old Quarter Horse colt. He's doing very well by the way. I was looking through my notes that I wrote right after your clinic and found there was something about "three Ps." I'm drawing a blank here, probably because I was too interested and stopped writing. Would you mind refreshing my memory about the "three Ps" idea or formula, and how it works?

A The three Ps are Patience, Persistence, and Praise. It's not actually a formula; I just put those three concepts together and called them the "three Ps" because I thought it would be an easy way for people to remember them. (Well, perhaps it wasn't so easy after all.) It's just a quick way to remember that these three things should always be part of your riding and training.

Patience. You must give your horse time to understand what you want from him, time to learn how to do it, and time to do enough repetitions to make the action easy and then habitual. You must be able to break down each new skill into tiny, easily learned pieces, and then teach those pieces one at a time.

Persistence. Although you only ask for small things, small movements, small responses, you need to ask for them often. This helps your horse understand, confirms your communication with the horse, and begins to develop the habits that you want.

Praise. Whenever you ask for anything, no matter how tiny and seemingly trivial, you must praise the horse — there has to be a reward for every effort. Praise is a good reward for a horse: a kind word, a pat on the neck or a scratch on the withers, and a moment of stillness will be appreciated.

This is not a new concept — it's very, very old. "Ask little, ask often, reward generously" is an old dictum of classical dressage. "Reward the try" is something you'll hear at any good natural horsemanship clinic. It all comes down to making the training experience — which is, when you think about it, *all* of your contact with your horse — a pleasurable one.

Good trainers are never without these tools.

Will My Horse Forget Her Training?

Q I bought my 13-year-old Appy mare in November and after several months, she seems to be much more trusting of me. She doesn't pin her ears as much and comes to me more readily. I have never had a horse before and was quite nervous, but I feel more comfortable since I took a four-day clinic in natural horsemanship in April. We had a great time and both of us learned a lot. My problem is that I don't have a place to train her at home. She shares a paddock with a pony and a goat, but there are too many stumps for us to work or play there. We are building a barn for them, and I am also having a round pen put up, but it will be a few weeks before it will be ready.

My question is, will she remember what she has learned or will I have to start all over again? I continue to study the books so I don't forget. This includes riding her. The one time I was on her she took off with me, so I am afraid to get on her without having worked her for two-and-a-half months. I was waiting for the pen so I have a controlled area to ride in before going to the street and trails. Any words of wisdom would be appreciated. I'm keeping it natural but feeling very guilty.

A First, stop feeling guilty. It sounds as though your horse has a good situation — turnout, grass, and a companion! She's not going to be at all upset if she isn't ridden. She's probably not going to forget anything, either. Horses have incredibly good memories. I've known horses that recognized individuals after 20 years and horses that remembered tricks they'd learned 10 years after they were last asked to perform them.

My answers to your questions are first, no, she won't forget; and second, yes, you should start all over again.

What often causes problems when someone doesn't ride for a long time is a combination of horse and rider issues. If the horse is confined and/or overfed, she can become difficult to manage. You aren't confining your horse, and I hope you're not overfeeding her, so you shouldn't have a problem there. If the rider isn't riding and isn't doing any other exercise that helps with riding, then after a few months there can be a big loss of coordination and balance and strength. If you can't ride, be sure to do other exercises, even if it's just walking, yoga, and light weight lifting. This will help you stay in shape so that you won't be all over your horse when you finally get back to riding, and it will make it much easier for you to become comfortable in the saddle.

A round pen is a nice addition to any farm — just be sure that you make the diameter large enough if you plan to use the pen for riding. Seventy feet will do nicely; 75 feet or larger will be even better.

While you're waiting for your barn and round pen to be built, and you aren't riding your own horse, this is a perfect time for you to take a series of lessons with a good instructor. The more you understand about your horse's body, your own body, and the balance and coordination that you need to use your own body to influence your horse, the better off you and your horse will be when you start riding her again.

You'll find, as riders invariably do, that anything you do to improve your own riding will produce a matching improvement in your horse. Some riders have found that the biggest "leap" in the improvement of their riding skills has happened after they had to take a few months off for whatever reason, provided that they did other things during that time. A summer spent swimming and bicycling, a winter spent skating and skiing, or anything else along those lines, can improve your riding dramatically. If you do that *and* add riding lessons with a good instructor, you'll find that when you do get on your horse, she will "magically" have improved enormously, even if she's spent the last six months eating grass in her field.

Horses have excellent memories. Yours won't forget her training even if she has a year off.

Starting all over again is something that all good riders and trainers do every single time they begin working with a horse that has had some time off. Starting from the beginning each time isn't a bad thing, it's a very good one. Make it your permanent practice, and you won't regret it. By starting over, you identify any weak areas in your own training, in your horse's understanding, and in your horse's physical ability to perform.

Time off for the horse shouldn't be "time out" for you — and in your case, it obviously is not, as you are reading the books to be sure that *you* don't forget what you've learned. That's a good idea and a good start. If you'll add some physical exercise and some riding lessons, you'll be able to help your horse more, every single time you work with her or ride her. Keep on learning — the better your own education as a rider, the better your horse will be. The more time and effort you put into improving yourself, the easier it will be for you to ride and train your horse.

You don't even have to wait for "time off" to start a horse again. Starting all over again is something that good trainers do *every day*. A good trainer never tries to pick up exactly where she left off at the end of the last session. Instead, she will do a quick run-through of everything that the horse has been taught, starting at the beginning.

In riding terms, we call this a "warm-up" — in addition to physically warming the muscles and increasing the circulation, the first part of your ride should take your horse, gently and gradually, through everything he's been taught to do. This reinforces the horse's responses to your aids, it helps build the horse physically so that he can meet increased demands later, and it gives you a chance to notice how your horse is feeling that day, physically and mentally. In other words, you find out whether your horse is energetic or sluggish, supple or stiff, and also whether he is interested or bored, happy or worried. These are all things that the rider should be very clear about during *any* riding session, and these are things that you need to know before you even think of introducing anything new.

Starting all over again also protects your horse's health — you'll be very quick to notice any change in behavior or movement, and if your horse is becoming sore, you'll find out before any more damage occurs.

Remember that horses are like humans — if you want them to learn something new, or to learn to do something in a different way, you need to give them very clear explanations and short practice sessions, and praise them for any effort that shows they're heading in the right direction. Once they've learned the new or revised skill, you need to do many *correct* repetitions over a long period — and by "long" I mean over the course of weeks or months, not over a marathon two-hour session! — so that the muscles can rebuild in a way that facilitates the work. Finally, you need to continue the correct work so that the new skill or the new way of performing the old skill will become a *habit*.

All of that takes time. You can go as slowly as you like, as long as you're making a little bit of progress and as long as neither you nor your horse becomes bored. Riders who cause injuries in their horses rarely do this by training slowly and carefully — it's impatience and hurry that create problems.

If you have to take time off at some point, don't worry about your horse forgetting anything. When you begin again, even after months, even after years, you *will* need to begin from the beginning, but that's for your horse's benefit and your own, so that both of you can rebuild and reinforce your strength, balance, coordination, and trust. A horse that is asked to perform some action he hasn't performed in years will usually try his best — he *does* remember what those signals mean, he *does* remember what you want, but he may not be physically capable of performing the action, or of performing it well, until he's had a chance to develop those muscles again.

You sound as if you plan to keep your mare for a long time. Relax, take your time, and enjoy correct, slow training — and get good help with your riding so that you can continue to improve your riding skills. The more competent you feel, the more confident you'll feel, and the more your horse will be able to relax and trust in your leadership. The great paradox of horse training is that the greater your hurry, the less you achieve; and the greater your patience, the more quickly and effectively you will get where you want to go.

CHAPTER 13

Training Equipment: Metal and Leather

EVEN A TINY TACK SHOP can have an intimidatingly large "bit wall." With so many bits and auxiliary equipment to choose from, it's easy to imagine that something on that wall could do some of your training for you. However, knowledge is your best defense against the purchase of useless or inappropriate equipment. If you understand how tack functions and how all the various gadgets and gizmos function, you'll keep things simple, and you're unlikely to come home with a bag full of "training accessories."

Never "buy" the idea that equipment can be a substitute for training. Before you use any piece of equipment, know what you're using, know why you're using it, and know what it can and can't do.

Confused by Different Types of Side Reins

Q I'm wondering if you could discuss the pros and cons of using various types of side reins, such as Vienna versus fixed, and whether you have an overall preference regarding type (assuming an experienced and knowledgeable person is doing the longeing). How do they all fit into a training program?

A First, let me offer a quick explanation of the difference between fixed and sliding (Vienna) side reins. Fixed side reins are primarily for use at the trot, although you can use them at the canter on an advanced horse that is truly in self-carriage. For young horses and all horses that are not yet advanced in their training, they should be used only at the trot. Even if a horse is very advanced, fixed side reins should *never* be used at the walk.

If you want to use side reins at all three gaits, use sliding or Vienna side reins. These resemble draw reins but are attached to the tack at both ends, the length of the reins running either from high on the saddle or surcingle through the bit rings to the billets or girth on the horse's sides, or from high on the saddle or surcingle through the bit rings and down between the horse's forelegs to the bottom of the girth. When properly adjusted, these reins allow the horse to seek contact and move forward correctly at walk, trot, and canter. The sliding side reins do not block the horse as does a fixed side rein. They allow the horse maximum comfort when he stretches his neck down and raises his back, limit him putting his head

No matter what kind of side reins you select, adjust them so that your horse can reach comfortably forward toward the bit.

up and hollowing his back, and allow a variation of positions in between for him to find his own most comfortable spot.

In the United States, riders tend to think in terms of fixed side reins, either the plain leather ones or the type with the rubber donuts. I would prefer to see sliding side reins or Vienna reins used whenever possible. Always be very careful with side reins. Although they can be helpful, they can also create enormous problems in a short time if they are too tight, too low, or incorrectly used. Poorly adjusted side reins will interfere with training and can cause damage (temporary or permanent) to the horse.

Ideally, the person longeing the horse understands the art of longeing, has found a suitable place with good footing, and is equipped with a proper longeing cavesson, a 35-foot or longer longeing tape, a longe whip, and either a longeing surcingle, a saddle, or a longeing surcingle *over* a saddle. A knowledgeable person always longes from the cavesson, never from the bit. And we'll assume that the person is taking into account the horse's age and training level, his conformation, his degree of fitness, and his mental and physical comfort level.

If the horse is young, unfit, or coming back from time off, the side reins would be adjusted long and low, but not so long that the horse is in danger of stepping over a Vienna side rein or being hit in the head by the rubber donut if the handler is using fixed side reins. On an older, more fit, or more experienced horse, the side reins could be adjusted slightly higher or shorter or both. On a younger horse, for schooling purposes, the side reins could be adjusted similarly for just a few minutes of trot-work toward the middle of the schooling session, then returned to their long and low position.

However the side reins are adjusted for a particular horse, they should offer a soft contact, and invite, not force, the horse to stretch forward and down with his head and neck, seeking the bit. This, in turn, should lead to a raised back and engaged hindquarters. When tight side reins *impose* contact and force the horse's head down and in, the horse will travel with a hollow back and his hind legs out behind him. The only individual to benefit from this arrangement will be the chiropractor who readjusts the horse's neck and back later on.

A typical session on the longe would begin with five minutes of playtime during which the horse trots freely with no side reins attached. Playtime does *not* involve running or bucking on the longe. Five minutes of trotting will allow most horses to begin to warm up and stretch, and will allow the handler to observe the horse and notice any hint of tension or lameness.

After this, the side reins would be attached and the horse would be worked for 15 or 20 minutes with the side reins attached. Most of this work would be at the trot, with brief periods of canter for more advanced horses, and with many transitions. The horse would be worked in both directions; more advanced horses would perform more changes of direction. The work would be followed by a warm-down of 5 or 10 minutes walking, either on the longe without side reins or under saddle on a loose rein.

When you longe, *watch your horse*. His posture and demeanor will tell you if the side reins are adjusted correctly. Adjust them so that your horse can stretch his neck forward and down comfortably, then watch him trot. When he trots with rhythm, looks cheerful, and carries himself well (his tail will be slightly raised and will be swinging in rhythm with his trot), and when he is working comfortably with a correct (small) bend to the inside, then he'll be ready to canter in (sliding) side reins. Keep the side reins adjusted evenly at all times; this bend in his body means that he is working more from the outside rein than the inside, which is exactly what you will want under saddle as well. Begin with the side reins comfortably loose, but not loose enough to allow him to catch a leg in the rein. If he canters smoothly and regularly and maintains his bend, take up the side reins gradually, but keep sending him forward and down to them — never adjust them so that they pull him back.

Cantering in Side Reins

Q My trainer longes my horse in either Vienna reins or side reins in both the trot and the canter. I seem to remember reading that a horse should only be longed at the trot with side reins and never at the canter, and also that a horse shouldn't be ridden with side reins on. I'm wondering about this, as I've seen many horses longed and ridden with side reins at the trot and canter and not seen any problems. I also think I remember seeing vaulting horses with side reins. Can you shed any light on this? My trainer said he learned to longe from a vaulter and they always used side reins. He said the Vienna reins shouldn't be used while riding, though.

A Vaulting horses generally wear side reins, but most vaulting horses are experienced dressage horses, quite capable of cantering in the restricted frame imposed by side reins. It takes about four years to train and condition a vaulting horse. By the time a horse is ready for work with vaulters, he can canter smoothly and evenly on a circle with steady balance, and the side reins don't present a problem. Even in a vaulting competition, though, the side reins must not be *too* restrictive, and the horse's poll should be the highest point of his body. If a vaulting horse becomes sore in the back or neck, one of the first things his trainer will do is put him back on a program of longeing *without* side reins, so that he can stretch his topline.

Young horses and unfit horses can be harmed by cantering on the longe with side reins; their training should involve longeing *without* side reins until they have found their balance and learned to engage and lift their backs. When a horse first learns to canter on the longe, or returns to cantering on the longe after a layoff, he lacks balance and finds the work on a circle to be quite difficult in and of itself. The horse must learn to move rhythmically and calmly and in balance at the canter on the longe before side reins are introduced. Young horses tend to go on the forehand, which is normal. They must learn to use their hind legs, stretch their toplines thoroughly, then slowly begin to use their hind legs even more, relax and lift their backs, and *finally* carry their front ends in a more elevated fashion. This takes time. Adding side reins too early will cramp the horse and restrict his movement.

Use side reins at canter only when the horse is able to maintain the gait easily without using his head and neck for balance.

If you watch two horses being ridden — a young, green, lower-level horse and an older, experienced, upper-level horse — you will notice that the young horse has a longer, flatter silhouette, and that at the canter he moves his head and neck, sometimes quite dramatically, to balance himself. The upper-level horse moves his head and neck much less, because he is strong, fit, and balanced, and he has learned to shift his weight and balance somewhat backward. This horse no longer needs to make large balancing gestures with his head and neck; he can be longed at the canter in correctly adjusted side reins.

Riding with side reins makes no sense, since the rider's communication with the horse's mouth would be interrupted constantly by pressure from the side reins. I've seen it done, but it's incorrect and does nothing to improve the horse, the horse's training, or the rider's ability to communicate with the horse through the reins. Riders can be longed on an experienced horse wearing side reins, but the purpose of that is to improve the rider's position, and the handler, not the rider, is in charge of the horse.

If your horse is sufficiently balanced and strong to canter comfortably on a circle with minimal use of his head and neck, and if the side reins are adjusted correctly so that the horse reaches for and moves out to the bit, then he is probably working at a level that allows him to canter on the longe wearing side reins.

Okay to Longe a Three Year Old in Side Reins?

Q I've been reading a book about starting a three-year-old horse. It recommends starting the horse in side reins (the author cautions against making the reins too tight). But he says that by using side reins you can teach a horse balance and collection. From your writing, I gather that they have to be taught this first and then they can be longed with side reins. Should I use them to start my young horse, or is this a bad way to train? Under what circumstances would you recommend using them?

A A young horse should not be started on the longe in side reins, because the youngster will have to work hard enough just to learn to carry himself naturally on a circle. A young horse needs someone experienced to longe him, a *long* longe line (preferably 35 feet or longer) so that he can work on the largest possible circle, and a proper longeing cavesson.

After the horse has learned to carry himself well, and when he understands and responds promptly to your asking him to walk, trot, canter, halt, and reverse on the longe line, you can add a bradoon carrier with a bit and perform his normal longeing routine while he wears it. After a week or two of this, you could introduce sliding side reins adjusted loosely, so that he can carry his head in its normal position and doesn't have to pull his neck up or his head in to "find" the bit.

You can encourage an already balanced horse to *stay* balanced by the careful, occasional use of side reins while longeing. If a horse is working in collection under a rider, you can encourage that horse to work in collection on the longe. Side reins can remind a horse to carry himself the way he has been trained and developed to carry himself, but they can only *remind*. They can *confirm* collection on the longe, but they cannot *create* it. If a horse is only three, collection should be the *last* thing on the trainer's mind.

I'll sometimes use side reins on an older, well-trained horse, during part of a longeing session, to reinforce whatever the horse has been doing under saddle. The horse is warmed up without side reins, then trots with them for perhaps 10 minutes in each direction, with breaks to stretch down, and then works the final 10 minutes without them, to stretch again. On a young horse, I'll use them just to introduce the horse to the concept of moving forward while feeling pressure

Side reins can remind a balanced horse to remain balanced
but cannot teach him to balance.

in his mouth from the bit. I'll adjust them long, so that they will be taut (but not tight!) when he carries his head normally, and I'll just trot him — no walk or canter, just trot and halt. He'll wear them for two or three minutes, no more. The next day, he might wear them for five minutes. It's a way to introduce a young horse to bit pressure without adding the stress of a rider.

Look for an instructor or clinician who is good at longeing and can help you work with a young horse *and* teach you how to longe a well-trained, experienced horse. Then, when you decide to try side reins, ask that person to help you.

Using a Neck Stretcher

Q My question is about neck stretchers. Apparently you can get them from tack catalogues. They are bungee-type cords that go over the poll, then through the bit rings. They invite the horse to stretch his neck down and not move as hollow. My instructor got one last spring for a lesson horse. From what I could see, it really helped. And it's not used as much any more, because the horse has been shown how much more comfortable it is to carry himself better. Or that's my conclusion. What is your view on them? Do you think they are more effective than martingales or are they a waste of money?

A Before I begin discussing devices, I want to make this point: Stretching the neck down doesn't make a horse less hollow. The way to make a horse less hollow is to strengthen and engage the horse's hindquarters and the joints of his hind legs, encourage the horse to tighten his belly muscles and lift and stretch his back, and *then* his neck will stretch. You can't achieve those results by starting with the neck, although many people try, and put horses through all kinds of hell by forcing their heads down and in.

I dislike head-setting devices, but it sounds as though your instructor used this one in the only way such a thing should ever be used: briefly, loosely, and temporarily to show the horse that it *could* move in a certain way. Devices are only good if they are *educational* rather than merely restrictive or forceful. When you consider using any device, ask yourself two questions:

1. Does this gadget *teach* the horse something or does it just force the horse into (or out of) a particular position?

2. Does this gadget help to develop the correct muscles in the horse, helping him become more capable of doing what is wanted, or does it develop the opposite muscles by creating pain and setting up a series of resistances?

Like a chambon, the neck stretcher can be helpful if used loosely and for a brief period of time, just long enough to show the horse (a) what you would like him to do, and (b) that he *can* do it comfortably. A few minutes on the longe wearing a loosely adjusted neck stretcher or chambon is unlikely to cause harm, as the horse's head will not be forced into any particular position. As soon as the horse discovers that he *can* move with his neck reaching forward and down, the gadget should be removed and put back into the tack box, and the trainer should take over the job of building up the horse until he *chooses* to move like that. Most people who use gadgets will point to the horse wearing the gadget and say, "Look, see how well it works, see how much better he carries himself, he's obviously more comfortable!" Sometimes this is true, but the proof of the gadget's effectiveness, and of the horse's comfort, is whether the horse remembers the lesson and continues to work comfortably in the same way after the equipment has been removed — and not just in the first few moments immediately after the equipment has been removed, but weeks and months later.

Many devices aren't actually *training* aids but rather are forms of coercion and shouldn't be used at all. Even so called "training aids" are often used wrongly, becoming coercive because they are adjusted incorrectly, used for too long, or used on a horse that isn't sufficiently developed or trained. If a horse is physically incapable of achieving or sustaining a particular position or way of moving on his own, and a device is used to force an approximation of the desired position, that is simply *wrong*. It's bad training, and it's not horsemanship.

Martingales are restrictive. They limit the degree to which a horse can toss his head (standing martingale), or they provide additional leverage to the rider (running martingale). They don't *teach* the horse anything, so they would not be considered training equipment.

The innocent-sounding "neck stretcher" puts pressure on the horse's mouth and poll.

Chains for Handling and Training?

Q We keep having the same discussion with one of our boarders. She maintains that using a chain over the nose or under the chin is okay, because if the handler knows what they are doing, the pressure is released "instantly." We maintain that using a stud chain on a horse is unnecessary and cruel, and even when used "properly" has debilitating effects on the horse, both mentally and physically.

Would you please settle this, if you can, once and for all? Does a horse anticipate pain from the chain resting on his nose (or under), even if the chain is not pulled on or snatched at? Does the use of stud chains cause rearing? What are the long-term effects of this "tool," both mentally and physically?

A Your boarder has missed some important basics in her education as a horse owner, and I hope that you, by teaching and by example, will be able to help her fill those gaps. The use of a stud chain — over the nose or under the jaw — should not be routine for anyone. Sadly, many horse owners and trainers are very fond of chains. The use of the term "stud chain" is revealing — it tells you that the person using the term sees stallions as wild, dangerous, unpredictable animals that need to be controlled by pain.

A chain, like a twitch, is for restraint, *not* for training. Chains have their uses — for example, a severe method of restraint might be necessary to control a horse in an emergency because the short-term pain used to immobilize the horse might help that horse survive or avoid greater and longer-lasting pain. Think of a horse caught in a barbed wire fence, for example, with one person holding it while another person cuts the wires or runs for help. Any method of restraint (using a chain or rope over the horse's nose, applying a twitch, tying up a leg, or sitting on the horse's head) would be appropriate in this situation in order to prevent the horse kicking, struggling, and tearing itself to shreds on the barbed wire. There would be no time to train

Chains should be used for emergencies, not for routine handling.

the horse to stand quietly, and the handler would have to do whatever was necessary to keep the horse immobilized until the wires were no longer a threat.

Routine handling, such as taking a horse from a stall to the cross-ties, or from a stall to a pasture, should not require a chain. If a horse doesn't stand quietly or doesn't lead well, the solution is to *train* the horse.

In answer to your questions: Yes, a horse can anticipate pain from a chain over his nose or under his jaw. If the chain has been used to inflict pain in the past, the horse *will* anticipate pain. Why wouldn't he?

Yes, the use of a chain can provoke a rear. Some horses will rear in response to a sudden, sharp pain across the nose or under the jaw; other horses will eventually rear in an attempt to escape the steady pain of constant pressure.

Training and coercing have very different effects, both physical and mental, on horses. A horse that cooperates out of understanding is a much safer horse to handle than one that gives in out of fear or pain. Sudden sharp pain can cause a horse to react violently. Jerking a chain over the delicate nose or under the jaw of a horse can cause the horse to rear or even to flip over. Putting steady pressure — even light pressure — on a chain will cause a different problem: Constant, dull pain can cause a horse to become less reactive to the pain over time. A horse that expects and accepts pain will typically become either dull or sullen, whereas one that anticipates and does *not* accept pain will become reluctant to cooperate, or even actively resentful.

Any handler who knows exactly when to use a chain, how hard, and for how long, and is sufficiently skilled to release it instantly and completely should not need to use pain and restraint as a substitute for training. Inflicting pain on a horse can produce a *reaction*, but not a *learned behavior*. Teaching a horse to stand quietly or to lead calmly requires time and effort, but it's worth doing. Causing pain to a horse does not mean being in control of that horse. Your boarder needs to learn how to control her horse; when she can do that, she will no longer believe that control comes from a length of chain.

Several years ago, someone sent me a comic strip that I loved and kept until it literally fell apart. It was such a perfect description of the "want control, must use chain" philosophy, and it showed the thought bubbles of the individuals at *both* ends of the lead rope. As I recall, a dog and his owner were walking along with the dog in front, pulling hard on a chain lead. The first caption, apparently quoting a dog-training book or a trainer, reads: "All it takes to assert control is a little jerk at the end of the chain." The second caption is the dog's thought bubble: "Yes, a little jerk — that's how I'd describe him."

Crank Cavesson

Q I was at a clinic of yours (my wife was riding) and saw you take off all of the flash attachments. I understand why you didn't like the idea of tying horses' mouths shut. But one of the other riders in the clinic had a crank cavesson, the kind that fastens back on itself so that you can pull it really tight, and you didn't take that away. In fact, you told the rider that you liked that cavesson. This seems inconsistent with your approach to riding. Am I missing something here? I trust you and your approach to horses and people, but my wife has been burned several times by instructors who talked about "softness" and then used cable tie-downs, twisted wire snaffles, and the like to get a dressage headset. I want to think that you had a good reason for allowing someone to use a piece of equipment that you yourself admit is designed to be abusive.

A I'm glad you asked this question, because it gives me a chance to clarify what I said. A flash noseband combines an ordinary cavesson with a dropped noseband and is commonly used to prevent the horse from opening his mouth. It's true, I do always loosen the cavesson and remove the lower strap, which the rider is then free to use elsewhere, perhaps attached to the front of the saddle as an "SOS strap" or "grab strap." I often tell riders that every bridle with a flash noseband "comes with a free grab strap." Crank cavessons *are* designed to allow the rider to pull them so tight that a horse's mouth is effectively tied shut. Of course I don't approve of that — it's miserably uncomfortable for the horse and makes it impossible for the horse to relax his jaw (and neck, back, and so on).

A crank cavesson doesn't have to be cranked.

However, you don't *have* to make use of the leverage-tightening opportunity presented by a crank cavesson. At that clinic, the rider you mention owned an older, gray mare with a tumor developing under her jaw. A cavesson is required in dressage competition, and this rider had been looking for one that would not chafe, rub, or put painful pressure on her mare's jaw. This "crank" cavesson was designed to be tightened severely without creating open wounds so the strap under the jaw was wide, soft, and padded with gel. There would be no "crank" effect unless the rider pulled the strap *tight*. This rider did not tighten the cavesson at all, and probably could not have overtightened it if she tried, as it was designed for a horse with a larger muzzle. It was ideal for this mare; the loose, soft, wide strap protected the mare's jaw (and tumor) from coming into contact with the buckle. In fact, she would have had to yawn *very* wide just to bring her jaw into contact with the strap.

Sometimes there are good ways to use even "bad" equipment. In this case, a cavesson that was designed for an unacceptable purpose (locking the horse's jaw closed) could be adjusted loosely and thus become *more* comfortable than an ordinary cavesson.

I hope you're reassured. And no, there is no good way to use either a cable tie-down or a twisted wire bit, and there is also no such thing as a "dressage headset." If any instructor tells your wife there *is* such a thing, please tell your wife to run away from that instructor as fast as she possibly can.

Draw Reins

Q I have some questions for you about draw reins. I ride dressage, and I don't use them. Partially because I've always been taught to push the horse into the bridle correctly, and partially because I just don't think that I have enough experience and education to use them properly. So, I would like to ask:

1. What type of situation would you recommend using draw reins in?
2. What is the correct use of draw reins?
3. Why do people use draw reins so much?
4. Why don't some people ever lose their dependence on draw reins?
5. What can you do with a horse that hasn't been worked *without* draw reins in a long, long time?
6. What is the difference between riding with draw reins and longeing with side reins?

A That's an interesting collection of questions, and I'll be glad to tell you what I think. There is no situation in which I would recommend using draw reins. These reins do not educate; they merely coerce, and their effect on the horse's body and brain is not a good one. Used wrongly or for too long — as is almost invariably the case — they create pain and muscle cramping, promote evasions, and teach the horse that the bit should be feared and avoided.

I don't think there *is* a correct use, that is, a use that is compatible with correct, systematic, progressive training. The only *marginally* acceptable use of draw reins is completely hypothetical. It would be the brief (no more than a few minutes) deployment of draw reins on a sound, comfortable, sensible, well-muscled but poorly trained horse, after a full warm-up, by an extremely adept and sensitive rider, to show the horse that he *could* carry his head and neck in a position other than the one he habitually adopted. The draw reins would be used for a few minutes and then removed. This would be repeated on the following day, and again on the third and final day, after which the draw reins would be put back into the tack trunk for the next 50 years or more.

I say "hypothetical" because I have never met a truly good rider or trainer who routinely used draw reins, and I have never met any horse that could not be taught what the rider wanted *without* the use of additional force and leverage. On the other hand, I have met hundreds of riders and trainers who use draw reins to substitute equipment and force for training and understanding, or to achieve a "look," or to dominate a horse they believe to be "challenging them." I have never seen a situation in which a horse was *helped* by draw reins, while I've seen thou-

The uncomfortable horse does not share his rider's enthusiasm for his arched neck.

sands of instances of horses being harmed by them. The example of marginally acceptable use that I've given above is not something you are ever likely to see, because any rider/trainer who could meet those conditions would also be entirely capable of achieving his or her goal without using the draw reins — and it's very unlikely that such a person would even *think* of using them.

Draw reins are popular with certain types of riders. Some riders believe that it's the horse's job to submit to the rider's demands, no matter how inept the rider, and no matter how physically impossible a particular demand may be for a horse to meet.

Some riders are lazy. They want their horses to look like highly trained, upper-level horses but are unwilling to work to develop the horses' bodies and minds. They see draw reins as a way to achieve a desired "look." They don't understand that although it's possible to force a horse's head down and pull his neck into a tight curve, this doesn't achieve anything useful, and the "look" thus created won't fool a real horseman, or a horse, for a single second.

Some riders are uneducated or inexperienced and allow themselves to be fooled by fast-talking, self-styled "trainers" who promote the use of such gadgetry. Others lack an understanding of leverage and imagine that adding draw reins and/or a more severe bit allows them to be "lighter" or "softer" with their hands. They don't understand that the leverage simply allows them to cause more pain to the horse with less effort.

Riders depend on draw reins for a number of reasons. Some feel that this is the only way they have "control" over their horses, and they are unwilling to allow a horse more freedom with his head and neck in case he does something they might not be able to handle.

It's difficult for people in any context, not just riding, to recognize that they've been badly taught and have learned to do something the wrong way. It's even more difficult for people in that situation to make the change and begin learning to do it the right way. If they've been badly taught, they probably have no idea of how to start over. If they *do* have an idea of what's involved in starting over, they may be daunted by the amount of work and time it will take.

Draw reins can hurt your horse. A horse that has been worked in draw reins for a long time is probably extremely sore, if not actually lame, and will require remedial work before he can be put back on a more reasonable training program. A good trainer can take one look at a horse and tell you whether he's been worked in draw reins, because the horse's musculature develops incorrectly through the neck and back. It takes a long time; a great deal of slow, patient, correct work; and usually a lot of trigger-point massage and passive stretching before such a horse can be brought back to a point from which correct training will *allow* him to develop properly. I've worked with horses like this and found 50 or more trigger points in their necks alone, each one of which requires attention and regular massage over a period of months, before the horse can be put into proper work.

Fortunately, some good-hearted, patient riders are willing to learn the skills and put in the time and effort necessary to salvage such horses. It takes infinite patience and can be heartbreaking to spend months coaxing a horse to begin to uncurl his neck, use his back and hindquarters again, and reach toward the bit. It's even more heartbreaking to reach this point at last, only to realize that the long-term effect of the draw reins damaged not only the horse's mind and jaw and neck and back, but his hocks as well, and that he will never be really sound again.

> *I have never seen a situation in which a horse was helped by draw reins.*

Side reins are almost as abused as draw reins; the difference is that side reins can actually serve a useful purpose if used correctly. Side reins, loosely adjusted, can help a horse learn to move forward while accepting light contact from the bit, but again, they must be used correctly, which means at a trot *only*, because the horse's head and neck remain steady at that gait, not at walk or at canter, when a horse in training will need to make balancing gestures with his head and neck. Too-loose, flapping side reins will only annoy the horse; too-tight side reins will restrict the horse's head and neck movement and set the horse up for lameness by causing tension in the horse's jaw, neck, and back.

In an advanced horse, working correctly at the upper levels of dressage, side reins can also be used at canter, because the horse at that level has learned to use his entire body and back, which makes large balancing gestures unnecessary, and so minimizes the movement of his head and neck. But side reins can be used only to confirm collection — *never* to create it!

Although side reins can be useful — you can teach a young horse to trot confidently, in balance on the longe line, in contact with the bit, without the weight of a rider — they are usually not correctly used or even correctly adjusted. Both side reins and draw reins create a great deal of work for equine chiropractors and massage therapists.

Adjusting Bit Position

Q It seems like a lot of people determine where to position the bit based on the number of wrinkles in the side of the horse's mouth. I don't see why every horse with all the different types of teeth, mouths, noses, and so on should have the bit placement based solely on the number of wrinkles that appear in the corner of his mouth. What am I missing here?

I have also seen some people let the bit hang loosely, almost where it can hit the front teeth, to see where the horse naturally would want to carry the bit himself, and then adjust the bridle so the bit is carried in that place. That seems more logical to me. Do you see any problems with this method? Would it matter as to where you would want the bit if you wanted contact, such as in dressage, versus noncontact, such as in Western pleasure?

A Your instincts are very good, and you are right. It makes perfect sense to adjust the bit to make the horse able to carry it in comfort. When it comes to snaffles, I'm not sure where this "wrinkle" idea came from — it's recent, and we would do well to be rid of it. It's a fad and a fashion, like the bearing rein (remember *Black Beauty?*), and just as (non)functional. You see those wrinkles everywhere, especially in some tack catalogues where every full-color photo of a bridled horse shows the bit far too high and the noseband far too tight.

Not so long ago, a snaffle was said to be positioned correctly if it just touched the corners of the horse's mouth. This is where I would begin with any bit, moving it up or down slightly according to the horse's comfort level and response.

When a young horse is first learning to carry a bit, it's preferable to adjust it a little high in his mouth, so that he doesn't manage to get his tongue over it. This typically happens with young horses when their bits are adjusted too low. Not only is this uncomfortable for the horse, but putting the tongue over the bit can become a habit — and a difficult one to break — so it's better to prevent it.

Begin with the bit just touching the corner of the horse's mouth.

Horses should have their teeth checked before they ever have a bit put into their mouths, and they should continue to have their teeth checked at regular intervals for the rest of their lives. Eating can be compromised if a horse's teeth need floating, and so can a horse's mouth comfort. A bit and noseband can become instruments of torture in the mouth of a horse with sharp edges and hooks on his teeth, and even the simplest, gentlest, mullen-mouth or French-link snaffle can cause acute pain to a horse who still has his wolf teeth.

A bit can rub the skin in the corners of the horse's mouth (you can prevent this by applying a little Vaseline). A bit can pinch the horse's lips against the outside edges of his teeth — no problem if the teeth are smooth but terribly painful if the edges are sharp. When you check your horse's teeth, check the inside of his mouth as well, including the insides of his lips and cheeks. Sores and ulcers are all too common.

The bit itself can be at fault, either in terms of design, suitability, or condition. Some bits are inherently unkind and should not be used. Inexpensive ones can be rough or pitted and can cause sores. Perfectly good-quality bits may be too small or too large for a particular horse or may simply be unsuitable for that horse's mouth conformation.

The French-link snaffle is the closest I have found to a "one-style-suits-all" bit, and even so, some horses are happier in a KK snaffle with the curved cannons and the center lump! In order to choose the right bit, you need to know about bits, and you need to know your horse — does he have a short mouth, a long mouth, a high palate, a low palate, a thin tongue, a thick tongue? How old is he; does he need his wolf teeth removed? Is he retaining tooth caps? Do his teeth need floating? It's really just consideration and common sense. Choose a bit for your horse with the same attention and care that you would give to selecting a comfortable pair of shoes for yourself.

What Is the Purpose of a Flash Noseband?

Q I am wondering if you can explain why the flash is used. I know it is meant to keep the horse's mouth shut and, I guess, so the horse won't try to get his tongue over the bit or anything like that. But if the horse is trying to do that, isn't something usually bugging him, like the way the rider is riding, or the way the bit feels, or maybe he needs dental work?

A Anyone who has ridden with me in a clinic knows that the first thing I generally do is remove the flash attachment and loosen the cavesson. Riders need to distinguish between goals of schooling and goals of showing: Schooling means looking for problems and fixing them, whereas showing means finessing the problems and showcasing what you do best.

The flash noseband was invented by someone who wanted to combine the features of the dropped noseband,

A flash attachment will make a lovely "grab strap" for your saddle.

which ties the horse's mouth shut, with the ordinary cavesson, which serves as an anchor for a standing martingale. You can't put a standing martingale on a dropped noseband. The flash (named for a horse called "Flash") is a very popular item just now, and in some areas it's difficult to find a bridle that doesn't come with a flash. But this is a fad, and as soon as everyone has bought a bridle with a flash, the style will almost certainly change just in time for all of us to buy new bridles.

Tying a horse's mouth shut is definitely counterproductive when schooling. If the horse is opening his mouth, he is almost certainly *reacting* to something: the bit, the rider's hands, or discomfort elsewhere. A good rider will want to know what the horse is feeling and will want to make the horse comfortable. A horse can't learn when he is tense or in pain, and he certainly can't enjoy himself. And whether you are schooling or showing or just hacking out, you should care, very much, whether your horse is comfortable and enjoying himself.

A tight noseband of any kind is uncomfortable for the horse. It makes no sense to tie a horse's mouth tightly shut if one of your riding goals is to get the horse to relax his jaw! Horses can't relax in the jaw, at the poll, or in the neck or back when they are stiffening their jaw in reaction to a tight noseband. Try this yourself: As you read this, tense your jaw, then while *maintaining* that tension, try to relax your neck muscles and those in your upper back. You can't do it. No human can, and no horse can, either.

Use your common sense. If you know that your horse can't relax his neck and back when his jaw is stiff, don't adjust your bridle in a way that causes your horse to have a stiff jaw. If your bit is riding comfortably in the horse's mouth and the cavesson is adjusted loosely so that the horse can flex comfortably, you will both enjoy your rides much more.

Will a Noseband Help My Horse Collect?

Q I've been riding my four-year-old Quarter Horse for about six months now, and he is unwilling to flex his poll and engage his rear end. My bit is just a snaffle and I use a plain noseband fastened not very tight. I have to wear long spurs to get him to use his hindquarters at all. I can get him to collect into a good frame at the walk but not at the trot and the canter is all strung out. He just won't round up. My friend Barry watched me schooling yesterday and said that my horse's frame was totally wrong — that it was upside down.

How can I make this horse go in the right frame (round) and collect at the trot and canter? Barry said that changing to a flash noseband and making it really tight might help, but that doesn't sound right to me. I think I'm a pretty good rider — I try to do all of my trot sitting — but this horse has some problems that I'm just not able to handle. Do you think changing the noseband would make all that much difference, and if it wouldn't, then what should I do instead?

A First, I don't think that your horse has problems other than the fact that he can't possibly meet your current expectations. Going in an inverted (upside-down) frame is a horse's normal reaction to a rider who is too active with the hands and is asking for too much, too soon. It's also a very typical reaction to a sore back, which would be the expected result of a young, green horse without much back strength trying to carry a rider who tries to sit when she should be up and off the horse's back.

Your horse *is* young, green, and simply doesn't have the musculature to do what you want him to do — *yet*. If you are willing to put in a few years of work, slowly and systematically helping your horse become strong and balanced and supple, he will be able to carry you easily and work in a much rounder frame by the time he is a mature horse. At four, he is still several years away from full growth and physical maturity. Quarter Horses can look large, sturdy, and muscular at very young ages, but that doesn't change the reality, which is that your horse's skeleton is still developing.

Second, you are right to question the advice you got about changing your noseband for a tight flash. Equipment problems should be solved with equipment, but this is not an equipment problem. Keep your loosely fastened simple noseband. There is no equipment that will change what your horse is doing or his reasons for doing it.

Your horse's frame is long because he is green. His frame is inverted or "upside down" because he lacks strength in his back. He needs to move more freely, stretch out, and begin to develop a topline, starting with his rear end. Instead of trying to "collect" him, which, by the way, isn't even in the realm of physical possibility now, here's what I'd like you to do.

Spend a lot of time at the walk — not a short walk, and certainly not a walk with you grinding your seat into his back and/or trying to hold his head up with the reins, but a relaxed walk on a long rein (at first, perhaps even on a loose rein). Keep your seat light, and alternate five or six steps of "normal" walk with five or six steps of asking him (with your calves, *not* your spurs — leave the spurs in the tack room!) to take longer steps at the walk. This will help your horse begin to stretch his back. When he can work consistently and well at the walk, lengthening his stride easily whenever you ask, powering from behind, lifting his back and the base of his neck, and reaching forward and down with his head and neck into light, steady contact, *then* it will be time for you to begin to do some trotting.

When your horse can do a powerful, stretchy walk on a long rein with light contact, you will be able to ask for a similar trot — powerful, stretchy, on a long rein with light contact. At the trot, your goals for yourself will be to stay light on his back — don't try to sit! This is an excellent time to practice your half-seat (two-point position), while maintaining a soft, steady contact with his mouth. When his trot is just as powerful, swinging, stretchy, and regular as the walk, and he is taking long steps easily, you can begin working on transitions from walk to trot and back to walk again, and so forth.

The time to begin sitting the trot — for just a few strides at a time — will be when your horse is lifting his back into your seat and making it easy for you to sit. This can't happen until he is much more developed in his gaits, balance, musculature, and understanding. For now, keep your seat light whether you are sitting the walk or canter; and when you trot, either rise or ride in a half-seat.

All of the above sounds easy and simple, and it is, but it's not fast. Plan to take *at least* several months to accomplish all of the above. It's not enough just to do it, you want to do it right, then do it well, and then do it really well so many times and for so long that "really well" becomes your default way of walking and trotting and doing walk-trot transitions. This will take time, which shouldn't be a problem, and during all that time, your horse will be getting steadily stronger and more coordinated and better looking. And remember, he will also be growing up. A four year old is a very young animal. Relax, ask for things he can do well, help him develop so that he can do other things and do those things better — and remember that he's only a little more than halfway to maturity.

If I were you, I wouldn't even ask him to canter for the next few months — or the next six months, for that matter. If you can, try to wait until the walk and trot work are going very well and have been going consistently well for a month or two. Then, when you do allow him to canter, deal with the first canters the same way that you dealt with the first walks and trots under your new program. Your priorities and goals should be the same: for you, a light seat and light following contact; for the horse, active use of the hindquarters, a lifted back, and head and neck reaching forward and down into your hand.

If the first few canters are less than wonderful, don't worry — that's entirely normal for a youngster. If you can, try to do your first canters in a large arena or field where you can work on long straight lines and not worry about which lead he is on. As long as he canters reasonably promptly (within a few strides) when you ask, comes back to trot reasonably promptly (within, say, 10 or 15 strides) when you ask, and responds reasonably well to your light steering, that's really all you can expect of a youngster just starting to canter under saddle.

I haven't mentioned flexing at the poll, because that's not something you should be concerned with at this time, and in any case it is something that your horse will need to offer. The flexion at the poll will come about naturally as your horse grows up, becomes stronger and better balanced, and finds it easier to work in a shorter frame. At the point where your horse rounds and reaches for the bit in response to a soft "ask" from your leg, you'll find that he will begin to flex slightly as he lifts the base of his neck and stretches into your hand.

As for why he won't collect, I hope you now understand that he won't collect because he *can't* collect. He doesn't understand what you want, and even if he did understand, he doesn't have the power, the balance, or the musculature that would enable him to collect under saddle. He shouldn't be asked to collect until he is older and better developed, so for now, don't think about collection. Your horse isn't failing you by *not* collecting, and you're not failing him by *not* collecting him — it's just that the entire idea of collection is premature.

What would help you more than anything, now and later, would be a good instructor to give you feedback and help you work with your horse. It's very difficult to work on your own riding progress and the training of a young horse at the same time. Having good help in the form of good instruction can make all the difference in the world.

Persistent Leg Pressure or Whip?

Q I am able to ride my horse only two or three times a week (we do regular training-level dressage type work, mixed in with some outdoor arena and trail riding). Lately he has been giving me refusals, halting dead and backing up if I use any leg pressure. Sometimes he gives little bitty bucks, too. This has so far only occurred (1) when asked to enter the outdoor arena under a rider, and (2) when asked to pick up again after a break at free walk.

I don't carry a whip when I ride because I find the weight makes my hand bounce a little. Should I carry one anyway until he moves forward when I ask him to with legs alone? Or should I just stay persistent with leg pressure until he goes forward? (He will eventually, but I'm not sure I'm winning this battle since he's still refusing.) Might it be that I'm working him too hard each time given the relatively low frequency at which I ride? I'm a little worried about the backing part. I have to release all rein pressure while using leg pressure, or he backs instead of going forward (with no rein contact he stays halted). So far he'll stop when I stop using my legs (as opposed to continuing to back).

Carrying a whip to enforce your leg allows you to use your leg lightly, briefly, and more effectively.

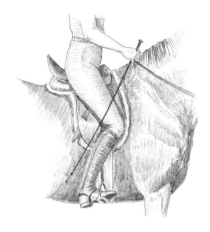

A I'll talk about the whip at the end of my answer, but first let's be sure that you are using your aids correctly. Sometimes horses will do things that we *ask* them to do, even though we really *want* them to do something else, so we need to be sure that we know exactly what we are asking.

When you ask for a halt, you should be riding your horse forward into a fixed hand. "Fixed" doesn't mean "pulling," it just means that when your elbows come back to your sides, they stay there, and your arms no longer move with the horse. Once your horse halts, your legs should relax too, and your fingers should relax from a tight fist to a soft one. Nothing else should change — you should still be sitting up and looking over your horse's ears. If you ask for a halt and your horse stops, but you forget to relax your legs and hands, he will begin to back — not because he is resistant, but because he is *obedient*, and that's exactly what you have asked him to do. It may help if you take a deep breath as you ask for the halt, then exhale completely as he halts. Breathing *out* can remind you to relax your legs, seat, and hands.

From what you've described, it sounds to me as though your horse is doing exactly what you've asked him to do: stop, and then back. The fact that he stands still when you release the pressure on his sides and mouth means that he *is* listening to you.

When you signal your horse to walk forward from a halt, there are two things you should think about. The first is that your leg aids should be a quiet, brief squeeze, not a kick or a long, drawn-out squeeze (constant leg pressure). You have to ask, back off, and give him a chance to respond. If he doesn't respond, you have to ask again with the same squeeze-and-relax movement of the legs. If he doesn't respond the second time, follow it up *instantly* with a sharp smack with the whip behind your leg. Be ready in case you need to do this — have your whip in one hand and your reins bridged in your other hand. If he leaps forward when you smack him, push your rein hand forward, too, so that you don't catch him in the mouth with the bit. If you ask him to go forward, he shouldn't be punished for obeying, even if he leaps into a canter.

The second thing you need to think about is that he won't want to go forward unless there is somewhere for him to go. When you want him to halt, you don't allow your arms to follow his motion. When you want him to back, you keep that steady contact, and squeeze your fingers once for each stride back that you want. When what you want is *forward* motion, are you remembering to *give* with your arms, so that your horse can reach forward? When you ask for that first step from halt into walk, you must relax your hands and arms forward, so that your horse can move off. To begin walking, he must reach down and forward with his head and neck. If you give with your arms while you ask with your legs, he will understand that you want him to move forward. If your legs say "go" while your hands say "no," he will do what he thinks you want him to do — walk backward!

If you practice these things and find that he truly is *not* listening to your legs, and ignores the leg aids unless you back them up with the whip, then you may need to carry a whip for a while, so that you can use it to reinforce your leg aids. I understand what you mean about your hand position deteriorating if you carry a whip — it's a common problem. But if you can put your whip down in your boot, or down the back of your breeches or jeans (this will keep it out of your way until you need it), it won't bother you as much to carry it. And once your horse understands the lesson, you may not have to carry it at all.

The idea is that you use the whip to reinforce the leg aid, so that the horse learns to listen to the leg aid, at which point the whip is no longer necessary. So don't worry, using it as a training aid doesn't mean that you will have to use it forever, or even for very long. But if you squeeze and then squeeze harder and then kick, or squeeze and keep squeezing and never stop, you won't make yourself clear, and you *will* make your horse uncomfortable and resentful and yourself frustrated and tired.

Since you're teaching your horse something and learning something yourself, but it's all your idea really, keep smiling and saying "please" and "thank you" to your horse. It really does help your riding — it's hard to be angry and pull or kick if you're saying "please back" and "thank you, that was nice" instead of "Back!" and "Finally! Why didn't you do that before!"

Baroque Horse Resists Bit and Bend

Q I have worked with my trainer for a couple of years and am very pleased overall with the progress my horse and I have made, although I have not followed *all* of her advice. She has long encouraged me to put my coming six-year-old Lusitano in either a double bridle or at least a Pelham or a Kimberwicke. I have kept him in a full-cheek snaffle on the basis of your comments on the subject, and I clearly understand that neither my horse nor I are ready for a double bridle. I know you are opposed to using a Pelham for early dressage training, and I'd like to understand why. Is the Kimberwicke equally as bad? I'm a relative novice having only been riding for about three years.

Fino is extremely strong and it is a struggle to keep him on the proper bend on the bit to the left. Would it be a mistake to follow my trainer's advice and use a Kimberwicke every once in a while to teach him to stop resisting on the left? This seems to be more a mental issue with him than a physical one as he can perform a lovely collected canter volte to the left if he feels like it with just a snaffle.

A   Using a curb can certainly teach your horse to back away from the bit, but if you want to do dressage, it would not be a good idea to teach your horse to work behind the bit like a Western horse. In Western riding, curbs are meant for one-handed riding and looped reins, not for two-handed riding with the horse on contact at all times. In English riding, the curb is used only as part of a double bridle, to enhance and refine the longitudinal flexion of a highly trained horse. In dressage, the curb is not used to train a green horse, to provide a set of brakes for a horse that's too large or too energetic for his rider, or to control the "made" horse (although you will probably meet people who will argue that one or another of those *is* the reason for the use of the curb).

Even a naturally "up-headed" horse needs to be trained to reach confidently forward into contact.

The curb is used to achieve great refinement in the rider's rein aids, in the same way that the spur is used to fine-tune the signals from the rider's boot. In competition, the real aim is to display the highly educated *rider's* ability to use both snaffle and curb to make tiny adjustments to the horse's head carriage without corrupting the horse's movement or impulsion. And indeed the curb, like the spur, is a very revealing piece of equipment — if the rider has insufficiently educated hands or legs, the horse's reactions to the curb or spur will make this fact instantly apparent to all observers. You can safely put the curb aside for now.

Early dressage training is aimed at relaxing the young horse, then helping him develop a clear, consistent, accurate rhythm in his gaits. Once the relaxation and rhythm are established, the rider will work at gymnasticizing the young horse, helping him become supple, flexible, and evenly developed on both sides. Bending and flexing exercises are part of this education and development, but they're not goals in themselves, and in dressage, there *cannot* be any question of "the end justifying the means." In any case, it won't work! If the rider's or trainer's focus is on creating a certain "look," rather than on building a horse that will eventually have the strength and flexibility and understanding to be able to carry himself and move correctly, then that horse is not ever going to be able to develop or move correctly, because there is no way to fake the necessary strength *or* flexibility *or* understanding.

If you've ever taken ballet lessons, you'll know that it takes time and effort and a systematic program of progressive exercises to create in a young dancer the ability to (for example) lift her leg straight up over her head. If anyone suggested to you that it would be much faster and more effective to use a rope and pulley to force her leg into that position and hold it there, you would be horrified and you would know that any such action would probably cause enough damage to the dancer's body to put an end to any hope of a career in dance. But this is very much what happens to a horse when someone says, in effect, "I can't be bothered to take the time to do this properly; I think I know what it's supposed to *look* like when the horse is fully trained, so I'll just use these ropes and pulleys (or a bit and reins and the rider's biceps) to *force* him into what I think the position should be."

Your horse's difficulty to the left is almost certainly *not* related to "attitude." It's far more likely to be related to ability — his and yours. Most horses that "resist to the left" are horses with a strong, well-developed right side and a less-strong, less-developed left side. Bending and turning to the right is easy for such horses, as the stronger, tighter muscles on their right sides will tighten and compress readily, while the weaker, looser muscles on the left side will readily stretch to accommodate the bend to the right. But when the rider asks for the equivalent bend to the left, the horse becomes very uncomfortable, because *now* it's the longer, looser, weaker side that's being asked to tighten and compress, and the shorter, tighter, stronger side that's being asked to stretch and lengthen. If he doesn't do it as quickly or as well, or if he can't do it for as long a time, he's not being "resistant," he's just being honest.

Dropping the contact is not the solution, but it's certainly a better choice than pulling. By removing the contact completely, you can accomplish two things. First, you won't be pulling against your horse; second, you won't give him any reason to push against the bit. I generally find that riders with "pulling" horses have overdeveloped biceps, which leads me to ask those riders to think about just which end of the reins is being pulled, and by whom! (Hint: It's not the bit end, and it's not the horse.)

The way to encourage a horse to bend around a turn is to take all the time that's needed, over weeks and months, to develop the horse's body so that he finds it easy to bend, and *then* to ask the horse to bend, and allow him to respond correctly to your request. The inside rein may be used to remind him, ever so gently, that you would like his nose to tip slightly to the inside. "Remind" means "a *brief* tightening of the fingers, and a *brief*, slight increase in the contact." It doesn't mean pulling or exerting heavy pressure. Imagine that you break a rubber band. Now, hold each end of that strip of stretchy rubber between the thumb and forefinger of your hands, and begin to separate your hands slowly. When the rubber is no longer looped, and you can make a tiny movement with one hand and feel it in the other hand, you've got the sort of contact that should be your ordinary, everyday riding contact. Now, move your hands another quarter of an inch apart. Feel the very slight increase in pressure, then move your hands closer together again, back to that straight-and-just-barely-stretched default contact position. The moment when your hands were slightly more apart created as much pressure as you'll need to remind your horse of where you would like his nose to go.

Contact is not about a *frame* — it's about a *connection*. Too much focus on the horse's head and neck, too much pressure on the reins, too-strong contact and trying to pull the horse's head into the turn will never have the effect you're hoping to achieve. The secret of good head and neck position on turns is simple: The horse's head and neck complete the curve established by the horse's body and legs in response to the rider's position and aids. A good rider will never try to pull the horse through a turn but will set him up for, and ride him through, a balanced turn. Keep your rein contact gentle and *even*, and let your horse keep his neck in front of his shoulders and his head in front of his neck.

The question is not "Can you pull his neck around?" but "Is he bending through his body?" A horse bent through the body should remain bent on a turn and through a circle, whether your contact is even on both reins or whether you drop the inside rein entirely. Dropping the inside rein is a very useful test: If a horse immediately counterbends when you drop the inside rein, he wasn't bent correctly in the first place. A correctly bent horse won't lose the bend just because you float the inside rein loose for a moment, eliminating the tiny amount of pressure that he was feeling before — if his head whips around to the outside when you drop the inside rein, you can be sure that the only reason his head was tipped to the inside was that the inside rein was forcing it there. That's not what the reins are for.

If dropping the contact entirely is a reward for the horse, what does that say about the nature of the contact and about the rider's attitude toward contact?

Now, let's look at bits. The reason I don't like to see a horse wearing a Pelham or a Kimberwicke for early training is that this is precisely the time when the horse needs to learn to move confidently forward and stretch toward the bit, into the rider's hand, *into* contact. The use of a curb, either alone (the Kimberwicke) or alternating with a snaffle (the Pelham — the effect depends on whether you are using the upper or lower rein), tends to discourage the horse from reaching forward and accepting contact calmly.

The rider who does trade her snaffle for a curb often finds that the horse "feels lighter," and thinks "Oh, my, he goes so much better in the curb, the contact is so much lighter!" Well, no, not really — what's actually happening is that the horse now feels a much *stronger* bit action from every movement of the rider's fingers and becomes understandably wary of the bit and cautious in his movement. This can reassure a nervous rider and make her feel "more in control," but it's bad for the horse's development and training. Once a horse has learned "bit bad, bit painful, must avoid bit" and developed the habit of tucking his chin and avoiding contact with the rider's hands, it's a difficult and often lengthy process to retrain him.

It's not always easy to tell which issues are mental and which are physical, but if you're willing to assume that horses are generous and willing — which they *are* — then you'll find that 99 percent of the time (or more), looking for and solving *physical* issues will eliminate the problem. Pain is a physical issue. Most mental issues, with horses, come down to fear or confusion. The former is caused by pain (physical) and the latter by inconsistency and/or lack of clarity on the rider's part.

Getting different responses on different days is usually the result of asking differently or under different conditions, not the result of the horse being in a "mood." For example, if you feel secure on a horse that is balanced and bent, you might ask for a pirouette with your seat and legs and get it without ever touching the rein. On another day, when your horse is neither balanced nor correctly bent, and you are worried about him going too fast, you might ask for another pirouette (although you shouldn't, under those conditions!) and find that he is suddenly "resistant." If you are honest with yourself and compare all the conditions of a successful pirouette with all the conditions of this attempt, you will probably realize that you were trying to pull your horse into a movement that he simply was in no position to perform.

I think that you need more time riding your horse in the gentlest possible snaffle, and perhaps even some time riding with no bit at all, so you can see for yourself that there is no need to put heavy pressure on the bit. Invest a few months working on relaxation, rhythm, energetic forward movement, and straightness, with the clear idea that your hands will only *ask* (gently), and that any insisting will be done with your legs. When your horse moves with confidence and balance, you'll find that riding a turn will require only that you look through the turn, indicate the bend with your legs, and ride your horse forward in rhythm. The reins will lie against his neck and provide reassurance, but you won't need to shorten your inside rein at all, and you certainly won't need to pull.

A thoughtful rider like yourself, especially one with a Baroque horse, needs to spend some time considering the more philosophical aspects of riding, including the purpose and meaning of contact. If dropping the contact entirely is a reward for the horse, what does that say about the nature of the contact and about the rider's attitude toward contact?

Whenever I hear someone say, "Don't drop the contact, you're rewarding the horse," I have to wonder how the horse perceives the contact. Classical training requires that the rider maintain the horse's comfort, respect his feelings and needs, and develop his physique and understanding so that he learns first to accept, then to welcome, and finally to *seek* contact with the rider's hands. In the process, the young horse must learn to balance himself under the rider, and so the horse may, early in his training, take a heavier contact than the rider finds enjoyable. When this happens, the rider must allow the horse to determine the amount of contact, and try to help the horse become lighter, *not* by pulling or tweaking or dropping the reins but by lengthening them just a little and using transitions, bending, and elementary lateral work to help the horse develop his strength, balance, and the beginnings of carrying power.

From the very beginning, think in terms of seat, legs, and hands — and use them in that order. Try to keep your contact even on both reins, and as light as your horse can comfortably accept, but don't "throw the horse away." The contact between his mouth and your hand is the connection that completes the circle of the aids: Your legs activate his belly and hind legs, your seat allows his back to lift and stretch, your hands accept the contact that he creates when he lifts the base of his neck and reaches forward. If you force or take away any part of this, either through heavy, forcible gripping with seat, legs, or hands, or through dropping the contact with seat, legs, or hands, the circle is broken and the connection is lost.

Be especially careful because your horse is a Lusitano. When dealing with horses that seem to be "born collected" and appear upright and balanced, it's often tempting for riders, instructors, and clinicians to ignore or "skip" the early stages of training. Be careful not to give in to that temptation — your horse will last much longer and enjoy his work much more if you take the time to go through the entire training process with him. He'll be stronger for it, and there will be no gaps in his skills or his understanding, or in your own.

CHAPTER 14

Longeing, Warm-ups, and Your Horse's Training

LONGEING AND WARMING UP are sharing this chapter for two reasons. First, because both subjects are frequently misunderstood. Second, because they are also often thought of as interchangeable. To train your horse effectively, you need to understand that although longeing can certainly be part of your horse's warm-up, it can also be an important part of your horse's training. It's also important to realize that longeing won't automatically warm up your horse, and that you can warm up your horse without longeing him. Warming up can and should be another important part of your horse's training.

Longeing and Warming Up — What's What?

Q I'm confused about the difference between longeing and warming up. At my old barn, we never just got on and rode, we would all longe our horses (or if you were the only one there you could turn your horse loose in the indoor and chase it around instead) for five or ten minutes before riding, to warm them up and let off some steam. Then I moved from that barn (hunter/jumper) to another barn where most people do dressage and eventing. The dressage teacher here is really good. The other day she saw me longeing my horse and told me that I shouldn't ever take a horse out of its stall and race it around like that, but should warm him up first before I "worked him" on the longe. I don't work my horse on the longe, I longe him to warm him up so I can get on and work him! Why isn't longeing a good warm-up? Isn't that what longeing is for? Also, how could I warm him up before I longed him? He always runs like crazy for about five minutes when I start longeing him. Isn't it better for him to use up his excess energy on the longe line?

A There are many different ways to warm up a horse. You don't have to longe him — you can warm him up under saddle — but it's very easy to use longeing as part or all of a preride warm-up. Longeing can be a fine warm-up for dressage horses and eventers, and a fine way to work a horse, as well. You can make this be true for your horse, too, if you'll learn to think about longeing, and do it, in a different way.

The dressage instructor gave you good advice. It's not a good idea to ask — or allow — a horse to race around when he's just come out of his stall. Racing around in a circle puts far too much stress on a horse's joints. One of the goals of a warm-up (on or off the longe) is to make the horse's muscles warm enough to stretch instead of tear; another goal is to lubricate the horse's joints by working them gently. The more careful you are about warming up your horse, the longer he is likely to remain sound. Even if he's been playing in a pasture all day, he shouldn't spend his first minutes on the longe galloping about like a mad thing. That can damage his muscles, tendons, and ligaments, and it certainly doesn't qualify as gentle work for his joints.

If he's been standing in a stall, it's even more important for him to move slowly and gently at first. Teach him to *walk* on the longe for the first 10 or 15 minutes after he leaves his stall, so that he can warm up safely before you ask him for anything more. If you're worried about your ability to control him, and you're sure that he will explode into action instead of walking calmly, begin by hand-walking him around the arena for 10 or 15 minutes and then put him on the longe and ask him to trot. That instructor was right about something else, too: Longeing should not be used to tire out a horse so that he'll be too exhausted to misbehave under saddle. Longe your horse so that he will be warmed up, flexible, relaxed, and attentive under saddle. Longeing is a very useful skill and serves many different purposes. You can longe a horse to:

- ▸ Start a young horse of two or three who isn't yet old enough for ridden work
- ▸ Help a green horse learn to move in rhythm and balance without the burden of a rider's weight
- ▸ Build up a horse to prepare it for ridden work
- ▸ Teach a horse to go forward with controlled energy
- ▸ Teach a horse to relax, bend, and stretch at all three gaits
- ▸ Get to know a new horse and help him get to know you
- ▸ Identify a new horse's "problem areas" and begin to solve those problems
- ▸ Build a horse's confidence in himself and in you
- ▸ Help a horse become more responsive to your body language and your voice
- ▸ Warm up a horse before you work him and warm him down afterward
- ▸ Evaluate his condition
- ▸ Notice and deal with any stiffness or soreness
- ▸ Improve his strength and flexibility
- ▸ Develop his musculature
- ▸ Help him develop a stronger back, and the habit of using his back
- ▸ Improve the rhythm, energy, and reach of his gaits
- ▸ Give him exercise when you can't ride
- ▸ Give him exercise if he has been injured and shouldn't carry a rider
- ▸ Introduce new equipment such as the bit, the saddle, and the stirrups
- ▸ Learn how your horse is moving and feeling today
- ▸ See how your training is affecting your horse's body and balance
- ▸ Develop your powers of observation and improve your understanding of gaits and movement
- ▸ Develop your responsiveness and your hand-eye coordination
- ▸ Develop your ability to "read" your horse and anticipate his actions
- ▸ Develop your ability to direct your horse's actions at the right time
- ▸ Develop your focus, communication, timing, and feel

And, of course, longeing allows you to warm up, exercise, and train a horse while his entire body is visible to you, something that isn't possible when you're riding the horse unless your arena has mirrors on every wall.

Warming Up — What to Do

Q I've been noticing that my friends and I always seem to be doing the same old exercises with our horses. This is probably because we don't have riding teachers in my area to teach us new ones. Do you have any easy, basic exercises we could do in our warm-ups? And are there some exercises that should always be used to warm up a horse?

A There are many different exercises you can do in warm-up. Your warm-up is very important — it's not mindless exercise; it's an important part of your horse's training. On many days, if your riding time is limited, a complete warm-up may be all that you have time to do. For a rider, warming up a horse is very similar to the process a pilot goes through when performing a preflight check. The pilot checks all the systems, gauges, and dials; he needs to be certain that everything is in good working condition before he takes off. Similarly, a rider

Hand walking your horse will help both of you warm up before your ride.

should check her horse's coordination, flexibility, and responsiveness, so that she can discover any problems (not just lameness or injury but any slight stiffness or soreness) and deal with them before beginning work.

Walking is always the best way to begin a warm-up. It's a good habit to adopt for other reasons, too — if your horse is ever lame or recovering from a lameness or an illness and needs exercise but only at a walk, he will be quite accustomed to coming out of his stall and *walking*, even if he is full of energy, whereas a horse that has always been allowed to come out and explode for the first five minutes will not understand why he is being "punished" by being made to walk — and may attempt to buck and run in protest.

Walking shouldn't be boring. Incorporate changes of direction, straight lines, turns, large circles — use all the large ring figures, make up patterns of your own, and make the figures smaller as the warm-up progresses. Do transitions within the walk — shortened strides and lengthened strides, just a few strides at a time, alternating with your horse's normal working walk.

When the horse is warmed up at walk, you can do the same thing at trot, but this time incorporating transitions within the trot as well as transitions between walk and trot. Again, turns, circles, patterns, large and small ring figures, spiraling in and out on circles, a little leg-yielding if your horse is new to lateral work, a little shoulder-in if your horse is more sophisticated.

After the trot work, you can do much the same thing at the canter. One caveat: If your horse is older or stiff from age, arthritis, or any other condition, he may find it easier to warm up effectively and comfortably if you allow him to follow the walk portion of the warm-up with a few minutes of canter (you in a half-seat, your horse on a loose rein). Horses are quite good at knowing what activity will stretch their warmed-up muscles — if your horse always volunteers a canter after his walk warm-up, allow him to stretch that way; his trot will be the better for it.

At all gaits, transitions are key: the better the transition, the better the gymnastic benefit to the horse. Tempo is also important, so do a lot of work on lengthening and shortening your horse's stride at each gait while maintaining the rhythm and tempo.

And now, I know what you're thinking: If you do all that, you may not have any time left to *ride*! Even so, you won't lose anything by taking the time to do a complete warm-up. If each system is "go," you'll know that you can, in fairness, ask your horse to work on something new or difficult. If your warm-up reveals a problem — a stiffness in one direction, an inability to stretch one way, a tight muscle somewhere, or a short stride in one leg — then you'll return to earlier exercises until you've solved the problem. If this takes your entire session, don't worry — it's giving your horse an increasingly solid foundation while promoting and preserving his soundness.

Throughout your warm-up and the rest of your ride, be sure to allow your horse to stretch his head and neck every three minutes — all the way to the ground if he finds this comfortable. This regular stretching needn't take long, a few seconds may be enough, but it's an important part of the warm-up and of the work for *every* horse.

Double Warm-up?

Q I hope to start eventing this year, and I am learning a lot about all three phases with help from my trainer. I have a friend who has already been eventing for two years. She has shown unrated and Beginner Novice, and she hopes to move up to Novice this year. She is a thoughtful and great rider, but I have a question. Every time before she schools dressage, she longes her horse for about 15 to 20 minutes, both directions at walk-trot-canter. After that, she mounts and does *another* warm-up and then does her ride.

I don't understand this. She canters her horse on the longe. But then when she mounts, she goes through a whole long loose-rein walk and the whole warm-up as if she hadn't longed the horse. I know that warm-ups are very important, but wouldn't it make more sense to warm up and *then* longe and then mount for further schooling? Otherwise it seems that the riding warm-up would be useless, and that the longeing would mean that the horse would be worked without being warmed up first.

A I can't tell you the specific reasons why your friend warms up her horse this way; why not ask her directly? Most riders are happy to explain their reasons for

Always take the time to warm up your horse.

doing things in a particular way. Meanwhile, perhaps it would help if I explain why *I* might warm a horse up this way. Think of the horse as an athlete, and the whole focus on warming up will make sense. Human athletes who want to stay sound and continue to perform for many years will put a lot of thought and time into a really thorough warm-up before each workout, and make the big efforts (lifting very heavy weights or running very fast) only briefly and only at the point when every system is fully prepared to take the strain.

If the horse walks in both directions and trots in both directions before cantering at all, it can help ensure that the horse is happy, comfortable, and sound. Without the weight of the rider to carry and balance, and with contact only through the cavesson, not the bit, the horse can relax and move freely. The rider, watching the horse, can ask herself "What horse do I have today?" and then plan the riding session (that is, select the exercises and level of intensity) based on her perception of the horse's strength, flexibility, and comfort level (physical, mental, and emotional) *at that time.*

If longeing is used this way, it's not part of the training/work session, but part of the warm-up. There are two parts to any good warm-up: First, the large muscles of the horse's body are exercised gently to make them warm and stretchy and improve the horse's circulation. Then the horse is "put through his paces," that is, asked to do all of the things he knows how to do, from the easiest and simplest to the most difficult and complex. After that, the "teaching session" or "work session" can begin. This is the comparatively brief period during which the horse is taught something new or asked to do something familiar in a different way (e.g., longer strides, more thrust from behind, more elevation or suspension).

Riding a warm-up is *never* useless. A warm-up on the longe line makes an excellent prelude to a ridden warm-up, which serves as prelude to the intense work part of any riding session. Using longeing as part of a warm-up is an excellent way to maintain a horse's soundness and a rider's keen awareness of the horse's condition, ability, and emotional state.

Are there times when it would make more sense to do the ridden warm-up first, or dispense with the longeing part of the warm-up altogether? Certainly. If the space available for longeing doesn't accommodate *at least* a full 20-meter circle, or if the horse has an injury or condition (e.g., hocks fusing) that would make *any* work on a circle painful, it would be better to forego the longeing and just warm up the horse under saddle, working long and low and beginning with straight lines and wide turns.

If your friend's horse is an old campaigner that she is trying to keep sound, her warm-up makes even more sense, because eventing is a very demanding sport. The older a horse is, the longer and more careful a warm-up he needs before work, and the longer and more careful a warm-down he will need after work. The better and more experienced a rider is, the greater the proportion of any riding session will be taken up by the warm-up and warm-down. A brief warm-up, a long, intense work session, and a brief or nonexistent warm-down are sure signs of an inexperienced, impatient rider whose horse is likely to have a short career.

No veterinarian is ever going to say to any rider "You shouldn't spend so much time warming up your horse, you're too careful," and no horse is ever going to become injured because his rider performed a gentle, thorough warm-up before every workout.

Warming Up: Effect on Training

Q Is there any actual training benefit to warming up? I know better than to take a horse right out of a stall and start working him hard, but my horse lives outdoors in a field and I know he moves around a lot, so I have never understood the point of warming up forever. Usually I bring him in, groom him, tack him up, and then we walk around the arena two times, trot around two times, and then we begin working. It just doesn't seem to me that he would need that much of a warm-up. I do dressage and I start with easier exercises before doing the harder ones anyway, so is there any reason that I should do more of a warm-up than what I do now?

A That's a great question. Many people these days are in a hurry and just want to come out to the barn and ride. They see warming up as an extra chore — something that isn't really necessary. In fact, a good warm-up can have a huge effect on the rest of your ride, making it good instead of frustrating, or perhaps great instead of good.

Every ride should begin with a warm-up, for several reasons. First, a systematic warm-up will help your horse work better in the short run and help him stay sound in the long run. Second, while you're warming up his body, you should also be warming up his mind and getting him focused and attentive and responsive. Third, if you pay attention during your warm-up, you will know what your horse can fairly be asked to do when you begin working him, because you'll know how he feels that day — how energetic, flexible, and forward he is.

A good warm-up will incorporate everything your horse knows how to do, from "go" and "whoa" to the highest-level work your horse is capable of doing. The warm-up will also tell you how *you* are feeling and how energetic, flexible, and forward *you* are.

Warming up lets you begin the "work" part of the session with a horse that is physically warm — circulation working effectively, muscles warm and stretched, joint cartilage lubricated, entire system oxygenated — and comfortable, therefore able to relax and give his full attention to you and the work. By warming him up gradually, you'll have made him ready to work and learn. And in the process, if something is wrong, you'll have noticed it and you'll know what to do about it, whether you need to back off a little on the work, change your plan for that session, or — heaven forbid — dismount and call the veterinarian. In fact, warm-ups can help you avoid extra vet calls. By doing a good warm-up every time you ride, you can save a lot of time that might otherwise be needed for treatments, hand walking, or some other form of rehabilitation.

The old adage "Walk the first mile out and the last mile back" was, and is, very sensible advice. Try it — it will do wonderful things for your horse, whether your "first mile out" involves a trail ride before work or a series of easy gymnastic exercises in the arena before beginning "real" work. A warmed-up horse will be more comfortable, cheerful, and attentive, and will demonstrate a better range of motion.

Using the Arena

Q After years of riding other people's horses, mostly on trails and outdoors in fields, I will finally be riding a horse of my own this winter. To make everything perfect, he will live at a stable where there is an indoor arena. I've had very little experience riding indoors, just enough to know that there are squares with letters on them on walls all the way around the school, and that those are supposed to be useful for training. You have probably already guessed that I don't do dressage, so would the letters still be useful for me? How can I make the best use of the indoor arena when I am training my horse? In the spring, when I bring my horse home, I will only be able to ride in a field, so what are some things that I can do in the arena that I can't do in the field?

A Given a choice between an indoor arena and a field, I must admit that I generally prefer the field, but it's true that there are any number of things that are much easier to do, or to do better, in an indoor arena. Learning the arena — knowing its dimensions and where the letters are — will let you ride school figures accurately. Riding from marker to marker across and around the arena will help you keep your lines straight, and also help when you are training your horse to lengthen and shorten his strides. In a field, you can tell yourself "I'm sure his strides were longer that time around" but in an indoor arena, you can count the strides between two letters and *know* whether you managed to leave a stride out.

In a field, it's not always possible to determine whether your transitions are as accurate as you would like them to be. Trees and fence posts are easily confused with other trees and fence posts, but there is only one of each letter in an indoor arena, and if you intended to halt at E and ended up halting at H, you won't be confused. Annoyed, perhaps, but not confused.

You can take advantage of the fact that the arena has actual corners to ride deep into them, which is gymnastically demanding, instead of cutting them off. Many fields have corners, but it's not always wise or even possible to ride deeply into them.

In an arena, you can more easily ride proper circles — the round variety. In a field, it's easy to ride in other shapes — eggs, potatoes, stars, and squiggles — and imagine them to be circles; in an arena with known dimensions, limited space, and markers on the walls, you can ride a real circle that touches the walls and the center of the arena only at single points.

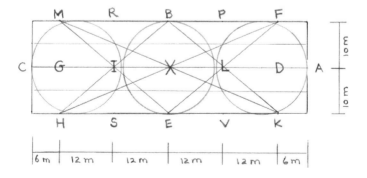

An arena offers many possibilities — there are so many ways to make use of that space.

If you are the first person to ride in the arena after it has been raked and/or watered, your horse's tracks will be visible. This gives you the perfect opportunity to test your ability to ride a round circle, a straight centerline, or a straight diagonal.

Enjoy the arena while you can. When you go back to training your horse in the field, remember that fields have some great advantages: fresh air, scenery, and a much smaller probability that another rider will cut across directly in front of your horse's nose.

Learning to Longe My Horse

Q I am currently leasing (maybe buying soon) a 10-year-old Thoroughbred gelding who used to race. He knows how to longe, but whenever I try him to the right, he stops and backs up. When I longe Jack to his left, he behaves perfectly and is willing, then I change sides to work him to the right, and he won't move out in a circle. I have tried having my dad lead him in a circle to get him trotting, and he will, until my dad lets go; then Jack stops and I am no longer behind him as I was when I was longeing him. I try to come near him and he backs up. I have tried tons of things to make him go on his right side, but no can do.

The longeing triangle consists of horse, longe line, and whip.

My friend suggests that when he used to race, he was longed in only one direction, so he is confused about what I want him to do. Please help! Are there any other things I should do when longeing besides standing behind him (in his range of sight, just farther back)? He seems to be afraid of the whip, as someone had abused him with one before, but I need a whip to keep him in a circle and to get him going (by slapping it on the ground).

A Your friend is right about Jack — in the United States, racehorses run around the track counterclockwise (to the left) as well as being handled from the left. Jack probably has no idea what you want. He may not have been taught to longe at all, although he's obviously used to being handled from the left. Most horses need to be taught everything twice, once from each side.

You need lessons from someone who can teach you how to longe a trained horse and teach Jack how to be longed, and then teach you and Jack together.

In the meantime, I can lay out the basics of longeing for you here.

Create a triangle so that Jack's body is one side, the longe line and your arm are the second side, and the longe whip and your other arm are the third side. Think of the triangle as a piece of pie, with Jack as the edge with the pie crust, and the longe line and whip as the two long sides, with you at the tip of the slice of pie. Position yourself just across from where your leg would be if you were riding Jack: a little behind his elbow.

The longe whip is an extension of your arm. Don't use it to hit him, to make noises, to wave, or to slap on the ground behind him. The whip simply makes your arm longer, so that when *you* point at his hind heel or his hock the whip points there, too. If he is afraid of the whip, he will either run away or turn suddenly to *face* the whip, and neither of those reactions is useful. Your position is important for Jack's training and for your own safety. Always stand to one side, never directly behind a horse — that's extremely dangerous. Wear your helmet when you longe, as a safety precaution: A frightened horse may kick.

Wear gloves. A longe line can burn your fingers if your horse runs off.

Fold the part of the longe line that you hold — don't loop it or wrap it around your hand. Hold the folds so that you can drop them if you need to.

Use clear verbal signals — the ones you use when you handle him in his stall, groom him, and lead him. Jack can learn to understand "walk" and "trot" and "whoa" and "stand." If you teach him to understand these when he's moving to the left and to the right it will be easier for him to learn to longe to the right.

Use a proper longeing cavesson. Longeing from the bit is *always* a bad idea, and it's an even worse idea when the horse is an ex-racehorse. If he's going to the left and you pull on the bit, he'll think he's supposed to run faster because that's how racehorses are trained. If he's going to the right and you pull on the bit, he'll think you want him to turn around, stop, or both.

Work on a circle is very stressful. Too much of it, and work on too-small circles, can damage a horse's legs. For at least the first month or two, keep the longeing sessions brief (10 or 15 minutes every other day, perhaps) and slow (walk, trot, and halt; no cantering). Keep the circle as large as possible — at *least* 20 meters (roughly 66 feet) across. If your longe line is too short, you won't be able to do this. Your longe line should be at least 35 feet long and should not have a chain on it.

Books are not a substitute for the personal instruction you need, but if there is no one in your area who can teach you to longe correctly, these two books will provide helpful information: *The United States Pony Club Manual of Horsemanship: Volume 1,* and *The USPC Guide to Longeing and Ground Training.*

Horse Turns to Face Me on the Longe Line

Q I'm not new to horse ownership, but it has been about 25 years since I owned a horse. My kids have graduated college, my husband will retire soon, and I grabbed the chance to have a horse of my own again. During my horse-less years I read horse books and horse magazines, and I'm overwhelmed with all that I don't know, including things that I used to think I knew how to do.

Take longeing: I used to longe my hunter before I rode him, back when I was young, but now I realize that (a) he was trained and knew what to do, and (b) I had no idea what to do except make him change directions and gaits once in a while! So, I'm reading books about how to longe, but my new horse has obviously not been reading the same books. I have a theoretical understanding of what to do, but in practice I can't make it work; it's like reading a book about how to play the violin.

In our round pen, Austin will walk in a circle fairly quietly, no acting up or bucking, but he has no interest in trotting on the longe line, much less cantering. In fact, he's mostly interested in stopping. If I get after him by waving or cracking the whip, or by hitting the ground behind him, he turns in and stops and stares at me. If I try to get after him before he stops, he walks right up to me until his head is practically on my chest. And that's to the left. I can't get him to go to the right *at all*. The man who sold him to me did a lot of round-penning with him, but only to the left. He said you didn't have to go both ways because round-penning is for creating respect and one direction is enough for that. I'm frustrated!

A Take Austin out of the round pen and longe him in an arena or in a field. He thinks that he knows what's expected of him whenever he's in the round pen, and he's on autopilot instead of listening to you. Also, unless your round pen is at least 66 feet across, he may find it physically difficult to work on the longe and physically impossible to canter, especially to the right. Change his environment by working him elsewhere; change his expectations by working him differently.

Practice your longeing technique. Your whip is not for cracking, hitting the ground, or hitting your horse. It's a visible extension of your arm, so that your signals can be seen easily. Unless you are walking a large circle within your horse's circle (which should be 20 meters or larger), you couldn't possibly touch your horse with even the tip of the longe whip's popper.

Your longe line should establish and maintain a soft contact through the longeing cavesson. The contact should be like rein contact: lightly stretched, elastic and alive, not hard and tight or loose and sagging.

To longe correctly, you need to use both the line and the whip.

Longeing is a great multipurpose exercise and training method. On the longe line, you can exercise your horse and watch him at the same time — something you can't do when you're riding. You can also teach him new skills, refine and improve his existing skills, help him become straight, help him stretch in both directions, improve his gaits and transitions, and work on just about everything from rhythm on up.

You didn't mention Austin's age, but remember that circles, whether on the longe line or in a round pen, put uneven strain on a horse's legs and can damage the joints, especially if the horse is young, working at speed, or working on a small circle. If Austin is at least three years old and your longeing circle is at least 66 feet across, it should be safe for you to give him 20 minutes on the longe every day, mostly at walk and trot. The circle size is important, so remember that every extra foot of longe line you hold folded in your hand makes the circle two feet smaller. If your longe line is a full 35 feet long (check yours — many items sold as "longe lines" are considerably shorter), holding the longe line 2 feet from its end will diminish the size of your horse's circle from 70 feet across to 66 feet across — that is, to a 20-meter circle. If your longe line is 35 feet long but you are holding 8 or 10 feet of it folded in your hand, Austin's circle will be only 54 or 50 feet across, and he may be unable to do the things you ask him to do.

Keep yourself positioned properly relative to your horse. Imagine that Austin is carrying a saddle and a rider. The rider's lower leg and foot should be just behind Austin's elbow, and your default position should be directly across from that imaginary rider's leg. By shifting your own position slightly — taking a small step toward Austin's shoulder or his hip — you'll shift Austin's position and tell him to move out or move on.

If Austin learned his round-pen lessons well, he will move *away* if you step toward the middle of his body, move *forward* if you step toward his hip, and *stop* or *come in* to you if you move toward his neck or head. This is what he is doing — whenever your body gets "ahead of him" (that is, out of position and too near his shoulder and neck), he tries to stop and come to you.

To change this behavior, you'll need to do a lot of longeing with hundreds of walk-halt-walk transitions. At first you'll ask for the walk *before* Austin has fully halted, because if you wait for him to stop, he will turn in as he has been taught to do. Don't give him the opportunity. When he is walking well, ask him to halt and watch him carefully. As he begins to halt, you'll see his nose begin to shift in toward you: that means he is about to turn in. Quickly use your voice

("Austin, walk on!") and body language (looking and stepping slightly toward his hip, lifting your whip smoothly forward and upward, toward his hock) to send him forward again, then praise him. Let him circle a few times, then ask him to halt again and be ready to send him forward as soon as he thinks about turning in. Do this in both directions, often, every day, until Austin understands that when you ask him to halt, he must halt *on the circle* and be ready to walk forward again instantly. As long as you're absolutely clear about what you want — walk, halt, walk! — and as long as you praise him each time he obeys, he'll be comfortable, not confused.

Young Horses and Longe Training

Q I will be training my yearling fillies to longe but not until they are at least eighteen months or even two years old. In the meantime, I'll work with them on ground driving and free longeing. They already know the signals for "walk," "trot," and "whoa." I know there are concerns about longeing young horses, so when I start them on the longe line, what guidelines should I follow?

Yearlings belong in the pasture, not on the longe line or the round pen.

A   You are wise to postpone longeing until your fillies are at least two. Working on a circle is too stressful for young legs and can cause permanent damage. When the time comes, put boots on all four legs (young horses, especially on a circle, find it easy to hit themselves), use a longeing cavesson and a longe line that's at least 35 feet long (no side reins), and ask them for walk, trot, and whoa. Since they already know the words, that will be easy. If they do it for ten minutes (five each direction) every other day for a few months, they should be off to a good start.

Longeing and Voice Commands

Q   I will begin to longe my three-year-old filly in the spring, and I was wondering what you use for voice commands. With my older gelding, I have always used "waaalk," "trot," "can-*ter*," for upward transitions, and "whoa" to do progressive downward transitions. Other people use those commands for upward *and* downward transitions. They say "whoa" should be reserved for *stop*. I'm wondering if using the same command, for example for an upward trot transition and a downward trot transition, would confuse the horse. It would confuse me if I were a horse!

My coach uses noises for walk-trot-canter cues, because she says that when you go to take your Canadian Equestrian Federation (CEF) level one coaching certification (she's got her level three), and you do your longe line testing, you'll be in a big indoor arena with three or four other horses, all longeing at the same time. The horses get confused and don't differentiate between their neighbor's commands and their longer's commands. She uses unique noises so that her horses never get mixed up. This filly is the horse I plan on getting my level one on (I'm 16 now, so by the time I'm 18, old enough to get my certification, she'll be going nicely under saddle).

A   You're very sensible to wait until your filly is more mature before starting her longe work. Your voice commands are perfectly fine; your horse will be able to distinguish quite easily among "waaalk" and "trot!" and "can-*ter*."

I don't use "whoa" for downward transitions, for several reasons. First, "whoa" has a specific meaning: *stop*. That association must be clear, so that when my horse hears me say "Whoa!" he knows that I mean *stop* and not *slow down* or *change gait.* If your horse is ever tangled in a longe line, puts a foot over a side rein, or steps into wire, or if a piece of equipment breaks or another horse is suddenly loose in the arena — I've seen all of these things happen — you need to be able to stop your horse instantly, *on voice command.* If you save "whoa" for "stop," you'll be able to do this effectively; your horse won't have to wonder what it means *this* time.

Second, when you ask your horse for a trot, "trot" or "tee-rot" should mean trot, whether the horse is standing still, walking, or cantering. It doesn't matter whether the transition is upward or downward; when you ask for a trot, that's what your horse should understand and that's what he should do. Similarly, "walk" means walk whether your horse is standing still or whether he is trotting or cantering. If a horse knows the meaning of the word, he won't be confused — what *would* confuse him would be your saying "whoa" when you mean "go from canter to trot" or "go from trot to walk."

Try this: Use either his name or "And . . ." as a preparatory command *before* you tell him "walk" or "trot" or "canter" or "whoa." A preparatory command ("Silver — waaalk" or say "And — waaalk") is like a verbal half halt. It warns the horse that something is coming and allows him to rebalance himself and prepare before you ask him to *do* something.

Third, I don't use "whoa" for transitions or anything other than "stop" because I want prompt, accurate, *forward* transitions whether they are upward or downward ones. When I'm longeing a horse at canter and I ask him to trot, I want him to move smoothly forward into trot, using his joints and muscles — I don't want him to canter more and more slowly until he *falls* into a trot. Similarly, I don't want a horse trotting, then jogging, then collapsing into a slow walk; I want a trot-to-walk transition that is forward and prompt and active.

If I want to modify the horse's gait or activity level on the longe while he's working in a particular gait, I'll ask him to trot (or canter, or walk) more slowly by saying "*easy* trot" or "*easy* canter" or "*easy* walk." Horses figure out almost instantly, especially if you say it on a long exhale and match it to your body language (lift your rein hand slightly, stand taller, and allow the tip of the whip to drop). If I want the horse to move more actively, I'll say, "walk *on*," or "trot *on*" — and match this to my body language by lifting the tip of the whip so that it points at the horse's hocks rather than at his hind heels, and by stepping slightly back toward his hip, getting myself in a position to drive him forward.

Don't worry about longeing in a large arena with several other horses. If your horse is focused on you (which is a matter of training, habit, and *your* focus), he will listen to you, not to anyone else. Even if another person is cracking a whip every few seconds, you can keep your horse focused on you and responding to your commands. Make it your habit to use ordinary commands, not "unique noises." If you have your horse's attention, he will be working from your voice *and your body language*, both of which will be clear to him. A well-trained horse with one ear and one eye on his handler is not easily distracted by another person's commands to another horse.

A horse can learn to respond to almost any signal, but keeping things simple will be good for you when you work with other horses, and will also be good for your horse when someone else works with him. People who teach their horses distinctive or unique signals aren't doing them a favor — other handlers, who don't know that the horse has been taught this way, may punish the horse for responding "wrong" to some signal or for "not responding" to a conventional signal. Unique noises may create problems if you are asked to longe another horse as part of the exam, or if your own horse is ill or lame when you go for your test, and you have to take a different horse with you.

Persuading Horse to Stretch on the Longe

Q I'm reschooling an ex-racer for dressage. He's been on the longe line in side reins for about three to four months, and we've just started him under saddle. My question is: How do you show a horse it's okay to stretch his back and reach forward, down and out while on the longe? I haven't even attempted this under saddle yet, but I feel if I can't get it on the longe then I won't be able to get it under saddle. I've flexed and released to the inside with a driving aid, tried the trotting poles, checked saddle and other things to see if he's in pain, and everything checked out. The most I ever get is wither height or a little lower. I always hear about how the horse should "automatically" stretch out when he's relaxed and working through. I'm still waiting for this "automatic" response! His tail has that relaxed swinging look and it's not clamped, and his gaits are fine. He looks as if he wants to but he's nervous about it.

He'll stick his head down to stretch but then instantly pops it back up. I feel the key to progressing is his total relaxation and learning to reach forward, down and out. How do you go about showing or letting a horse know that it's okay to do this? I'm at my wits' end, and I'm going to be missing some hair soon. Please help!

A   Stop worrying, it sounds as though everything is going well. You're in too much of a hurry, that's all. Take a deep breath, get rid of some equipment, and continue to do what you're doing. After only a few months of work on the longe, your ex-racehorse has good gaits, a relaxed and swinging tail, and is reaching forward with his head at withers height or lower. You can't really ask for much more at this point.

Remove both side reins and return them to the tack trunk. Then use very simple equipment: a longe whip, a 35-foot longe line, and a longeing cavesson (correctly designed, with a jowl strap and hinged metal nosepiece). Send your horse out on the largest possible circle and concentrate on his balance, the use of his hindquarters, and the quality and purity of his gaits, *not* on how far below the level of his withers he is carrying his head. When he is working energetically forward, reaching with his hind legs, in balance, in rhythm, at a suitable tempo, and with a relaxed and lifted back, I promise you that his head will be in the very best place he can manage to carry it *at this time*.

You'll know when his back is relaxed, lifted, and swinging, because his feet won't make as much noise or kick up as much arena dust. You'll also notice that his tail is swinging gently and carried a little away from his body — just as you described it.

A horse being longed correctly will have a reaching neck,
a lifted back, and a gently swinging tail.

When you've achieved this, you should ask for transitions within and between gaits and spiral the horse in and out of the 20-meter circle (from about an 18-meter circle — certainly no smaller for the first few months — to a 25-meter circle). If your longe line isn't long enough to allow the larger circle, you can still ask for it and get it: Just walk a 5-meter circle while the horse circles *you* at 20 meters' distance.

As your horse becomes more balanced and comfortable, and as his muscles build and his reactions become more refined, he'll "use himself" better, and you'll see him reach farther down with his neck and head. *But you must give him time.*

He needs time to stretch the muscles that allow him to reach, and more time to build the muscles that will allow him to continue reaching all the way around the circle. Don't worry if he reaches for a second and then pops back up — as time goes by, and as his abilities and body develop, he'll stay down longer because he'll be more comfortable carrying his head and neck that way.

Your voice can help him understand what you want from him. When he's doing what you want, praise him, but be sure that you are praising *while* he's doing it, not after. If he drops his head, then lifts it, and then you praise him, you may think that you're praising him for dropping it, but he will understand that you're praising him for lifting it. Timing is all-important: If you say "good boy" while the head is dropping or while he's carrying it low, he'll learn that this is something you like.

It's possible to force a horse's head into a particular position, but you don't want to do that. You want relaxation and muscular development, which *cannot* be forced — they must be set up, encouraged, and allowed to develop over time. You've been willing to allow your horse some time already — now give him more time, and help him develop so that he will be able to do the things that you want him to do.

Remember, you're training a horse, which means there aren't any shortcuts and the only way to hurry up is to take your time. If you're willing to take months and even years to get what you want, you'll get it much more quickly. If you want something to happen quickly, it may never happen at all.

And please don't pull your hair out. You may need it for warmth when winter weather comes.

PART IV

Training and Retraining at All Ages

AGE-APPROPRIATE TRAINING

RETRAINING THE ABUSED
OR CONFUSED HORSE

RETRAINING THE EX-RACEHORSE

SPECIFIC SITUATIONS AND SOLUTIONS

CHAPTER 15

Age-Appropriate Training

As long as your horse has a pulse, he isn't too young or too old to learn. His training should be suitable for his conformation and energy level; the duration and intensity of his lessons should match his age, development, understanding, and attention span. Young horses — and all green horses — need shorter, more frequent training sessions; older, more developed horses can handle longer sessions. All horses need steady, systematic, progressive training over time. Whether you're training a dressage horse, a hunter, a competitive trail horse, or a working ranch horse, always keep his age, condition, and experience in mind. You'll train much more effectively, and your horse will enjoy it more.

Starting and Backing — What's the Difference?

Q I've heard different things about when it's right to start a horse and when to back a horse, and I realize that I don't know exactly what those two terms mean or how they relate to training. I have just bought a foal to bring up and train when he is old enough. What is the difference between starting and backing, and which one comes first?

A "Starting" comes first, and it's a process — it's everything that you do with your horse in preparation for backing. "Backing" is the specific event of putting a rider on the horse for the first time, and it's the culmination of all of the earlier work. Think of starting as a calendar, and backing as the red X marking one day on that calendar.

Starting is a long process: It's what you do from the time you acquire your foal until the day, several years later, when you first back him. Starting encompasses all of the things you can do with a foal, a weanling, a yearling, a two year old, a three year old; even a four year old, if your horse is large and "growthy" and won't begin real work under saddle until he is five.

Early lessons should take place with the foal being led rather than tied.

My New Foal: A Starting-to-Backing Program

Q I will soon be bringing home a beautiful young weanling foal to raise and train. I have heard horror stories about foals that are worked too hard too early, and I have no intention of stressing my young horse like that. I have heard other horror stories of the problems people have when breaking in a two- or three-year-old horse that has been allowed to run wild, and I want to avoid that as well. What sort of training can I do with a weanling, and what would be a good starting-to-backing program or training plan?

A Start your foal by teaching him that he can trust you. Handle him daily if you can, and handle him everywhere so that he accepts everything you do to him. Some people are content if they can run a foal into a corner, trap him there, and eventually get a halter on his head. That's not handling a horse, that's just overpowering him, and it's not a good beginning to the training process. Start your foal by doing what previous generations of horsemen have called "gentling."

Maintain a soft connection with the lead rope. Strong pulling can injure a foal's neck.

Handle your foal gently and kindly, but persistently, until he is equally unperturbed whether you are putting on or taking off his halter, picking up his feet, brushing his coat, taking his temperature, applying fly spray, or giving him deworming paste with a syringe. Starting a foal means teaching him that his default reaction to new and different forms of handling should be calm acceptance. Taking the time to bring your foal along like this will save you a lot of time and trouble later when he is taller and heavier and stronger. At that point, you'll be very glad that your young horse learned willing cooperation at a very young age and practiced it forever after.

Teach your foal to respect your space, so that although he may be afraid of something and leap in fright, he won't leap on top of you. He needs to learn to stand calmly while you enter his space to groom him, halter him, deworm him, or whatever you need to do, and he needs to learn *not* to enter your space to lick or nip or push you.

Teach him to lead, not just to follow, but to accompany you, to watch your body language, to walk where you walk, to turn with you when you turn, to stop when you stop, and to step back and step over — one step at a time — on your signal. The lead rope shouldn't hang loose, but it *should* sag gently so that it requires a movement from you or the horse to create any tension on the rope. You can test your foal's leading skills every time you turn him out into the field. He should maintain that soft, slightly saggy lead rope connection between you as you approach the gate, stop, open the gate, walk through, stop, turn, and close the gate.

Help him learn more by ponying him from another horse, perhaps his mother. Ponying is good exercise for young colts that have just been gelded — you don't want to ask a very young colt to trot on the longe line or in the round pen, since circle work is bad for young bones and joints. Straight line work with wide, sweeping turns at a trot is best, but that's not so easily done if you are on foot and your foal has even a moderately ground-covering trot. Ponying can give him the exercise he needs. It will also introduce him to something he'll be able to remember later when you back him: the concept of a human looming over him, taller and apparently stronger than he is, giving him direction but demonstrating no aggression toward him.

Teach him to ride in a horse trailer. His leading training will have taught him to load and unload calmly and quietly, stepping up calmly and walking forward when asked and taking one step after another backward and stepping down when asked. The rest is up to your trailer and your careful slow, steady driving.

Teach him to tie, but only after you teach him to lead. If you teach him to lead and pony, he will understand tying when the time comes, because he'll be familiar with being attached to something and staying a certain distance from it, not trying to charge past it or pull back against it.

By the time he is two you can teach him to carry a simple bit — ideally a not-too-thick mullen-mouth or straight bar snaffle (assuming that you know that his teeth are in good shape and that any wolf teeth have been removed). He shouldn't wear a bit for long periods, and there should be no reins attached and no pressure placed on the bit, but he can learn to open his mouth and accept a bit, carry it quietly, and then open his mouth again to drop the bit when the headstall is removed.

You can familiarize him with a surcingle and longeing cavesson at this age, as well as with the longe line and the longe whip. Learning to longe means learning that the whip is an extension of your arm — nothing threatening or alarming, just a source of clear signals. Once he is longeing well you can introduce him to the saddle, saddle pad, and girth, and let him wear that gear occasionally for longeing. When the saddle and girth are no longer unusual or interesting to him, you can add the stirrups — adjusted so that they cannot hit his elbows — and let him work on the longe while the stirrups hang and bounce.

By the time your young horse is three years old, he should be well grown and well muscled. He should be proficient at walk, trot, halt, back, and reverse on the longe line. He should longe well, with understanding, quietly and confidently moving with his belly muscles engaged and his back lifted and stretched. He should trust you entirely and understand and follow both your body language and your verbal signals — these signals will be important when you back him.

Haltering and Leading a Foal

Q I'm so excited because my mare has just had the most beautiful foal! He is the sweetest and most adorable little fellow anyone could imagine. Unfortunately I was not there to imprint him at birth, so I am a little bit worried about some of the things I will need to do with him soon. I have never taught a young foal to wear a halter or be led with a lead rope, and I was counting on the imprinting to make all those things easy and automatic. But now it seems I will have to do them "the hard way." This is my first foal; my mare and the gelding I owned before her were completely trained when I bought them. I have no idea where to begin.

A Teaching a foal to accept and wear a halter isn't terribly difficult; you'll just need a lot of patience and a very small, very soft leather halter. My personal preference is for a foal slip made from thin, soft leather. If you buy one that comes with two crownpieces, a shorter one and a longer one, it should fit your foal for the first three or four months. Most foal slips come equipped with a short (18- or 24-inch) catch strap that hangs from the foal's chin and will enable you to handle the foal in the field. I generally remove these, for two reasons. First, the presence of a strap sometimes encourages humans to attempt to catch and hold the foal by the strap, which will frighten a young foal and could badly injure its delicate neck. Foal restraint, for the first few months, shouldn't mean holding the foal by the catch strap or even by a lead rope — it should mean standing or squatting and "hugging" the foal, with one arm at its chest and the other around its backside. Tiny young foals can be lifted and held, even carried, in this way; larger foals can be held immobile, especially if they learned to accept and respect this form of restraint when they were tiny.

My second reason for removing the catch strap is that it often annoys the foal, who then flips and shakes its head in an attempt to rid itself of it. All that head-tossing and head-shaking can cause the end of the catch strap to hit the foal in the face, even in the eye. I've dealt with eye injuries, and I never want to create any situation that could lead to another one.

Maintain a soft connection with the lead rope. Strong pulling can injure a foal's neck.

In any case, the catch strap is of limited usefulness. When the time comes to teach the foal to lead, a proper halter and lead rope are best, together with a flat-braided "foal rope" to put behind the foal's backside and use as a "come-along." Buy a small, soft halter for your foal, then put it aside and focus on making your foal comfortable with people handling his head and face. Foals generally welcome rubs and scratches in key areas: the underside of the jaw, the ridges over the eyes, and the area just behind and below the ears.

When your foal accepts having his face, ears, and head rubbed and scratched by you, you can begin to focus more on those areas that will be in contact with the halter — across the top of the foal's nose, for example. You can then begin to carry the halter with you, let the foal see it and smell it, and occasionally use it to rub one of those itchy areas. When you are doing this one fine morning and the foal is calm, stand next to the foal's shoulder and neck, reach forward to pet it, rub his jaw gently with the halter, and then slip the halter onto his head. If he shakes his head and the halter goes flying, don't worry; there's no harm done. You would do far more harm by trying to get the halter on and fastened by main force. Keep trying, calmly, until the foal stands quietly long enough for you to fasten the halter. Then let the foal run in circles, toss his head, and rub against his mother if that's what he wants to do. Some foals don't object to the halter at all; others seem to think that enough head-tossing and rearing and running will persuade the halter to go somewhere else.

Foals are often thought to have learned to lead when all they actually know is how to follow.

Eventually, you will find yourself standing next to the foal, talking to him and rubbing and scratching his itchy bits. At that point, you can gently unfasten the halter and let it drop. Continue to talk to the foal and rub and scratch him, and eventually pick up the halter and have a repeat. For the first day or so, you may feel that the foal will never relax and stand quietly while you put on his halter, but you will be wrong — it just needs time and practice. There's no need to pester the foal incessantly and halter him every 10 minutes, which will annoy and intimidate him. Your foal will learn his lesson well if you repeat it several times a day, always taking the necessary time to ensure that the experience is a calm and pleasant one.

Don't leave the halter on the foal unless you are there, because even a thin, lightweight "breakaway" leather halter may *not* break when breaking would be appropriate. Instead, put the halter on the foal, let him wear it for a little while, and remove it before you leave.

Leading is another lesson, and again, the delicacy of a young foal's neck cannot be overemphasized. Any "leading" should be done using both the butt rope or come-along *and* the halter and lead rope, but with the lightest possible touch on the lead rope, and all the real pressure on the come-along.

Foals are often thought to have learned to lead when all they actually know is how to *follow*. It's comparatively easy to keep a young foal moving in a certain direction when the mare is being led just ahead of the foal. When you can do this easily, every time, try leading the foal *next* to his mother. When that becomes easy and habitual, try leading him in a different direction — just for a few steps at a time. As the foal becomes older and more independent, spending more time away from his mother and spending time at a greater distance from his mother, you can practice leading him this way and that, turning, stopping, starting, even backing a step or two. You can also begin teaching the foal to move away from pressure by poking him near the hip and saying "over." When he steps or even sways sideways away from your poking finger, stop poking and praise and pet him. Do a little of this every day, and soon you will be able to tap, touch, or even just point at the foal's hip or chest and say "over" or "back" and see him step over or back instantly.

Don't worry about having lost the opportunity to imprint your foal. It doesn't matter. For one thing, imprinting is not really the correct word — certain animals do imprint, but horses do not. If you handle a foal extensively very soon after his birth, you will simply be desensitizing him to your various forms of handling. The idea is that by touching him all over, putting your fingers into all of his orifices, putting a rope around his belly, running clippers, banging on his hooves, and so on, you will teach him to accept those sensations forever after. This doesn't actually work — at least, not in the sense that a foal so handled will always remember and accept those actions. It does provide a foundation that you can build on by doing some or all of those things regularly as your foal grows and matures, but you can do that *anyway,* whether you begin handling the foal when he is two hours old or whether you first meet him when he is a weanling. It can be fun to handle a new foal, but don't worry that his training, his personality, or your relationship with him will suffer because you weren't there during his first few hours. That's just not so.

What's the Problem with My Yearling?

Q We have an 18-month-old Quarter Horse that seems to have a slow or lazy personality. She is a sweet horse that is learning lots of things and mostly behaves quite well. She was introduced to walking on the road just recently and has grown comfortable. We are able to sit on her and she is getting used to that also. But she just seems slow and lazy. She walks slowly, does not want to trot, and generally has no energy. She seems to be in good health, no feet problems, and eats well. How can we get her to enjoy walking, trotting, running, or simply moving? I am quite surprised that a one year old is this slow. My 25 year old runs circles around the young one. What's the problem?

A Your yearling is probably quite normal, but it would be a good idea for you to have your veterinarian come out and take a look at her, just in case something is wrong. She may be okay now — I certainly hope so — but something will be badly wrong with her soon if you don't change her program right away. The biggest problem your filly has is not laziness — it's youth. She is a very young baby and needs her energy to grow. This is not the time to be asking her to carry a rider, walk on the road, longe, or do anything else like that. Those activities are not suitable for yearlings.

A very young horse may be docile enough to saddle and ride, but don't let yourself be tempted.

Growing takes a great deal of energy. Why not let your filly grow up a little more? Wait until she's older before you try to work her on the ground, and wait until she is *much* older before you ride her. Right now, work can only damage her. Her body is still developing, and if you want her to develop normally, you'll have to provide her with an environment and activities that will allow that normal growth. Since you have a 25-year-old horse, you must have some experience looking after older horses; looking after a very young filly is somewhat different. She would probably have plenty of energy for normal activities such as eating, sleeping, and running around with other youngsters in a large field.

Your filly's skeleton is only partly bone right now — much of it is cartilage. Her skeleton won't be fully developed for a very long time. She won't be physically mature until she's six or older, so you have plenty of time. Starting a horse too young and asking too much is a quick way to ruin a horse. Give your filly a chance to grow and develop the coordination and balance and the strong bones that she'll need as a riding horse.

As for her energy problem — if she doesn't enjoy walking, trotting, and running *on her own* or with her friends, there could be a real problem somewhere, and that's why you should consult with your vet. But if she just doesn't seem to enjoy doing work under saddle, that's normal. Her reluctance to carry a person and trot down the road is like a four-year-old girl being reluctant to run track while carrying a backpack full of rocks. You'd never ask such a young child to do anything like that, because it wouldn't make sense; asking your filly to carry a rider doesn't make any sense either, for all the same reasons.

What to Do with My Two Year Olds?

Q I've realized that I was trying to teach my young horses to longe too early in their lives. Could you tell me what I should be teaching them and at what age of life to teach it? I have been around horses a long time and have owned several horses but I have never trained one at a young age. My babies will be two in July. They both have good manners and lead well. They know the voice commands for whoa, walk, and trot. I will be riding Western. Where do I go next in their training?

A Your babies may be too young to longe until next summer, but there's a lot you can do with them in the meantime, and it sounds as though you are off to a good start. Do their good manners include standing, stopping, turning, backing, and lifting their feet when asked? All of this will make them very popular with vets and farriers forever! Other useful lessons would involve teaching them to stand calmly and quietly while being clipped, and to walk into and step out of different sorts of trailers without becoming agitated.

Do they really lead well — staying balanced, remaining attentive, keeping a polite distance from you, not getting behind you or ahead of you? If you practice leading them from *both* sides (including turning, stopping, standing, and walking and trotting), they'll be ahead of the game when you teach them to longe next year.

You can teach them about tack. Provided that their teeth have no sharp edges or points or retained caps, two-year-old horses can wear a smooth bit for 15 minutes or so at a time under supervision. They can learn what it feels like to have boots and pads and saddles put on and taken off. They can wear tack while you walk them around, and if you already have a good longeing cavesson, that's great — put it on, add a bradoon strap and a simple, straight snaffle, and let them be directed by the cavesson and just carry the bit.

Take them for walks; lead them *everywhere*, up and down hills, over and through ditches, in mud and sand, on grass and gravel. Lead them over ground rails and through pole mazes (look at older books on TTouch by Linda Tellington-Jones' for ideas). And keep doing all of this *from both sides.*

Young horses can benefit from learning to walk over ground poles.

If anyone in your area trains driving horses, you might consider doing some of this with your two year olds. Driving is less problematic than longeing or round-penning, because the horses are worked on long straight lines and gentle, wide turns, which is much less stressful to their legs than work on a circle.

I think you'll stay busy and keep those babies busy as well. Keep their lessons short and friendly, and let them spend most of their time exercising themselves in their field. Short, friendly sessions will ensure that your horses will run *toward* you, not away from you, when you walk into their field to get them for lessons.

None of this is a waste of time. People may point and stare and wonder (loudly) why you aren't *riding* those big strong horses yet, but just smile and keep on keeping on. You'll have happy, strong, educated young horses, and by the time you begin their actual longeing and eventual riding program, they'll behave as though *nothing* is new, and they knew everything all along. Then those same people will tell you "You're so lucky" — and again, you can just smile.

What Is "Light Riding" for a Three Year Old?

Q I am purchasing a three-year-old Warmblood-Arab cross gelding who has not been backed but has been handled extensively. It is my #1 goal to keep him sound in body and mind and to let him develop so that I have a healthy, strong partner for the future. I understand that light riding of a three year old is generally okay. My question is, what is "light riding"? Does it mean walk only? Walk/trot? About how long should each riding session be? How many days a week? I am 5'4" with good (though not professional) riding skills, soft hands, and an understanding of riding from back to front.

Since I have an 18-month-old daughter, a much-reduced riding schedule for a year is ideal for me. And of course, my interaction with this youngster will be general handling, progressing to longeing, and *then* light riding.

A If you have a healthy, happy three year old, he'll benefit from your time limitations over the next year or so. Too much riding is detrimental to a youngster, as you know. The wrong kind of riding is also very detrimental. At three, your gelding will still be changing physically, and he will continue to change for the next three to five years (depending on which side he takes after more — the Arabian or the Warmblood).

You are wise to think in terms of longeing, then long-lining, and finally riding — by the time your youngster is rising four, he will be much better able to deal with the weight of a rider. Longeing, done correctly, will help him develop his musculature and balance. Long-lining will make a wonderful transition from longeing to ridden work. Both activities will help prepare him physically for the demands of carrying a rider and will help prepare him mentally for the change-over from "The person I can see is asking me to do things" to "The person I can't see, but can feel, is asking me to do things." Long-lining is a sort of middle ground: "The person I can sometimes see and sometimes *not* see is asking me to do things." It takes time and education for a horse to make the shift from reading the handler's body language to correctly interpreting the invisible handler's use of leg pressure, weight shifts, hands, and voice.

I prefer maximum turnout and a light training schedule for youngsters, with perhaps half an hour of work every other day. If you have the time, two 15-minute sessions are even better than one 30-minute session, especially in the early stages. The advantages of the every-other-day schedule are twofold.

First, the horse has time to relax and reflect on what he has learned. Horses are very good at latent learning, and I've found that you can make the same progress (and sometimes better progress!) with an every-other-day training schedule as with an every-single-day schedule.

Second, if the horse is overstressed physically, the extra day between sessions gives him a chance to recover or to show some lameness or unevenness, which will warn you to change your program. A horse that is overstressed on Monday may not show the effect until Wednesday; if that horse is worked every day, the effect of the stress will be *more* than doubled by Wednesday if the horse is stressed again on Tuesday.

Walking is ideal; after a few weeks of walking, add a little bit of trotting so that the horse (and you) can get in the habit of coming back from the trot to the walk, calmly. If you *only* walk, then when the horse startles or becomes unbalanced and begins trotting, as will inevitably happen at some point, it will be more difficult to effect a calm transition back to the walk. If you practice trotting, you'll have a background of quiet, calm returns from trot to walk, and you'll be able to deal more easily with the transitions down from those sudden, energetic, spontaneous trots.

All you're really going to be asking from your youngster is that he move forward when you ask, turn when you ask, and stop when you ask. Take him out in the biggest space available, with the safest footing, and work on long straight lines and wide, gradual curves. No tight turns or small ring figures for the young horse just starting under saddle, please! And don't ask him to move backward under saddle — to do that correctly, in a gymnastically beneficial way, requires more development and coordination than he will have at three. If, late in his three-year-old year, your youngster can stay calm and move with relaxation and rhythm while carrying you on straight lines, on large circles (at least 20 meters in diameter) and on figure eights consisting of *two* 20-meter circles, if he can make wide turns and go back the other way, if he can stop and start without hesitation, moving confidently forward from light leg pressure and calmly into whatever amount of contact makes him feel secure, then you'll have done an excellent job.

"Light" means not too often, not too long, not too demanding, and with not too much pressure from the rider — both mentally (keep the questions simple) and physically (sit lightly, with more of your weight on your thighs than on your seat bones).

Congratulations on the way you plan to bring him up. The slow and careful way will be more than convenient for your schedule — it's going to create a wonderful equine partner and companion that should live a very long, sound life. This may well be the horse that, eight years from now, gives your daughter her first ride.

Quiet walking and trotting outdoors is an ideal introduction to ridden work.

Bits, Signals, and a Three-Year-Old Horse

Q About three years ago, my wife bought me an Arabian colt for Christmas. Although I've been around horses all my life, I haven't really learned how to ride "properly," and in fact, I probably never will have the time, interest, or money to join the "horsey set," but after retraining several spoilt horses, we thought it would be nice to have one who only had faults that were his, or mine, or ours, rather than fixing someone else's mistakes. Besides, one more good horse will probably see me out.

So far we are getting on fine and enjoying the process, but I am starting to realize how much I don't know. At just over three, Shasta is being ridden gently. We had a rider (me) in the saddle for his third birthday but he had done a lot of work before that. There were no traumas, just a progression from his previous lessons. In fact I am sure he looked up at me and thought, "Oh so that is what it was all about. Why didn't you say so before?"

I have tried to be extra gentle on Shasta's mouth. I taught him to longe in the round yard on a light flat rein on a ring attached to his halter or a ring on a leather strap passed through the bit rings. This has been pretty much standard down here for bush horse-breakers for two hundred years. Before I read your articles I had never heard of, let alone seen, a cavesson, and although I would hate to have him tread on the longe rein we have not had any major difficulty so far. He normally has a good attitude that I could describe as "his bit, my hands, our reins." Lately, however, when we first start lessons, or when he is losing concentration (bored, I think, rather than distracted), he has started lifting the bit and chewing it . . . not the lovely "tasting the bit" that I read about, but a distracting "crunch crunch" (and then I think I hear him chuckle). Is this a common concern and what, if anything, should I do about it?

A Your Arabian colt sounds like a dream. Shasta's reaction to being backed is just what a well-started young horse's reaction *should* be: "Right, there you are then, now what do we do today?" When horses are handled properly, backing is just one more step in a calm, easy progression, and no one, horse or human, makes a great to-do about it. Well done! Now, in answer to your questions:

With regard to longeing, if possible, do try to procure a proper longeing cavesson with metal plates and rings on the noseband. I still don't recommend longeing in a halter, just because of the way halters are designed: If you make it sufficiently snug that it can't pull around on the horse's face and perhaps rub against an eye, it's bound to be too snug to permit the horse to flex at the poll and carry his head comfortably. That's one of the main advantages of a longeing cavesson, actually — there's a strap that fastens under the jowls and holds the cavesson steady on the horse's head.

However, here's what I'll suggest for you. Longeing from the bit is not something I recommend, but if you do it this way you can at least minimize the potential for damage. Instead of running the line through both bit rings, run it through the inside bit ring *and the noseband of the bridle*, and fasten it back on itself. This will accomplish several things. It will hold the bit a little more steady in Shasta's mouth. It will keep the bit from pulling through Shasta's mouth. It will distribute the pressure between mouth and nose, instead of putting all the pressure on his mouth. And, last but not least, it will prevent the bit from creating a "nutcracker" effect — the bit will be kept reasonably straight instead of bending (if a jointed snaffle) and pinching Shasta's tongue and bars.

You say that he seems to be losing concentration — that's possible, but I would suggest that you might be losing *your* focus first. In my experience, any horse, regardless of his age or training level, has an attention span that is precisely the same length as that of his rider/handler.

I can think of several reasons for him to be lifting and crunching the bit. Perhaps one of them will ring true:

Teeth. Have your vet take a close, careful look at his teeth, and do whatever needs doing. Points, edges, retained caps, emerging wolf teeth — all sorts of things can make a young horse uncomfortable and fussy with his bit. At age three, Shasta is losing baby teeth and growing new, adult teeth, and he will be doing this very actively for another year. I've always found it strange that we so often introduce young horses to bits and bitting at the very time when they are teething and most likely to be uncomfortable in their mouths.

Tack. Take a close look at Shasta's bridle. Begin with the bit. Look with your eyes, then "look" with your fingers (put a silk scarf over the bit if you want to be dead accurate about what you're feeling). Bits can develop sharp edges and rough spots, and riders need to know the condition of their horse's bit. If the bit is in lovely condition and Shasta's teeth are too, then look at the type and size and style of the bit. Most of the Arabians I've met have had smallish mouths — not just narrow but short, with thick tongues and low palates; this makes them excellent candidates for gentle bits. Most Arabians will go sweetly in a French-link snaffle, a mullen-mouth snaffle, or even a short-shanked curb such as a Kimberwicke provided that any curb you use has a mullen-mouth or low-port mouthpiece rather than a broken one.

Growth. Third, look closely at the corners of Shasta's mouth. Sometimes we forget that horses' heads grow along with the rest of their bodies, and consequently we may forget to lengthen the cheekpieces of the bridle. If this happens, then the horse will begin to experience more and more pressure from the bit, even when the rider isn't touching the reins. Adjust the bit so that it just barely touches the corners of Shasta's mouth, then let him tell you whether it is comfortable or whether he would prefer it adjusted a little higher or a little lower.

Loud, dramatic chewing can also be an expression of anxiety; if that's the case, backing off a little in the training and repeating things that the horse already does well can be sufficiently reassuring that the horse will stop chewing. In this case, though, I strongly suspect that you're dealing with three-year-old teething issues, nothing more.

Stopping from the Bit

Q I've just started riding my three-year-old colt. He is used to longeing and responds well to voice commands, but I need to know how to change the signal for stopping from voice to the bit. We are getting this together a bit better now, but to achieve it we had to use walk-trot-whoa rather than just walk-whoa. We can do transitions from whoa-walk-trot-whoa and walk-whoa now, but I fear you may not approve of the trot at his age and my 200-pound weight. How do we normally achieve light responsive stops from a bit when the horse doesn't know that this is what we want?

A I don't disapprove of trotting a recently backed three year old, provided that most of the trotting is done on straight lines and the rest on wide, gentle curves. Your weight shouldn't be much of an issue if the saddle fits well, you're a balanced rider, and the riding sessions are brief. The key to light riding, especially with youngsters, is partly balance and partly the ability to lighten your seat and carry less weight on your seat bones and more weight in your stirrups and in your thighs.

Stopping will become easier as your colt becomes stronger and more coordinated. At three, he's like a young boy who can't trust his constantly changing body to obey him on Tuesday in the same way that it did on Monday. For now, your stops are likely to be through quite a few steps of walk. That shouldn't be a problem for either of you.

The answer to changing the signal from voice to bit is simple: *don't*. The signal for "stop" should not be from the bit. The signal for "stop" is you stopping your body and letting your colt stop with you. Light responsive stops will be your goal — light responsive *everything* will be your goal, really, but all of that will take time.

Here's what's involved when you ask your horse to stop:

1. You think "I'd like to stop."
2. You signal the horse to stop.
3. The horse's body receives input from *your* body (and perhaps voice as well).
4. The horse filters this input through his brain and says to himself, "Right, I remember, that means he wants me to stop."
5. The horse's body responds to the signals from his brain, and he stops.

Your body, not the bit, gives your horse the signal to stop.

When a horse is fully grown and fully trained and ridden by an excellent rider, this process can all take place in the space of a heartbeat. When a horse is half-grown and at the start of his training and under-saddle work, *even* when he is ridden by an excellent rider, this process may take five seconds, ten, fifteen, or longer.

What's your signal for "stop," then, if it's not pulling the reins? Try thinking about stopping and putting your body in a position to ride a stop: Your eyes are still looking ahead, over your horse's ears; your back remains straight; your lower back and hips stop moving with the saddle and the horse; your legs cease to ask for forward movement; your elbows come back to your sides and remain there; and your fingers remain closed on the reins. You hold yourself immobile *and you wait*. When the horse begins to slow and stop, you relax your fingers very slightly as a reward, but you change nothing else; when the horse stops, you praise him immediately.

As to how to achieve any action/reaction from *any* signal when the horse doesn't know what you want, you do what you've done throughout the horse's training: You ask, relax, wait for a response, and if you get even the merest *hint* of what you're after, you praise, take a break, and then begin again. If you don't get even a hint, you ask again, being sure that you are asking in a way that makes it physically easy for the horse to respond, that you are asking at the right time (when he *can* respond), and that you aren't interfering with the answer by your movements or your balance shifts.

Eventually, the horse begins to understand that you appreciate certain responses, and then you can begin to ask for more response and for a more refined response. After a good deal of this, you'll be well on your way to eliciting those generous, quick, balanced responses, and you'll have the lovely feeling that riders experience when they whisper a question and the horse immediately shouts the answer.

Four-Year-Old Horses: Different Breeds, Different Expectations

Q I went to see a four-year-old, 16-hand Quarter Horse, trained hunter style and still a bit green; he just started fence work. He has a nice temperament but lacks impulsion. He does not want to carry himself and wants the rider to pick him up. I mentioned to the owner that I would like to see him use himself more, and the owner was very offended and said I wasn't squeezing with my legs to support the horse.

My impression is based on two things. First, I learned from a coach in Germany that I should ride from my seat and keep my legs quiet. The legs should be "on" but I never reapply leg aids; that is, once the horse is doing what I ask, my legs should be "passive" until I ask for something else. I should not keep squeezing or tapping with my legs.

Second, my coach here has been letting me ride her four-year-old Warmblood. He is very big yet very light and responsive; he carries himself and has lots of suspension. When I close my legs, he immediately opens up/extends. My coach said the first thing a horse must learn is to be forward with impulsion and to "think" forward. When riding a young horse, it is especially important to ride as quietly as possible so the horse is tuned to small aids.

Going back to my horse-shopping experience, I feel I needed to help the QH too much with every step. He seemed to have only one gear; with much support, he produced an average trot, but I believe he (like any horse) is capable of a finer response and more self-carriage. Is it the discipline that makes us think differently? Am I wrong in my expectation or assumption on how a horse should go? Who is right?

A The quick answer to your question is that both of you are right. There are some key differences between the horses you have ridden and between the disciplines in which they are being trained, and there may be some key differences among the styles, methods, and beliefs of the trainers, as well. Here are some general points to consider.

First, the Quarter Horse is still a baby. At four, he is not full-grown, much less fully trained. Young horses typically need more support from the rider's legs, hands, and sometimes seat. Young horses also are typically on the forehand to a much greater extent than older, fully developed, and fully trained horses.

There's nothing wrong with your expectation/assumption about how a horse should go, as long as you understand there are a number of factors that affect the way a horse goes, including his breeding, his build, his training, and his rider. Hunters are taught to expect a certain style of riding and certain aids; Western horses are taught to expect a certain style of riding and certain cues; dressage horses — more about this further on!

Young horses don't have the balance, experience, or neuromuscular development to offer an instant, impulsive, balanced response to light aids. Neither do older horses *unless* they have been trained to offer such a response and have been developed in such a way that they (a) *understand* what the rider is asking, (b) are physically *capable* of offering such a response consistently, and (c) have been ridden so consistently and so well that they are in the *habit* of offering such a response.

Hunter training, like dressage training, is a variable commodity. Just as there are dressage trainers who produce light, soft horses that go on a gentle, stretchy contact and respond to even the slightest shift of the rider's boot, there are dressage trainers who produce horses that push against the bit and need constant reminders to keep going forward. The difference is in the trainer, not the horses. There are huge Warmbloods that are very sensitive to light aids; there are thin-skinned, sensitive Thoroughbreds and Arabians that respond only to strong, crude aids because that's how they've been trained.

Can any horse feel a rider's light aids? Certainly. Does every horse understand what light aids mean? *No.* Can any horse learn? My experience says *yes*, but with one caveat: A horse can be trained to respond instantly to light aids *only* if he is trained by a good trainer who is constantly striving to use lighter and lighter aids. The rider's leg staying passive until/unless she wants to change something is an ideal, but ideals must be worked toward, and horses must be trained to understand what you want. A couple of points to consider here:

1. A young horse that is built for and has begun to be trained for one sport may seem unresponsive to you if all of your expectations and aids are based on riding horses that are built for and have been or are being trained for another sport entirely.

2. A horse of any age that has been taught one set of signals, or taught to expect a certain type or level or frequency of signals, will always tend to wait for those signals — in other words, a horse will tend to go as he has been trained to go. A horse that has been taught to listen for and respond instantly to a slight tightening of the rider's calf is not necessarily more sensitive or attentive than another horse that has been taught to ignore the rider's leg and wait for a push from the seat, he just has a different set of expectations.

Don't underestimate the combined effects of breeding and training. Your coach's four-year-old WB was bred for dressage, and it sounds as though the horse's training has begun very well. Your coach seems to be focused on keeping this horse forward, energetic, and "uphill," and all of the training is taking place within this context.

The four-year-old QH was not bred for this sort of movement and has not been trained for this sort of response. You could probably buy the horse, take him home, give him a month off, and then start him again from the ground up with your coach's help, and you would eventually have a very nice, willing QH that would be much more responsive and energetic and forward than he was when you first tried him. He would still be a QH, though, not a WB, and he would still be bred and built for a different style of movement, so it's unlikely that he would ever become a WB clone.

The horses that do well in QH hunter classes are very unlike the WBs that do well in dressage, both because of their training and because of their physical attributes. While some of these things are amenable to change — horses can be retrained and bodies can be remodeled to some extent — it would be cruel to ask a long-backed, long-necked horse with daisy-cutting gaits and a naturally low

Both the young Quarter Horse and the young Warmblood are potential athletes, but their different conformation makes them suitable for different sports.

head carriage to become round and elevated, with more knee and hock action, a lifted back and neck, and a much higher head carriage. You could, over time, develop such a horse so that he was as round and as elevated as his conformation would allow, but years of work might just bring your QH to the point at which your coach's young WB is starting out.

Similarly, it would be unkind to take a free-moving "uphill" horse with natural knee and hock action and a naturally higher head carriage, and try to train him to move like the horses that do well in QH hunter classes. Even after years of careful training and the redevelopment of his muscles, the WB *might* reach the point at which that young QH is starting out.

Attitude and energy level (equine and human) can be quite different, too. A naturally quiet horse that needs to be encouraged forward may make a wonderful mount for QH hunter classes and would be an ideal choice for a timid rider whose greatest fear is being run away with by an excitable animal; however, the horse may simply lack the energy level, boldness, and presence that would help make him successful in the world of competitive dressage. A horse that is naturally "full of himself" and perhaps a bit fizzy would be quite unsuitable for those hunter classes, where a quiet, calm demeanor is all-important; but (assuming appropriate movement and conformation) he might be very well-suited to dressage competition under a secure, confident rider.

When you said that you would have liked the QH to use himself more, you were comparing him with the dressage horses you have ridden, and to your coach's undoubtedly talented, well-built, and well-started young WB. The comment made perfect sense to you but probably made no sense at all to the QH's owner, whose experience is based on a different way of going and different standards.

Disciplines vary widely, and an energetic, rounded horse that "uses himself well" in dressage terms would be unlikely to be placed in the ribbons in a QH hunter class, just as a well-trained Western horse that stayed obediently behind the bit on a loose rein would be unlikely to earn a ribbon at a dressage show. Could those horses learn to carry themselves and use their bodies in the ways that would let them win ribbons in those classes? Yes, it's possible, but it would take a considerable amount of time to retrain their bodies and change their understanding and their expectations.

Now, as to whether you ride more from the seat or from the legs — that's a very individual question, and the answer is "both and neither." The legs initiate movement; the seat either allows the movement to happen or blocks the horse's back and prevents the movement from happening. Riders who do too much with the seat — who attempt to create movement from the seat or amplify movement by pushing or grinding with the seat — will simply create sore-backed horses. There has been a regrettable tendency to overuse and misuse the seat, particularly in certain styles of dressage, because of a misunderstanding about equine biomechanics and also, I think, because of mismatches between horses and riders.

For many years, the "typical" competitive dressage rider was a small woman, generally hopelessly overmounted on an enormous, Brontosaurus-like Warmblood. Such riders couldn't hope to use their legs effectively; like tiny children on large ponies, they were riding their saddles more than their horses. When a rider's leg only comes halfway down her horse's side, and her calf is in contact with the upper part of the horse's barrel, that rider can't possibly use her lower legs correctly, because they aren't in the right place. A rider in this position will probably learn to push and grind with her seat instead, even though this is bad for the horse and bad for the rider's riding. Fortunately, we seem to be getting away from this trend. Today's WBs are smaller and lighter, more athletic, and easier for smaller riders to ride correctly.

Don't underestimate the combined effects of breeding and training.

It sounds to me as though you have an excellent coach. What did your coach think about this young QH? If she felt that the horse had the physique, temperament, and movement to do what you would like it to do, then don't worry about having to support the horse too much — that is something that will change to some extent as the horse matures and can be changed by you and your coach to a much greater extent through training, as the horse learns what *your* expectations are. If your coach thought the horse would be suitable for you, or wouldn't be suitable for you, then you just need to ask "Why?" and you'll have the information you need.

One last point: I find that whether you are looking for a young horse for a particular sport or whether you are trying to determine which sport will best suit a particular horse, it's always better to observe the young horse, analyze his build and movement and temperament, and figure out what's going to be easiest and most natural and most fun for that horse to do — then help him develop in that direction.

I am a firm believer in the usefulness of basic dressage for every sort of horse, as correct training and development promotes health and soundness and self-confidence and happiness. But if you are looking for a horse that can perform and succeed competitively at high levels, why spend years trying to teach a horse to do, with difficulty, what another sort of horse could do easily and happily and well? Life is much simpler and more pleasant for everyone when trainers and riders work *with* the nature and abilities of their horses instead of trying to change what they have into something else.

Big Green Horse Insensitive?

Q I'm working with a Percheron, and this guy's a member of the one-ton club. He's very sweet and doesn't have any bad habits. He doesn't mind being saddled, nor does he mind it when I put his bit on. He just stands there, like the big impressive monolith he is. He's so big that the amount of force I use on my Saddlebred doesn't seem to be "heard" by him. I can elicit a response but I have to be very forceful.

This horse does not lack sensitivity — he lacks education.

I usually longe him for a few turns to warm him up. When I mount up he just stands there, quite unlike my Saddlebred who starts "circle-dancing." At this point I feel like I'm doing well because he hasn't showed any sign of discomfort. Then again, he hasn't showed any sign of being awake either. For the first minute or two I gently nudge him and give him a mild "go" kick, looking for the start button. After he starts off, I begin asking him to make turns, and he's pretty good at turning either way. I find that if I have to pull a rein to either side to get his attention, then I really have to crank on him to get him to turn. He can walk a straight line with his nose touching his ribs. After about five minutes of asking and rewarding him for good performance, I try asking him to stop. Either he won't stop, or he won't start up again afterward.

Sometimes I'm wondering if he feels my "limit" of using too much force and is walking all over me for it. I'm not using any devices to assist me mechanically. Just a straight bar 6-inch snaffle bit on a fairly typical bridle and headstall. No martingale, no cable halter, no crop or quirt. Maybe I'm a bit gun-shy about really cranking on him in part because I've been accused of overdoing it with my Saddlebred, Blondie. Then again, I'm used to her and the way we communicate.

A Yes, you can train a Percheron as a riding horse. It will help if you try to avoid comparing this horse with your trained Saddlebred, though. That's really an "apples and oranges" situation, though less because of the breeding than the training.

What you've described isn't just a typical draft horse, it's a typical *green* horse that hasn't been frightened by anyone yet: quiet, gentle, willing, and perfectly happy just to stand there all day. You've made a good start with him, but I think you can do better. You should have a whip with you whenever you're in the saddle, and you should use it appropriately, to reinforce your leg. If you find that you have to escalate your aids, something is wrong, and your horse is learning the wrong lesson.

Horses are not born understanding the aids — that's why you are working so hard to teach your horse what you want when you say "turn please" or "walk please" or "trot please." Horses *are* born with an innate ability to learn what we want, if we'll just explain it to them clearly and spend a little time listening to them, too.

There is nothing *seemingly* less sensitive than a green horse. You can put a bit in his mouth and pull until the horse bleeds, and he won't know what you want. You can kick until you're exhausted and the horse is badly bruised or has cracked ribs, and he won't know what you want. You're teaching the horse a new language, and until he understands the basic concepts and words, you won't be able to communicate effectively and you certainly won't increase the effectiveness of your communication by yelling more and more loudly. Imagine someone giving you orders in a language you don't know. If you don't know what the command is, will it help if the person repeats it over and over or screams louder and louder?

Your young Percheron needs to learn what you want from him. That's where your whip is going to help both of you! From work on the longe line, he should already know what a light tap on the hindquarters means. Now you can apply this under saddle, by using your legs *gently* to indicate "forward please" and then by repeating the signal — no escalation — and immediately following up with a smack on the hindquarters. Then, as soon as the horse moves forward at all, praise him. This will help him learn that the signal for "forward please" is a small, quiet leg squeeze, not a small squeeze followed by a larger squeeze followed by an even larger squeeze followed by a sharp kick. If you teach him the former, you'll be able to use lighter and lighter aids as his understanding increases and his reaction time decreases. If you teach him the latter, you'll be one very tired rider.

Similarly, he can feel the rein pulling him, but he obviously doesn't interpret this as a turn signal. Stop thinking of him as a Percheron and start thinking of him as a horse: Turn him as you would any other horse, from your seat and legs and upper body *first*, with the reins acting last and least of all the aids. "Big" isn't the same as "insensitive" or "dim." Use visual aids to remind both of you to turn: Cones or buckets can serve to mark the centers of circles or the insides or the outsides of curves, and poles on the ground can help you work on your steering *and* help him learn to pick up his feet.

Don't give up on your big green horse. Percherons can make wonderful riding horses and terrific hunters, so if this horse has the conformation and movement to make a good riding horse, you're going to have a lot of fun.

Does Time Off Hurt a Horse's Training?

Q I bought Hope as a five-year-old, unbroken broodmare at the end of last year. I started her myself, and she is showing all the signs of an amazing future. Her paces and conformation are wonderful (trust me on that, because I'm *really* fussy), but that's not as important as her personality. She's the best horse to ride and to be around. She's everything I ever wanted in a horse and more!

Unfortunately, life threw a curveball as life often does. For the last six months, Hope has been turned out without one bit of work. This doesn't look like it will be changing anytime soon. I know that *she* is perfectly happy. She's in a big field with shady trees, with her gelding friend, wearing no blanket, doing no work, and eating grass all day. It's a perfect life for a horse! I, on the other hand, am very concerned about all that talent going to waste. Selling her is really a last resort for me, as I've been waiting all my life for this horse, and I don't want to send my best friend away. I guess what I want to know is, do you think Hope can still have a successful show career if she doesn't start serious work until she is eight or nine years old?

She was going walk-trot-canter when I was riding her and doing ground poles and very tiny jumps. I would love to take her to some dressage and jumper shows one day. I'm worried because she wasn't saddle broke until she was five and a half, then she was only in light work for a year before I had to turn her out. Part of me says not to worry about how many prizes my best friend can win, but the other part knows that she really could go places if she has the chance. Is it bad to start a horse under saddle, work her lightly, and then put her out in the field? Should I just enjoy my wonderful mare and not worry about trying to compete with her? If (or should I say, when) I am able to ride her again, do you think she'll take a long time to remember her training? As far as I know, horses can keep on learning throughout their adult life but that doesn't stop me from worrying about *my* horse! If you can offer any insights into the effects of interrupted training, they would be most appreciated.

A"Time off" is generally a good thing, provided that it's productive time off, building the horse's health and well-being and physical coordination, and provided that when the rider finally puts the horse back into work, she doesn't feel compelled to pick up exactly where she left off.

It sounds to me as though you've found the ideal horse for you, and (granted, owing to circumstances more than planning) she's been lucky enough to be started slowly and worked lightly and given the best possible sort of time off. Starting a horse at five or six and then turning her back out for half a year or a year isn't quite the norm, but in many ways, it's *better* than the norm.

Many horses are started too early, worked too hard, given insufficient time off, and consequently they burn out — mentally, emotionally, and physically — far too soon. Many riders use force and gadgetry to torque the horses' bodies into an "outline" without regard for the damage this causes to the horses' bodies and minds. Some horses are more fortunate; they're allowed to become more mature before beginning work, and they are worked gradually and systematically and given time off at regular intervals.

When "time off" means "time locked in a stall," it's bad for the horse's body and mind and spirit. But when "time off" means "time in a field, time to move freely and enjoy the company of others, and time to relax and just be a horse," it's good for the horse in every way. No, you won't be able to take an extremely fit, competition-ready horse, turn her out for six months, and then fetch her out of the field on a Friday and take her to a competition the following day! But the horse won't have forgotten anything she's learned, although it may take a few weeks or months to build her strength and reflexes back to their previous level of fitness and ability.

Life has a way of throwing those curveballs, and we just have to find ways to deal with them. You seem to have found a good way to deal with yours — your horse is in an environment that keeps her happy and healthy, and you work with her (sensibly) whenever you can. Training is a process of skills acquisition, so don't feel guilty for not being "on schedule." There *is* no schedule.

Enjoy your wonderful mare, work with her, build your partnership and your skills, and have *fun*. If you would like to compete, then you should plan ahead, work with your instructor to set long-term and short-term goals, work toward them, go to shows, and have fun competing. In my experience, horses can continue to learn as long as they have a pulse and as long as anyone is willing to teach them.

What you call "interrupted training" is what many other people would see as "the luxury of working slowly and doing things right." Horses are brilliant at latent learning — they continue to process their lessons during their time off. If you do the same thing and are systematic, patient, and observant when you work with your mare, I expect you'll do very well indeed and will still be actively enjoying each other's company 20 years from now.

CHAPTER 16

Retraining the Abused or Confused Horse

Whether the horse you are training or retraining is a neglected animal that you're rehabilitating, an abused horse that you're teaching to trust humans, a young horse that was badly started, or a mature horse that's had one career and is now embarking on another, you'll find that starting over is much like a slow-motion version of starting from the beginning. The main difference is that your horse won't be a blank slate. He'll have some learned habits, reactions, and expectations that you will need to replace with others, so it's likely that you'll need to proceed more slowly, especially at first.

Training Is Slow, Damage Is Fast

Q I have been going to horse camp for the last three years and every year we have had the same wonderful English trainer. This year we had a new trainer who seemed fairly good at first. My horse was in tip-top shape and was doing better than I have ever seen him — he was collected, using his haunches, had a lovely headset, and was relaxed in my hands. It was a sight! The trainer wanted to get on him and (I thought) put him through his paces, but instead she ripped his mouth off!

Everyone asked her to get off, but she just kept saying that the horse was hanging on the bit! He had a roller bit, and if I'm not mistaken that prevents him from hanging. When she dismounted after about 10 minutes, my horse was in a heavy sweat and his mouth was *very* tender. He is now afraid to put his head down and when he does, he tightens up and becomes very stiff. He is refusing to give me his head, and I'm not sure how to get him to collect anymore. He was coming along so nicely and blew the competition away in training levels.

A Your horse was doing very well until you made one simple (and, unfortunately, *big*) mistake: handing him over to an abusive rider. This "trainer" may have been on your horse for only a few minutes, but that was enough time for her to do damage that may take many months to repair.

Good training takes time. Good training is a matter of educating the horse and developing his body and mind while increasing his confidence and trust in the rider. Good training can't be done in an instant, or with the help of coercive equipment or harsh methods. But, sad to say, *bad* training and abusive riding can have an effect very quickly. That's what happened to your horse.

Horses don't hang on the bit. Riders hang on to horses' mouths, causing horses to push against the pain (it's an expression of their natural instinct to run away), and it sounds as if that's exactly what happened here. Abusive riders can pull and haul and rip away at horses' mouths until the horses drop their backs and twist their necks, sometimes damaging muscles in the process, in an effort to escape the pain. When a horse finally can't bear *any* contact with the bit, and puts his head straight up in the air or tucks it into his chest to avoid the contact, the abusive rider who caused this may say proudly, "Oh, look how light the horse is now, he's not hanging on the bit anymore." But the horse never *was* hanging on the bit, and he hasn't become "light," but instead hurt and frightened.

Your horse was certainly sore in the mouth; he may also have sustained neck and back damage. You had better forget about collection for the foreseeable future, and also forget about showing until you've restored your horse's health and soundness as well as his mental relaxation and his confidence in the rider. To do this, you'll need to go back several steps and work him gently on a long rein with the softest possible contact.

One reason your horse isn't relaxing his back and neck is because he is frightened. Another reason is because he is hurt and he *can't* do what you're asking him to do. Tell your vet exactly what happened, and ask him or her to come out and check your horse closely for both obvious and less obvious injuries. Your horse may need a few months of turnout; he may also need help from someone skilled at chiropractic, massage, or both. Sudden violent traumas can cause enough muscle tension and cramping to pull a horse's bones out of alignment. When that happens, the horse won't be physically able to relax and carry himself the way he did before, at least not until he's had some help.

When he is once again capable of normal movement and carriage, you'll be able to start over at a much lower level, working on relaxation and rhythm and *nothing else* for weeks (or months, depending on the severity of the injuries). Even so, you'll need to pay regular attention to the horse's tight muscles and trigger points.

You've had a hard lesson: You must be careful about whom you allow to ride your horse and careful about what you allow someone to *do* with your horse. If you allow someone to ride your horse, *you* are still responsible for the horse's welfare and well-being. If the person is a bad or abusive rider, your responsibility is to get that person *off* your horse immediately. Don't wait a few minutes because you hope the rider will stop hurting your horse. Even if the rider is your best friend, your boyfriend, a trainer, or your employer, don't wait! You must always protect your horse.

Abusive riding can cause damage very quickly.

Fixing Young Horse after a Traumatic Week

Q We have recently hit a stumbling block in our training and have a question about the Bitless Bridle and how it might relate to a young horse in training. We were wondering what the negatives/positives might be of introducing bitless riding at an early stage in a horse's riding life, before he's really had a chance to get used to a bit.

My mother, our trainer, and I have been training a young Percheron-cross for approximately a year and a half. We adopted him as an unhandled rising three year old and moved him into our trainer's backyard barn, where we spent the better part of a year on ground work. We felt honored to have a horse come our way who did not have training or experiences that we needed to "un-do," and working with him has been pure pleasure for all of us. He has become quite soft and responsive and willing to do anything asked of him.

Our problem is that he went through one week where he was "overcooked," and he hasn't quite come back around. A very capable 12-year-old rider (another student of our trainer) asked to ride Olie in the last schooling show of the season. She is nearly 6 feet tall, Olie is nearly 18 hands, and she had ridden him a few times before. We decided that she could show him walk/trot, provided that she took some lessons on him first.

The week before the show, she took one-hour lessons on him (he usually works in 20- to 30-minute sessions) on four consecutive days (we do not ride him on consecutive days). His last two sessions were group "show-prep" lessons with four other riders (his first time riding in a group). At first, Olie seemed to enjoy the new work. He went over his first cross-rail nicely, though we hadn't planned to jump him for another year or so. One would have been fine with me, just to see how he did; but he continued taking rails right along with the other older and more experienced horses, even though he was not going to show over fences. I believe that was a turning point.

It takes time to restore a horse's trust.

As the week progressed, he became visibly more resistant, and I was terribly worried and upset but didn't want to second-guess his trainer, for whom I had the utmost respect (this incident has changed that). They kept pushing Olie to do more in each lesson, which culminated in him refusing jumps, running out, and ultimately bolting out of the arena rather than be forced over one more jump. At the show, Olie did beautifully. He looked and acted magnificent and took home the blues in several classes, but he has been a different horse since that horrible week. He is noticeably agitated when his riding tack is introduced. He has bolted out of the arena (twice with riders, once while longeing, once while long-lining). Our trainer has, very uncharacteristically, said that the solution is a stronger bit (she requested a mullen-mouth Pelham instead of his usual simple snaffle) and a pair of spurs.

This really threw my mother and me. We have not ridden Olie since these events, but have instead returned to his familiar ground work. We would like to resume his saddle training, but we are hesitant to ride him in the arena where he's had his bad experiences, as we don't want to reinforce his bolting behavior. We are thinking of riding him under completely different, unfamiliar circumstances so that we don't somehow "trigger" his bolting memory/patterns. Perhaps just some relaxed walking in a fenced pasture, in his rope halter, which he associates with the ground work that he likes. We would like to give him some pleasant, positive riding experiences before he takes the winter off, and we were thinking that perhaps if we changed enough elements and somehow made it an entirely different experience, that would be a way to start overcoming his "riding is bad" association. That's where the Bitless Bridle comes in.

I called my former riding coach, in tears, to tell her about what had happened to Olie. She came and spent an afternoon with us, and her feeling was that Olie was definitely not too far gone and that he could come back from his four days of trauma, but that he should not be ridden for quite some time. In the meantime she recommended he return to ground work, long-lining, and such. Given that his ground work and long-lining are still excellent, and that he does not seem the least bit hesitant or upset by these familiar activities, I guess I don't understand how that will help his *riding* unease?

Please help. I'm guilt-ridden, heartbroken, and willing to do whatever it takes to get my horse back to where he was before that horrible week.

As we learn, we *all* make mistakes, and sadly it's often our horses that suffer for our errors. Now you must (a) do everything you can to put things right with this horse, and (b) remember what you've learned so that you'll never make *this* mistake again with *any* horse.

Your horse was unlucky to have been entrusted to two irresponsible people: the "trainer" who took advantage of your ignorance and goodwill, and the child who rode your horse aggressively and badly. Hindsight is not always perfect, but I'm sure you now realize that everything about this situation was wrong, and that no trainer worthy of the name would be willing to sacrifice a child's education in horsemanship and a horse's soundness and well-being for the chance to bring home a few strips of brightly colored nylon.

What matters now is that you restore your horse's soundness and his trust in you. From the first minute you begin to handle a horse, you begin to establish a "confidence account." Every interaction you have with the horse will either add to or subtract from that account. Your plan now should be to accumulate confidence-building interactions with Olie. This will rebuild your confidence account, and you'll be able to draw on the account when you need to ask Olie to do something frightening or unfamiliar. You will also be able to make occasional mistakes with your horse, because if he has enormous confidence in you, he will forgive and you will both get through the problem.

It's a good idea to change tack and give Olie a different set of associations, spend time with him on the ground, and then go for short, pleasant, undemanding rides. You could certainly use the Bitless Bridle for this. It's an excellent bridle for youngsters, even if you intend to put them into a conventional bridle with a bit when they are older and their education has progressed a little more. Alternatively, you could use his familiar rope halter and lead. If you need more subtle and precise control, the Bitless Bridle would be preferable. If you plan to ride him with a bit in the future, it would be helpful to use a bradoon carrier or a one-strap Western headstall to hang a bit in his mouth. Don't connect the reins to the bit; just let Olie carry it so that he can remember what it's like to be ridden *and* to carry a bit when no pulling or jerking is involved.

You can indeed do ground work and long-lining for years without him making any connection between those activities and ridden work unless *you* make the connection for him. *Use your voice.* Talk to him, teach him to respond to verbal commands, teach him to relax when he hears your approving tone of voice. The progression from ground work to long-lining to riding is a logical one:

1. On the ground, the horse learns to respond to your voice as well as your body language.
2. Then, on the long lines, the horse learns to respond to your voice even when he can't see you as well, and when he can't see you at all.
3. When he is thoroughly confirmed in his responses to your voice, you can use your voice from the saddle to slowly teach him to associate leg pressure, shifts of weight, changes in breathing and position, and rein actions with the verbal commands he already knows well.

Eventually, you'll depend less on verbal commands and more on the language of the aids. In the meantime, remember that your voice can be very calming and soothing to your horse; don't hesitate to use it.

Your horse came out of that experience sore in his back and mouth and probably in his legs as well. A sore horse is not a happy horse; a sore young horse that is doing his best and being pushed and punished is going to be an extremely unhappy horse. Olie is young and inexperienced; after being treated badly every day for a week by his rider and trainer, he has a right to be suspicious. Give him time off to let his sore back and legs and mouth become less sore. Then, *if* his saddle fits really well, let him wear it while you do ground work with him. Then put a rider on his back, but it must be someone he likes and trusts, and someone who is gentle, undemanding, and will listen to *you*. Give him some good experiences under saddle if you can.

Treat him as if he had never carried a rider before. Have the rider mount from a mounting block, sit quietly, praise the horse, and get off again. Walk him around, talk to him, and do it all again. At the end of 20 or so of these "one-minute rides," spread over several days, he'll remember that on many occasions, someone got on him and got off him, and nothing horrid happened to him while the person sat on him.

Once he's relaxed about this, let the rider walk him around the arena or ride him in large circles or figure eights, on soft contact on a long rein, keeping the focus on relaxation. Then have the rider dismount. Repeat several times, then do it all again the next day. If you do this for a week and it all goes well, you may want to add a brief, calm trot to the walking and turning activities.

The rider should *always* praise the horse for effort, not just for perfect performance. The rider should also concentrate on asking clearly, *waiting* for a response from the horse, and then staying out of the horse's way so that he can effect the response. If, for instance, the horse is trotting, the rider asks for a walk, and the horse takes 15 steps to make the transition, *fine*. No matter how well each ride goes, the ride should end *before* the horse begins to "lose interest" or "get tired" or (heaven forbid) "get cranky." Stop while it's still pleasant and fun, even if that means stopping after five minutes. Keep your horse listening to the rider, trusting the rider, and responding to everything the rider asks with "I can do that!"

After a few months of this, he'll have some pleasant, positive associations with the saddle, bridle, and rider; eventually, if the rest of his life is good, he'll view the whole unpleasant week as the horse equivalent of a bad dream. Above all, don't be in a hurry. He's *had* that. Take your time, ask for very little, praise him, and repeat. At the end of it all, you'll get your sweet, trusting horse back, and you'll have made yourself a better horseman in the process.

"Advanced" Horse Needs to Start Over

Q I board at a nice dressage barn where most of the people ride at a high level. Some of them show on the Florida circuit. It's usually a pleasant place, but everything has changed now that one woman brought a really fabulous Warmblood home from Europe and suddenly everyone is taking sides with her or the trainer. This horse is absolutely gorgeous and the best mover I've ever seen. He is only five years old and trained to a very high level. His owner hasn't been doing dressage for long, but she was planning to have the trainer show him in Grand Prix next year, and she wanted to show him herself a year or two after that.

Now everything is falling apart. She went to Europe by herself and bought this horse against her trainer's advice. Now the trainer says that he won't show the horse and won't teach her on the horse; the only thing he will do with the horse is to start him all over again, from the ground up, because the horse isn't correct or "through" — and this is a *very* fancy, expensive, advanced horse! He says that the horse is not advanced at all and doesn't even have correct basics. His advice was "put him in a pasture for six months and then teach him to walk on the longe line, then we'll see."

I know this trainer pretty well, and I really don't think he has such a big ego that he would refuse to ride a horse or give lessons on it just because he didn't get a commission on the sale. But none of us know what to think. If he's right, how could the owner have made such a big mistake, and if she's right, how could a good trainer be so wrong about a horse? If he's right that the horse needs to be started over, have you ever heard of starting over again from ground work with an advanced horse?

A This is a familiar story. It's not unusual for an ambitious rider, especially one who is relatively new to dressage and can't quite bring herself to believe how much actual work is going to be involved in her climb to the top, to purchase a horse that she believes will make it all happen for her immediately. It's also not unusual for a money-hungry broker (usually not a breeder) to use gadgets, gimmicks, and trick-training methods to force a talented young horse along so that he presents the superficial appearance of being "advanced."

After the proud new owner signs a huge check and returns home with her "advanced" horse, she is actually lucky if her trainer is honest enough to tell her the truth about her purchase. It takes courage to be honest, because "He might make a dressage horse someday, but he'll need to start over" is not what the owner wants to hear. In an ideal world, the horse's owner might answer, "I made a mistake, thank you for correcting me." In the real world, that is *not* likely to be the reaction of someone who has just poured a huge amount of money into what she believed would be her expensive but instant ticket to competitive success and fame.

It's easier to buy an advanced horse than to become an advanced rider.

The best hope for this horse is exactly what the trainer has proposed: several months of turnout followed by careful ground work and then a reintroduction to ridden work, all as if the horse were completely green, which is not far from the truth. You've said that he's a beautiful animal and a lovely mover; let's hope that his owner will provide him with the turnout and then the training that he needs, so that he will have a chance of someday becoming the horse that he was advertised to be.

Here's something else to consider: Training isn't static. It changes constantly, according to the way the horse is ridden. A horse that is trained to a high level but is then ridden at a low level or by a poor rider will quickly slide down the levels as he loses the mental and physical habits he once had. The horse's physical appearance will reflect that downward slide; just as correct, systematic, progressive training makes a horse more correctly muscled and more beautiful, a steady deterioration in the quality of a horse's riding and training will create a corresponding deterioration in the horse's physique. This is why an upper-level horse, whether he's a competition horse or a schoolmaster, should be schooled regularly by someone who can maintain the quality of his responses and his physique.

Restarting a Mature Horse

Q I am new to horse ownership. I fulfilled a lifetime dream by taking a year of riding lessons (Western) and now have an 11-year-old Appaloosa who was given to me by a man who owned him for eight years but needed someone who could care for him properly. This guy used Cash as a trail horse and did intense distance/terrain hunting and fishing expeditions on him, but I don't think that he ever taught him anything other than "stop" and "go." Cash was left in a field for the last two years with absolutely nothing done to him except feeding — no shots, no shoes, no trims, no riding, no other horses. My veterinarian has given him a complete checkup and has found him to be in good health. The farrier is coming this week to do the first of several trims to get his hooves in better shape to shoe.

My question: Is the training approach the same with a young "green" horse as with a mature horse with no manners? I have read copious amounts about training in general and am working on ground work with him, but he has had years to establish his patterns. He doesn't bite or kick, and after a week of work, I can walk right up to him in a two-acre pasture and halter him, but his ground manners are terrible. He walks into me while on the lead, walks on briskly without permission, and so on.

I have been working on "forward," "whoa," and "stand," and am wearing out my elbow on his shoulder to make him aware of my space. I want to develop his ground manners before riding him and am hoping that "old dogs can learn new tricks." Any special advice for working with the "mature" horse?

A This is a big project and not one that I would normally recommend to a novice horse owner. You seem to have maturity, patience, and a good attitude, but there's one more thing you'll need to make this work: a really good instructor who can help you teach your horse to be a good citizen and good riding horse. If you have that, it will take a long time, but at least it will be doable. If you don't have that, then you would do better with an educated horse. If you have room for another animal, you might consider adding an older (late teens or so), well-educated horse to your stable, so that it can be your second teacher while you teach your uneducated horse! You may be able to borrow or free-lease a suitable horse from someone for six months or a year — your instructor and your vet can help you investigate the possibilities.

Old horses can certainly learn new tricks, and your horse isn't at all old in any case — he's just coming into his prime. You should begin exactly as you would with an untrained youngster, and teach him everything he needs to know beginning with leading, standing for grooming, picking up feet for cleaning, and so forth.

This is always a good idea, even if you buy a fully educated animal. If your horse is wonderfully trained, you can go through everything he should know in a day or two, and be ready to begin work. But you will always discover some "gaps" in the horse's education, or areas in which the horse has learned to do things *this* way and you want him to do things *that* way. It's best to discover these things before you begin real work. You'll get to know your horse, learn what he's been taught (and how he's been taught), and fill in the gaps by teaching him the missing bits and pieces.

The fact that your horse is not a baby is an advantage rather than a disadvantage. Because he is physically mature, he'll be able to do much more, and more quickly, than would a youngster. Because he hasn't been worked with in two years, you'll be able to help him develop, physically and mentally, along the lines that you want. Because he was previously used as a trail horse for hunting/fishing expeditions, he knows about terrain, distance, trees, water, and various other things that a stall-raised, paddock-kept horse would require to be taught.

There is one drawback of training a mature horse: He may feel that he already knows what to do, because he has learned something *one* way. If you want him to do something in a different way, be patient and take the necessary time to teach him what you want him to know. You can't say to a horse: "Forget how you used to do this." You *can* say to a horse: "This is how I want you to do this now and forever," and you'll say that by showing him what you want, by being consistent, and by rewarding every hint that he is trying.

If he doesn't automatically do what you ask, don't imagine that he's being deliberately disobedient: He isn't! Begin teaching him to keep his distance, to maintain his position, and to respond to your body language, physical aids, and verbal cues. You'll be surprised at how quickly he'll learn, especially if you remember that every session should be fun for him and that learning should always be pleasant. Eventually, he'll see you as the antidote to boredom, and you won't have to walk up to him in that two-acre paddock. He'll see you coming and beat you to the gate.

Remember there is no "right" or "wrong" from the horse's point of view; there's only what comes naturally, what's been taught (and rewarded) by the handler, and what has been accepted (which is also a form of teaching) by the handler. If your horse steps on your feet, crowds you on turns, walks ahead of you on the lead rope, and puts his head high in the air when you try to bridle him, someone in his past has taught him to do those things, or has done things that provoked those reactions and then accepted his reactions, thus reinforcing them. So *never* get angry unless your horse does something that would never be acceptable to *any* trainer: biting or kicking at you. Those behaviors rate a loud "No!" and a strong slap to the neck or chest (in case of biting) or to the offending leg (in case of kicking).

No matter how well or how quickly his training goes, there are two things that you cannot take for granted: fitness level and attention span. You will need to develop both, and both will require time. Old horse, young horse, or in-between horse, it doesn't matter — muscles may develop in one month or four, but it takes much longer to develop a strong cardiovascular system. Bones will take a full year to remodel, and support structures (tendons and ligaments) will take almost as long. It's especially easy to be fooled by horses with a lot of Quarter Horse ancestry, because those horses will develop muscles very quickly and *appear* to be strong and fit. Never forget that those big lovely muscles are only a small part of the fitness story.

Retraining an Abused Mare

Q As I have gained more and more of my barn manager's trust, I have been given free rein of her stable and can ride any of the horses. Well, I went and fell in love with this mare that my manager bought from an abusive home. Aimee is a 15-hand, light gray, seven-year-old Arabian mare. She was bred on an Arabian racing farm, as far as I know to be a broodmare. I don't know when she was broken to ride, but whoever did it apparently didn't take enough time. She is horribly frightened of whips, spurs, and ropes; is head shy, hoof shy, and touchy about her underside and hindquarters; and freaks out whenever someone raises his or her voice.

She was purchased, unridden, on account of her bloodlines about three and a half years ago. Her previous owner worked her in the round pen for my manager, who described it as tyranny on the handler's part and terror on Aimee's. A rider at the barn worked with Aimee for a while after she got there, rode her twice, as far as I've heard, but soon quit with her. Aimee was put to pasture and bred once (her daughter is now two years old). Her racing lines are supposed to be impeccable, and my manager wants me to show her and calm her enough to breed to a racer — her baby is "wild" just like her.

Aimee had not been ridden in about three years when I started working with her, so I went about it like I was starting her for the first time. I have been doing a lot of ground work, bonding time, round pen, and longeing. She's made unbelievable progress — lets me put on her halter, accepts being tied, will let me pick her front hooves, lets me handle and quickly pick her back hooves (I was told that she was "thrown" by her previous owners to be trimmed), tolerates fly sprays, and has accepted the bit. I recently got on her in the round pen.

I was bareback, since the cinch we have that fits her rubs her raw. She jumped about 10 feet to the right, reared once really high, then bucked

Retraining an abused horse usually takes considerably more time than training a green horse.

a little, but calmed down after about 15 seconds. We walked around and worked on stopping and responding to the bit. She really prefers voice and weight shifting, as she's very sensitive to leg pressure and her mouth is extremely tender.

Now I admit I let my guard down as I dismounted 10 minutes later. As I slipped off, she leaped away from me, kicked at me, and tried to run off. For about five minutes, she acted like she thought I was going to kill her, but after a while, settled and let me run my hands over her neck and head. Every time I dismount now, she rears and jumps away and becomes horribly head shy again.

As to mounting, she's gotten much better as I spend lots of time leaning over her and putting my weight on her without actually getting on, and then I give her treats. She's learned that won't hurt her, but I don't know what to do about the dismounting part. I can't spend a lot of time getting her used to it since the second she feels me getting off she freaks. I have to be careful where I get off or she'll plow right through fences and hit walls or people, if I don't warn them and have them move.

A When I read your letter, I was reminded of a time when I injured my knee quite badly as I was dismounting from a mare in a similar situation. The problem, I figured out later, was that she had never been ridden bareback! And although she was perfectly at ease while I was riding, the feeling of someone slithering down her side at the end of the ride was just too much for her. She reared, and I hit the ground hard. Luckily I had dropped the reins on my way down, because if I had compounded the frightening effect of the slide by putting sudden, hard, painful pressure on her mouth, she would have been so terrified that it might have taken me months to convince her that a dismount didn't mean trouble.

You can't know exactly what has happened to her, but I've seen both racehorses and show horses jerked by the reins and hit in the head after failing to win a race or a class, and it's possible that your mare has had such an experience. I've also seen bad injuries caused to a horse's mouth when a rider has fallen and pulled hard on a rein attached to a severe bit, and other, only slightly less severe injuries caused by a rider pulling hard on a quite ordinary, everyday snaffle.

There are several things you can do in working with this mare. My first suggestion is that you begin using some TTouch massage on her. The following ideas should be helpful as well.

Move very slowly — take extra time with every movement, whether you are bridling her or brushing her. All horses are very sensitive and very aware of potential dangers in their surroundings — abused horses are 10 times as sensitive and aware, because to them, these are not *potential* dangers, they are *actual* dangers.

Speak very softly around her, and keep breathing deeply. Loud noises and harsh sounds are terrifying to abused animals, as is anger. Be sure that you remain calm at all times. If you hyperventilate and your voice becomes tense, the mare will probably panic.

If you fall off, do *not* hold on to the reins, for two reasons. First, this mare has obviously been punished with the bit, and a sudden painful yank on her mouth will convince her that you too are using the bit to hurt her. It won't be deliberate, but there is *no* way you'll be able to explain that to her. Just let her go. Ride her in an enclosed space so that you won't have to worry about where she goes.

Second, horses have a well-defined, easily provoked flight-or-fight response to fear and pain. A horse's first instinct will always be to escape, to get away. If you force the mare to remain in one place while she is being frightened or hurt (again, although it isn't deliberate, she won't understand that!), you will very likely invoke the "fight" part of that response. "Fight" happens when a frightened or hurt horse is cornered and cannot escape — and this is not a response you want to provoke.

Change *everything* about the equipment used on this mare. She needs to learn to trust you, and she is obviously terrified of her bridle. I suggest that you put her in a bitless bridle or a jumping hackamore, or that you remove the bit and cavesson on her bridle and replace them with a jumping hackamore noseband. This is *not* a mechanical hackamore — it's a circle: a piece of heavy rope covered with soft leather, fastening under the jaw (not too low) with a leather strap. There are two rings on it, for your reins. It goes where the cavesson would go, and you may find that your mare is much happier when her mouth is no longer threatened. Since she won't be familiar with this piece of equipment, it won't hold any horrible associations for her.

Enlist a friend to help with the dismounting issue. The two of you should take turns leading the mare, asking her to stop, start, turn, and back, and rewarding her. When she calmly accepts being handled by both of you, and when you have identified some treats that she enjoys, one of you can lead her in the pen with the other one sitting quietly on the mare's back for a few quiet minutes at a time. Since you know the mare may rear in fright, both people should wear helmets and exercise extreme caution.

The rider should reassure the mare while leaning forward, sliding her arms around the mare's neck, then dismounting gently by sliding down her side (a quiet, calm version of an emergency dismount) while the other person continues to reassure the mare and offer her treats. The rider should also pet the mare and offer her treats after dismounting. Many calm repetitions over many weeks should help this mare develop positive associations with the rider dismounting, but a change of venue, a windy day, or loss of focus on the rider's part may cause her to react with surprise and fear. Continue to work to build her trust, but never take her calm reaction for granted. When she is calm being mounted, ridden, and dismounted bareback, and when her sores are completely healed, you'll be able to begin again with a comfortable, well-fitted saddle, but for the first few months at least, the rider should keep her feet out of the stirrups and be prepared to do an emergency dismount at any time.

Don't even think about showing her — focus on calm time at home. It takes much longer to reclaim an abused horse than it does to train a green one, so be patient. Everything you are doing on the ground sounds good — just make an extra effort to be quiet and calm while you do it, and whenever she gets especially nervous, *back off* and let her have time to relax and think. It won't be a waste of time — she can't learn when she's tense, and it will give *you* a chance to relax as well.

Deal with her as though she were a wild filly right off the range — only *more* slowly, *more* gently, and with even *more* care. You're on the right track.

Retraining a Saddlebred for Dressage

Q How would you go about retraining an American Saddlebred show horse for dressage? I have found a nice Saddlebred, but his background has been as a five-gaited show horse and later as a Western pleasure show horse. His life at six years of age is to be kept inside in a dark barn most of the time, trained along the aisle of the barn, and then ridden in shows. Period.

He has good conformation and good bone, not the extreme swan neck or the other unfortunate parts of the ASB show-horse look. In his present Western pleasure training, he is trained *not* to extend his trot; I'm not sure how it will look if he is retrained, but considering what his training has been his gaits are pretty regular looking and feel smooth as silk. I would like to use him for lower level dressage, as well as hunter pace, trail, fun shows. His living situation with me would involve all-day turnout in a 30-acre pasture with five geldings, and he'd only be in a stall to eat and for a few hours at night. This may seem crazy, but he seems like such a nice fellow that I somehow think he could make the change both in terms of training and lifestyle.

If so, where would I begin? Since he's had so little exposure to the natural world, I thought first I'd send him to my very good friend, a cowgirl trainer, for a month or two, thinking he may benefit from some trail riding around her place, sacking out, and gentle exposure to some real-world experiences. She also has a small herd he could live outside with in a normal pasture.

As to his gaits and training him for dressage, I was told that he has been ridden with *no* leg, except as a "go faster" command; that he thinks that if your hands are low that means to raise his head; and that if you signal him to increase speed at a trot that means canter. So, where do we start and what do we do to retrain? A dressage trainer I talked to said her problem would be that she has no clue what Saddlebreds do in the Saddlebred training world, so she doesn't have the cross-over knowledge to know where to start.

A My question for you is this: Do you enjoy the long, slow process of developing and training horses? If you don't, then don't tackle this project. If you *do*, then read on.

A well-built, compact Saddlebred can make an outstanding dressage horse.

This horse will need to start over from the very beginning, more or less from square one. The best thing in the world for him would be to have six months in a field before *any* retraining begins — or three months in a field followed by another three months with very short training sessions every two or three days. It will be much more comfortable for him, and easier for you, if he can adjust to the new environment and management style before you add retraining to the mix.

In addition to the total change in his environment, he'll need to make equally dramatic changes in his body. Before he can begin constructive retraining, he will need to lose a lot of the muscle, posture, and movement that he's had to develop for show purposes. He'll also need to learn to use his body in a new way that is better for him in the long term but will make him sore at first.

If you let him relax in the pasture for a few months, his musculature will change in a way that will get him much nearer to a good starting point for his new training. Even for a horse with basically good conformation and nice natural gaits, it will be infinitely harder to go directly from "show-horse body and movement" to "dressage-horse body and movement" — those are *huge* changes and the attempt may not be successful. If you let him relax and be a pasture potato, and then start to rebuild his body, you're far more likely to achieve your goals.

Don't make too many changes in his life at once. Even the best adjusted, sanest, and sweetest animal can become disoriented and fearful when there's nothing familiar in his environment. I know you're eager to give him the life you've described, but please plan to take the time to introduce each new element singly.

Introduction to pasture turnout should take place gradually and incrementally. Begin by putting him in a stall. Just being in a normal stall in a normal barn, with plenty of air and light, will require a big adjustment on his part. Ideally, the stall should have an attached run, but keep that closed until he becomes comfortable in his new stall. Take one of the horses you plan to turn him out with, and put that horse in the next-door stall so that they can get to know each other.

When he's adjusted to that, open up the attached run and let him take a few weeks to become used to more space, assimilate the idea of being outdoors, and continue to get to know and be friendly with a neighbor. When they become buddies, you can eventually turn him out in a small area with that other horse; if that goes well, wait a few weeks before introducing the two buddies into a larger pasture with the other horses.

You'll have to pay close attention, though, and notice how your new horse interacts with his neighbor before you even think of putting him out with a group.

If he's been in show barns since he was a foal, which is entirely possible, he may not have the necessary socialization skills to be part of a herd — in other words, he may not "speak horse." If he has a relaxed, laid-back personality, he may be able to learn some of those skills, but it will take time and he will probably acquire some bites and bruises while he is learning how to behave around other horses.

He will also need to learn about uneven terrain, dust, mud, trees, and fences, all things that your other horses take for granted. He'll even need to learn about light and air and weather. Don't take *anything* for granted — given this horse's history, what you have is the equine equivalent of the "boy in the bubble." Nothing about normal farm life will be familiar to him. If he was raised for show, he may never have had anything approximating what most of us would consider a "normal" life for a horse.

When you begin working with him, focus on ground work at first. Simple handling comes first, then sacking out — that is, proper sacking out, which is simply gentle, progressive desensitization to certain things, and equally gentle, progressive sensitization to other things.

Before you ever ride him, he should learn how to use his belly and stretch his back, and he should understand what the rider's legs are for. You can teach all of this from the ground, using belly lifts to teach him to engage his belly muscles. This will be difficult and uncomfortable for him at first, so ask often and be satisfied with *any* result. Use your hand on his side to teach him to reach forward with each hind leg in turn and step away from pressure when asked.

Once you've achieved that, I would start him in dressage training in exactly the same way that you would start a horse that had spent his entire life in a pasture. Begin with leading, parallel leading, and longeing, then move on to longlining and riding. Take everything slowly, teach him in tiny increments with lots of repetition and lots of praise. If you're at all interested in clicker training, try that. Most Saddlebreds are intelligent and eager to please, and take to clicker training quickly and easily.

What I've just described will probably take up his first 18 months with you. It's possible that things will progress more quickly than this — it's also possible that the process of remaking this horse will take longer. Either way, it's for the horse's benefit, and as long as both of you enjoy the process, it will not be time wasted but time *invested*. With any luck, you'll end up with a balanced, confident, seven-and-a-half-year-old Saddlebred ready to begin serious dressage training. And don't forget that the first few levels of dressage are just a matter of developing basic riding-horse muscles, movement, understanding, and skills. Don't get caught up in the competition mystique. Dressage is a wonderful way to make your horse the best horse he can be, strong and supple and confident and happy, whether or not you ever set foot in a show ring.

If you do compete him, be aware that he will revert when you go to your first competition or an event of any kind. No matter how well you've taught and trained and conditioned him, the first time you go somewhere that "feels" like a show to him, his head will come up, his back will drop and stiffen, and his hocks will suddenly be behind him instead of under him. Plan to take him to a few events just as a spectator — walk him around the grounds, let him graze and eat some treats, maybe do a little quiet longeing in a corner, possibly hack around the grounds if he isn't too nervous. Make it pleasant and change his expectations.

There are many factors working *for* you here. You have a lovely attitude and, I think, the willingness to take the time to do the work correctly. You have a physical setup that will allow your horse to ease into a much more normal life. You have a helpful friend and an honest dressage trainer. With those assets and your own good attitude and good sense, there's no reason you can't take on this retraining project.

Saddlebreds can live a long time, and they're a real delight to work with. If yours has a good build for dressage, which many Saddlebreds *do* (it's one of the best-kept secrets in the dressage world that Saddlebreds can make *wonderful* dressage horses), a year or two of careful work now could create a beautiful, happy horse you'll be able to ride and enjoy for the next 15 years or longer. That would be lovely for both of you.

"Dead-Mouthed" Horse

Q I am considering buying a gelding that is apparently about nine years old. He has a nice placid temperament and is fairly comfortable to ride. Apart from being very one-sided, quite unfit, and down on the forehand, he is absolutely dead in the mouth! It goes way beyond hard mouthed — I had to haul his head around to change direction and again to halt. (I was using leg and body aids but he clearly doesn't know them.) I don't think I have ever ridden a horse quite so unresponsive in his mouth. I rode him in a jointed snaffle provided by the owner. He is unlikely to ever bolt (though one can never say that for sure!), but if he did, I can't see that he would be stoppable.

His history is that he has been ridden by all and sundry. Because of his nice temperament, anyone who is nervous or who has never ridden before is put onto his back. I suspect he has developed this dead mouth from having the reins used as a lifeline to keep people on his back.

My questions are (1) can a very hard-mouthed horse ever become a reasonably soft-mouthed horse? and (2) can you suggest anything to retrain this horse's mouth? A different bit? No bit? Special "mouth" exercises? I'm not hard on horses' mouths, normally riding English style with light pressure and occasionally going onto the buckle to let the horse stretch his neck and relax. I tried neck-reining this horse, and he was slightly better, but not much. I realize that the entire horse needs to be trained, and I think he may be worth the work, but I am concerned that he will stay hard-mouthed forever, regardless of all other improvements.

A This horse sounds very sweet, and I'm sure that you are right about him. He is clearly good-natured and has learned to tolerate constant random input (read: abuse) without complaint, but no one has bothered to teach him anything else.

All of the problems you've described — one-sided, unfit, heavy on the forehand, and "dead-mouthed" — fit neatly under the single heading "untrained." Horses are naturally one-sided and heavy on the forehand, which is why two of the main goals of training are to straighten and strengthen the horse, and to develop his ability to work in balance, carrying more of his weight behind. Horses are also naturally "dead-mouthed," in the sense that without an education in the language of the aids, *including the rein aids,* they simply don't understand what the bit and reins mean. If the rider pulls, they *do* feel it, and they experience the pain it causes. But as these horses have no idea what is wanted, they will typically react either by recoiling in shock and horror when approached with a bridle, or — does this sound like your horse? — by going obediently forward, leaning against the rider's hand, and putting enough pressure on the bit to (eventually) numb their mouths.

It's often best to begin the retraining process with a Bitless Bridle.

If you take on this horse, you will need to handle him as you would a completely green, unstarted three year old. The training process will take longer than it would with an unstarted three year old, however, because *this* horse has certain expectations of the rider and of what riding is all about, and it will take some time to teach him new expectations.

Unless your veterinarian tells you that the horse's bars and tongue are nothing but a mass of thick scar tissue, it's likely that the "dead mouth" can be put right. As I said earlier, horses feel the pressure and pain, but if a horse doesn't know what the rider wants, and/or has no expectation of relief from the pressure and pain when it *does* what the rider wants, it will learn to "tune out" what it has learned to think of as random input like constant radio static or a loud, endless lecture in an unfamiliar language. The "hard mouth" or "dead mouth" is rarely a physical reality; rather, it's almost always the result of an untrained or badly trained horse being ridden by a series of heavy-handed riders who spend their time in the saddle "water-skiing" off the horse's mouth. The horse has no idea that his mouth is anything but a "handle" for the people who sit on him, and he certainly has no notion that the bit and reins could be used for communication, let alone subtle, quiet communication. So far, his relationship with the bit and reins has been all pressure and no release. Constant, light contact, subtle brief pressure, and instant release are things about which this horse knows nothing at all — if he's going to learn, you'll have to teach him.

Can such a horse become a good riding horse, alert and responsive to the aids? Certainly, if someone is willing to invest the time and effort to train him from the ground up. Can a "hard mouth" become a soft, responsive mouth? Certainly, because the softness depends on the horse's ability to understand and interpret the rider's signals. If someone will take the time to teach the horse what the signals mean, what response is wanted, and what reward will follow, a "hard" mouth can become very soft indeed.

Something you need to consider is that this horse's "dead mouth" may be accompanied by "dead sides" and a "dead back," all of which are equally illusory. In other words, this "sweet and unflappable" nine-year-old horse may have achieved a state of all-over numbness that gives you no idea of his actual personality. Once he has become fit and balanced, and once he understands that riding is a matter of *two-way* communication between horse and rider — say in two years' time, if all goes well — he may prove to be a horse that can enjoy life instead of being resigned to it, and he may also "wake up" and prove to be vastly more energetic than he seems to be right now. This prospect may please you or alarm you, but either way, you should keep the possibility in mind.

I've found this to be true of many horses rescued from bad situations. They accept everything, never trying to run away or fight; they remain preternaturally calm at all times. If you look such horses in the eye, they don't seem to be "in there" looking back at you; their eyes look glazed. Too much sensory input has put them on hold mentally and emotionally: They have simply given up. When rescued and placed in the hands of horsemen, they can change — not quickly, but dramatically. After a time, often a year or more, the horses appear to shake themselves mentally and wake up from their previous state of near-autistic sleep-walking. At that point, they often prove to be cheerful, happy, energetic souls.

If you like this horse and are willing and able to take on what may be a long-term project, by all means have a go. He's only nine, which means that in two years' time you could have a very pleasant riding horse with many years ahead of him.

The best advice I can give you is to begin his training as though he were very young and very green, and work to develop his body, mind, and spirit in the way that someone should have done long ago. Since he has learned to lean on the bit, I would suggest that you not use one for some time, perhaps a year or longer.

Basic ground control and stable manners can be taught in a halter; longeing should be done using a proper longeing cavesson (find someplace with good footing, put him on the largest possible circle — at least 20 meters — and do hundreds and thousands of transitions).

For riding, I would recommend the Bitless Bridle, which works on the nose and poll rather than the mouth. When the horse has become fit and learned a new way of carrying himself and a new way of going, when he has learned to understand, listen to, and respond to the rider's legs and seat, and when he has learned to respond to gentle pressure from the bridle, you'll be able to add a bit to the ensemble. That is, if you *want* to — many riders find they enjoy riding without a bit.

If you do add a bit, let him carry it for a month or so while you continue to use the reins attached to the noseband. When you are sure that he accepts and is comfortable with the bit in his mouth, you can add a second rein (this one attached to the bit) and begin riding him with both reins. Gradually, over several months, you'll be able to shift the emphasis from the Bitless Bridle rein to the rein attached to the bit, and finally you'll be able to begin riding him with that rein only.

Begin with a very mild and obvious bit: a mullen-mouth snaffle, not too thick. When the horse responds easily and well to this, you can replace it with an equally mild but more subtle bit: a French-link snaffle. This will be easy on his bars and tongue while allowing both of you more subtlety in communication. If, at that point, the horse tells you that he is more comfortable in the mullen-mouth bit, or more comfortable without a bit, believe him.

The real questions here are not "Can the horse be retrained?" and "Can a hard mouth become soft?" but "Do you have the willingness, the experience, ability, patience, and time to begin and continue this horse's education from the ground up?" and "Are you comfortable with the idea of working this horse without a bit for a year or more, then gradually introducing a bit over a period of some months?" And, finally, "If, after all the time and effort, the horse makes it quite clear that he is much happier being ridden *without* a bit, will that be a viable option for you?"

The first two questions *must* be answered in the affirmative — the third is more speculative but is something to consider. Much will depend on your plans for the horse. If you're thinking in terms of hacking for pleasure, or of competitive trail riding or endurance riding, the presence of a bit won't be an issue. If your ambitions include competitions for which bits are required, this may be too much of a gamble, and require too much of an investment in your time and energy.

Retraining after Spur Abuse

Q I have a question concerning a horse at the ranch where I work. This horse (in his early teens) was ridden with spurs in order to get him to lope, even though I know for a fact that he generally only needs fairly light cues; however, spurs were still used. The rider used spurs harshly the first time in trying to get him to extend the lope and did not use them gradually, contradicting the ask, tell, command method. (This person has since stopped riding him.)

The horse is generally well behaved and has never given me any trouble in my eight years of experience with him. Now when this horse is cued for the lope (with the light aids he needs), he crow-hops, throws his head, and leaps into the lope, which is not like him. He also will not slow down on cue. I've checked my position and my equipment to make sure that it isn't me causing his discomfort or his actions, but I can't find anything. This reaction has continued to the point at which it is too dangerous for new riders to ride him at a lope (he was previously used for lessons for teaching riders to lope).

I am afraid that having spurs used on him is the cause of these reactions, and I'm not sure how to correct it. I believe he is just reacting to the pain he remembers from being cued at the lope with spurs. At any other gait he is fine. What is the best way to correct this? How do I get him to stop trying to take off when I cue him and how do I help him forget what happened this past summer?

A You've described an interesting problem. I'd say there are two components here: One is the physical pain caused by the spurs, which may have caused some bruising and tearing of the tissues, and two is confusion from the way he was handled. Spur damage is not always visible from the outside. Anyone can see blood on the horse's sides from a sharp spur, some people are astute enough to notice thinning hair from the constant use of blunt spurs, a few people will look closely enough to see localized swelling from bruising, but *nobody* can see internal tissue damage. Necropsies on some animals with intact, "normal" looking skin and hides show significant scarring and deep bruising *under* those hides from the constant use of spurs. So, just as an aside for all those of you who

The overuse of spurs can cause both physical and emotional damage.

wear spurs: Don't assume that riders need to be careful with *sharp* spurs but can just kick away with the blunt ones. Those blunt spurs can cause a lot of damage, and with no surface blood to tell you, you won't necessarily *know*.

When a horse has experienced physical pain of this kind, he is likely to be very sensitive to, and worried about, *any* contact with the injured area. This sensitivity can diminish over time, if the cause of the pain is removed, until there is no longer any real pain, and the horse is reacting to the memory of pain and the anticipation of pain.

This brings us to the *real* problem here, which is that the horse was already well trained. He knew how to lope, knew the cues, and responded to them promptly. He has now been *retrained* to know that "lope" means "pain and confusion." Look at the situation from his point of view: For years, he would lope when the rider cued him for a lope, and that was that. Suddenly, someone got on and used spurs hard, and what the horse understood was that the rider didn't want a lope, he wanted something else, and the horse was in trouble because he didn't know what the "something else" could be. He is crow-hopping and tossing his head and leaping into the lope because he now associates that particular transition with pain and confusion. Right now, this horse doesn't know *what* riders want from him; he just knows that "lope" means that he's in trouble. Horses don't deal well with pain and confusion — all they want to do is get away.

You can probably *re-retrain* him, but it will take time and patience, and you can't put anyone else on him while you do it. He needs a secure, balanced rider who understands his problem — not a nervous novice and certainly not another bad, brutal rider. You're going to have to convince this horse that the transition to lope is *not* a guaranteed bad experience, and that means that he will have to do hundreds and probably thousands of those transitions. And during each one, you will have to give him the lightest possible signal, and then *praise* him when he goes into the lope no matter how he does it. Praise him even if he crow-hops first, even if he tosses his head — these actions don't mean "Make me do it," they mean "I'm afraid it's going to hurt when I do it."

The only way to teach him that loping *won't* hurt, and that it *is* what you want him to do, is to make each transition a positive experience. Ignore the crow-hopping and head tossing; those things will go away in time as he learns to relax. If you reprimand him for doing them, he won't understand — he'll think that you didn't really want the lope after all, and he's doing something wrong, and he's in trouble again. A frightened, tense horse cannot learn anything except how to be *more* frightened and tense. You want him to relax, so you're going to need to sit up there, relax, and remain relaxed no matter how bad his first 5 or 10 or 50 transitions to lope may be. Leave the reins loose, sit up, pat him, praise him, and let him know that life is back to normal.

The good news: If you're careful, you can do this. The bad news: Even once you've resolved the problem, *one* bad incident involving another idiot with spurs can undo your good work and you'll need to begin all over again. So if you can, keep everyone else off this horse for as long as possible, and when you eventually put other riders on his back, limit them to walking and jogging. And don't let anyone with spurs on get anywhere *near* him.

I'm assuming that you don't own this horse, because I don't think you would have allowed the abuse if you had been in charge. Talk to the horse's owners, and explain how much time and effort will be needed to effect a real "fix." It may not be easy for you, but it's a very important part of the retraining — if you don't do this, and the owners just think that the horse is "misbehaving," there's a good chance the horse will be beaten for crow-hopping and put in a tight tie-down to keep him from throwing his head up during the transition, and then it will be too late, and instead of an abused horse being retrained, what they (and you!) will have on your hands is a recipe for disaster.

You mentioned the "ask, tell, command" method of giving cues. The problem with using a system based on escalation is that the horses learn to ignore the "ask" and the "tell" and respond only to the final, most powerful "command." This is something you can see every day, not just with horses but with dogs and children. We've all encountered the parent attempting to control a child at a supermarket: Even at several aisles' remove, it's possible to follow the sounds of escalation from "Honey, please put that down" to "Stop, put that down, do *not* open that box!" to "NO! Stop! Give me that RIGHT NOW!" (accompanied by the sound of parent's hand meeting child's backside). It would make far more sense to teach the child that the first request will be repeated once *and instantly reinforced.* This would help the child learn that the initial quiet, politely worded request actually requires a response. In the long term, everyone — child, parent, and all of the onlookers — would benefit from this.

But things being as they are, many children, like many horses, simply learn that the sequence is first "meaningless 'ask', no response necessary," then "meaningless 'tell', no response required," and finally "command — *this* is the one that will be reinforced! Response required, must act promptly!"

Is It Too Late to Retrain My Horse?

Q My horse, a 10-year-old Anglo-Arab, seems never to have learned the "proper" relationship between horse and person. He didn't go under saddle until he was six and was leased to several riders before I bought him about a year ago. Since I am a beginning (but older!) rider, I started working him slowly, but now I wonder if I should have gone even more slowly and spent more time on the ground with him. If so, the damage is done (I try to be gentle and steady with him, but firm at the same time). After a couple of months, he learned to trust me and will come to the front of the stall when I arrive at the barn, but he still won't come to me in the pasture, and sometimes he still invades my space. Is it too late to go back and do some natural horsemanship stuff with him? Will he ever learn to come to me in the pasture? I'm hoping that since it wasn't too late for this old broad to learn to ride, maybe it won't be too late for her steed to learn to trust and like her!

A Your horse is only 10; that's not very old. I'd give you the same answer if he were 5 or 15 or 25: If he has a pulse, he can learn. If you skipped over the basics, go back and begin training him again from the ground up. This is something that many, many people need to do with their horses; you're one of the rare ones who notice and acknowledge the need.

The damage is *not* necessarily permanent. Begin again, and treat him as if he were a new horse that had just arrived at your barn. Treat him as if he were a young stallion — you wouldn't hurry things or skip *any* steps if you were dealing with a young stallion, because you'd be thinking about safety and the importance of a complete education.

Riders who keep the same horses for many years often find themselves needing to start over, sometimes many times, with the same horse! The more you learn and the more you know, the more aware you will be of the gaps in your horse's education and your own. As you learn more, and wish that you'd taught your horse this or that, simply go back, begin again, and *teach* the horse. Each time you start over, everything will go a little more easily,

If your horse has a pulse, he can learn.

smoothly, and quickly. And each time, you'll be building a better relationship with the horse.

You know that this horse has had several years of inconsistent handling, and you know that you, as a beginner, were probably just another inconsistent human in the horse's life. Now that you know more, you're in a perfect position to take the horse back to the very beginning, and teach him every single thing you want him to know. After all, you're actually teaching him every time you handle him anyway — it's just that now you'll be aware of that, because you'll be doing the teaching deliberately.

This is the time to make your horse into the horse you want. Here are some very basic skills that every horse should know:

- Does he put his head down so that you can halter him easily? If not, begin with that.
- Does he lead on a slack rope, staying at your shoulder whether you speed up, slow down, turn right, turn left, or stop?
- Does he stand, go forward, stop, and back on voice cue?
- Does he step sideways in his box when you tap him gently on the side and say "over"?
- Does he tie and cross-tie easily?
- Does he pick up his feet for you to clean?
- Does he lower his head for the bridle and open his mouth for the bit?
- Does he stand quietly for you to mount and wait for you to signal him before moving off?

As for how to do it: a little at a time. Be clear, be consistent, and be kind.

Ask for very little, ask often, give the horse a chance to respond, and praise him for trying even if the effort is almost imperceptible. Keep your training sessions short, frequent, and pleasant. Your horse will learn to trust you, respect you, and view you as a source of friendly entertainment. After a few months of this, getting him to come to you in pasture will not be a problem; he'll probably appear as soon as he sees you.

If you're persistent with your ground work and consistent in your verbal cues and praise, you'll see good carryover to his under-saddle work. Take your time. Hurrying won't get you where you want to go, but going slowly *will* get you there.

CHAPTER 17

Retraining the Ex-Racehorse

WHETHER THE HORSE YOU ARE TRAINING or retraining is a neglected animal that you're rehabilitating, an abused horse that you're teaching to trust humans, a young horse that was badly started, or a mature horse that's had one career and is now embarking on another, you'll find that starting over is much like a slow-motion version of starting from the beginning. The main difference is that your horse won't be a blank slate. He'll have some learned habits, reactions, and expectations that you will need to replace with others, so it's likely that you'll need to proceed more slowly, especially at first.

Tips for Dealing with Ex-Racehorses

Q My wife and I have purchased a four-year-old ex-racehorse that has been turned out for five months and is ready to come to our barn and start his new career as a hunter. My wife is very small, and I am much newer to horses than she is, so I was hoping that you could offer us some more tips about handling our new horse. We thought that we would spend several months just taking him out on the trails with our other horse to keep him calm, since he wouldn't associate the trail riding with racing.

A Taking him on trails is a good idea. Take him out with other horses, but choose your company carefully. The other horses should be quiet and well behaved, not excitable and nervous. Your horse won't associate the trails with racing, but he may be surprisingly timid in a completely unfamiliar environment. You and your horse will both feel more secure if you can find a large, sedate horse to follow on the trail. Be sure that the other riders understand what kind of trail riding you will be doing with your new horse, so that they will *not* ride with you if they plan to canter and gallop. You'll want to ride at a walk and trot, calmly. If other horses begin to go fast, your horse will shift into racing mode, which isn't what you want.

You may find that your horse doesn't seem to lead very well. At the track, most grooms use a lead shank with a chain attached. The chain runs from the upper right halter ring through the lower right halter ring, across the nose, and through the left side halter ring. If your horse leads well, you won't need to use a chain. If he doesn't, you'll want to teach him to lead well without the chain, but you may have to wean him away from it gradually. Even if you have no intention of using this chain, you may want to use two leads at first, one attached in the way you are used to (to the ring under the halter's chinstrap) and the other, with chain, attached in the way the horse understands. You can then use the "normal" lead unless you need reinforcement and then you'll have the chain in place. Remember, the idea is to do without the chain, eventually if not immediately.

You probably pick out your other horses' hooves by walking around them and picking up each foot in turn to clean it out. At the track, horses have their hooves cleaned in a different way: The groom stands on the near side and cleans the near forefoot, then reaches across under the horse and lifts and cleans the off forefoot. Next, he cleans the near hind foot, and then reaches across and cleans the off hind foot. You don't have to do this, but you should know that this is what your horse expects, and he may be very surprised when you walk around him and stand on his off side to clean those two hooves.

Your farrier will probably want to change your new horse's angles. Racehorses are traditionally (and foolishly) trimmed and shod with long toes and low heels, which create angles that are both stressful and unhealthy. It may take several shoeings (or trimmings, if your horse has good feet and doesn't need shoes) to correct your horse's angles. Your horse will be much happier with less toe, more heel, and a more natural angle, but remember that any change, even a positive one, will have an effect on his musculature and movement. Ask any woman about the effect of changing into flat shoes or sneakers on the weekend after a week of wearing city shoes with heels. It always takes a little time to get used to the change and feel comfortable.

Here are some other tips for handling horses off the track:

▸ On the trail and elsewhere, footing that the other horses take for granted may be a complete surprise to your ex-racehorse. Uneven ground, deep mud, rough footing, rocks, and water may be new to him. Allow him to follow another horse over or through the terrain at a walk.

At the track, your horse had all four of his feet cleaned from the left side.

- When you want to pat and praise your horse, *especially* while he is in motion, talk to him and pat him slowly and gently on the neck or withers. If you have the habit of reaching back and patting your other horses behind your leg or on top of their rumps, don't do this with your ex-racehorse until you know each other better. Those are the areas where the jockey would use the whip, so your reaching back behind your body at a walk, trot, or canter may be interpreted by the horse as "Go, *go*, full speed ahead!"
- Be gentle, soft, and slow when you handle his head and especially his ears. Many racehorses are head shy because they have been "eared down" by handlers who grabbed and twisted an ear to keep the horse under control. If your horse seems extraordinarily nervous about having his ears handled, this is probably the reason.

If you keep his past experiences in mind, and make it your business to help him relax, feel at home, and trust you, you should do very well with your ex-racehorse.

Off-the-Track Mare with Bad Attitude

Q Two months ago I purchased a five-year-old mare off the track. I was hoping to school her, compete her, and then either keep her for myself or sell her. She has very nice gaits and a good jump, she is usually quiet and sensible, but she has a very bad attitude. She was all right for the first couple of weeks after I got her but then she started napping out on hacks — at certain places she stops determinedly, puts a foul expression on her face, and stamps with her hind leg. I try being firm by shouting and hitting, but that doesn't do any good. I try being nice and talking softly to her — no good either. When I try to school her, if she isn't in the mood she'll pop her shoulder and run into one particular corner near the gate; sometimes she bucks or rears and throws me off. I'm just about ready to give up on her, but it's such a shame as she is naturally quite talented and could be a good jumper. Is there any way I can turn her sullen attitude into a willing one?

A I love your last question. *Yes*, you can almost certainly turn your mare into a willing partner, but it will involve some thoughtful work on your part and a change of attitude for *you*. First, stop reproaching yourself for not being able to turn an ex-racehorse into a smooth, polished hack and jumper overnight.

You haven't failed; you're just trying to achieve the impossible. Ex-racehorses *can* become wonderful hacks, eventers, hunters, jumpers, dressage horses, and trail horses, but the process takes time.

Two months is a very short time. Your mare has had several years of training and racing, and she knows her job. Now you're giving her an entirely new job that involves new skills, new habits, and a new way of using her body.

Taking a horse from the racetrack and immediately putting her into training for a new career is risky. It's usually preferable to let the horse down as quickly as possible: This means taking the horse out of race training, removing her shoes, cutting back her high-octane diet, gradually increasing the amount of free exercise she gets, and increasing the size of the pen or paddock in which she is allowed to exercise herself. After that, turn her out in a field for several months, ideally six months or longer, so that she can learn to relax, lose some of her racing condition and fitness, and begin to cleanse her body of any residual drugs from her racing days. This time will also give the horse a chance to learn to know you and to watch your interactions with other horses.

Your mare seems to have been cooperating beautifully, up to a point. I believe that she is trying to tell you something. Here's what her actions and attitude may mean.

A horse fresh from the racetrack will need some down time before her retraining begins.

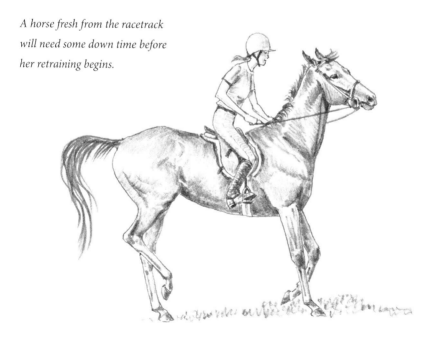

Many horses are unbelievably sweet and docile for the first couple of weeks in a new home or a new job — like a child at a new boarding school, they are confused, curious, a little worried, and willing to accept just about everything. But after they begin to feel a bit more familiar and a bit more secure, they may relax enough to complain about the things they don't like about their new environment.

A horse that has a dramatic change in her exercise type and routine, and behaves beautifully for a week or two, then begins to get more and more difficult, is very probably experiencing pain and confusion. Imagine how your body and mind would feel if you were suddenly taken out of your familiar environment and put into an unfamiliar sort of physical training — ballet, say, or gymnastics. After a week or two you would be physically sore after using your muscles in such different ways. You might try to avoid the exercises, or you might make faces while you were performing the exercises. Now imagine that the person in charge becomes angry with you and hits you.

Flat racing uses equine muscles in a very specific way; other types of riding put different demands on a horse's muscles. Your mare is probably *sore* and confused about what is expected of her. When she was racing, she understood her job; now, she does not.

You're quite right: Shouting and hitting won't elicit a horse's cooperation. Their use should be restricted to those occasions when a horse does something completely unacceptable such as biting or striking at a human. But as training techniques, shouting and hitting are no good at all. Discard them, please.

Talking softly works well; it will work even better if your horse understands what you're after. Imagine that someone screams at you in a language you don't know, then hits you — you'd be frightened and want to escape, you wouldn't understand what that person wanted you to do, and you would learn to fear that person. Now imagine that someone talks to you softly in a language you don't know, tells you something again and again, and then becomes frustrated and angry with you. In spite of the soft voice, you'd still be upset, and you still wouldn't understand what was being asked of you.

Spend time grooming and talking to your mare, then take her for 15- or 20-minute hacks, talking to her all the time. Give her little, simple tasks to do and reward even the slightest hint of an effort on her part. To create a willing attitude, show her that you notice and appreciate her smallest effort. Then turn her out and let her be a horse for the rest of the day. Don't jump her yet. She needs preparatory work on the ground and under saddle before she works over jumps.

Your mare has been racing for quite some time. If she is sound and has correct basic body structure, she can probably become a jumper, but not if you rush her. She's no longer expected to go as fast as she can for a short time, then return to doing nothing until her next exercise session. She's learning how to wear a different sort of saddle and carry a rider who *sits* on her. She's learning how to react to signals from a rider's leg and seat, and how to respond to a bit and reins that are sending her new and different signals. She's learning how to spend a long time under saddle, doing a great many transitions and turns. Her new skills must first be learned, then repeated many times before they become easy, and repeated many more times before they become habitual. And as her body slowly rebuilds along new lines, all of this puts new and different stress on her feet, her legs, her back, her neck, her joints, and her mind.

It takes a year or more to rebuild a horse's body. A good trainer can create dramatic changes in a horse's musculature in just a few months, but tendons and ligaments take a year to strengthen and bones take *at least* a year to remodel. Your mare is trying to figure out what you want her to do, and the combination of mental and physical stress is likely to make her emotionally insecure, mentally anxious, and physically sore. Add to this your own anger and annoyance, and you've got a sure formula for an unhappy horse.

There's no reason you can't school, compete, and sell this mare, but if you want to enjoy the schooling, compete successfully, and sell her at a tidy profit, you'll need to relax and work with the horse she is *now*. Give her the time she needs. If you sell her, you'll want her to be sane, sound, and successful at her new job; if you keep her, you'll want exactly the same things.

Am I Spoiling My Ex-Racehorse?

Q I am working on my own with a five-year-old ex-racehorse that I adopted three months ago. He had surgery for a condylar fracture in a foreleg (with three pins inserted) a year before I adopted him and has been checked by two vets, who x-rayed the leg and pronounced him sound for riding; he has apparently made a full recovery. His muscles were quite weak initially as he was on stall rest for several months and then turned out for a few more months in a field at a foster home before I brought him home.

I was afraid to longe him in case it stressed the formerly injured leg, so I started by just riding for 10 minutes at a time and now ride for up to 30 minutes at a time. I have been working for three months at a walk and slow trot, using *very* light contact and *very* light leg. I had to start by riding him with *no* contact as he wanted to lurch along quickly with his mouth open. He has learned to halt from a walk, to respond to a teeny bit of leg-yielding, and to maintain a steady pace at a walk and trot. He is beginning to soften through his body (at least that is what it feels like), accept the bit with a closed mouth, chew a little, and relax his neck (it was "upside down" and in my lap before).

In general he is beginning to feel relaxed and calm but is, of course, still quite crooked and stiff. I think this will resolve itself slowly given enough time and patience and large circles and turns. When he spooks occasionally, I just sit through it and slow him down gently to a walk. I used to ride daily as a teenager and in my early 20s but have not been able to ride consistently for a few years as I had no horse. I am very excited to be back in the saddle but a bit nervous (I am in my early 30s now).

My railside critics tell me I should try to put him on the bit, ride with a stronger contact or more forcefully, not allow him to spook, move on to more aggressive lateral work, and in general say that I am not asking him to do enough. (We seem to be wandering around in circles doing nothing.) They think I am a "chicken" because I have not cantered him yet. Should I be alarmed that we are still moving slowly in large circles after this amount of time? He seems quite happy during our riding sessions now and is a much softer, more willing, and more athletic horse than he was.

I love him dearly and do not want to push him too hard or ask him to do things he cannot physically accomplish yet. He is intelligent and eager to please. I worry about asking for too much too soon and destroying his trust in me and his enjoyment of our rides. Will I know by feel when he is ready to do more, or am I just not doing anything at all, as my critics suggest?

Kindness and consideration do not add up to "spoiling."

A You're not a "chicken" and you shouldn't be alarmed by anything, including the random squawkings of those railbirds. The only critic whose voice should matter to you is your horse. Listen to him. He is obviously happy and making good, steady, incremental, positive progress — well done, you!

Far from "not doing anything at all," you are doing an excellent job of retraining your horse according to his needs. What you're facing now is the trainer's dilemma: Whenever you train a horse in view of others, you will be given advice and suggestions — some sensible, many quite useless. If you are going to train horses, you must develop a thick skin and become selectively hard of hearing.

There will always be someone to tell you that you would achieve more with gadgets, an accelerated training schedule, or punishment. Ignore all such suggestions. *You* are retraining this horse, and *you* will be living with the results. Don't allow yourself to be pushed. The correct rate of training is the rate at which you are comfortable, the horse is comfortable, and you make progress. For the sake of politeness, respond to railbirds with a smile and say, "Thank you for sharing!" Then continue as you were. Don't change a thing.

If your horse is more relaxed, softer, more willing, and more athletic after only three months, then you are doing everything right. Training is education — there is no timetable or schedule. It's a matter of understanding and skills acquisition, not a race to see how quickly you can force a horse to do something. Relax and continue to do what's right for your horse.

When you train or retrain a horse, there are two processes going on simultaneously. One process involves building mutual trust and confidence so that you can train your horse correctively and effectively. You achieve this by taking your time and working gently and gradually with the horse, *not* by hurrying or pushing him. The other process involves slowly and systematically developing the horse's body and mind so that he is always physically able to achieve each new step, and so that he is comfortable and confident, reacting to each new training demand with a cheerful "I can do that!"

Continue to build your horse's trust in you and your mutual enjoyment of your rides. You've done a great deal in a short time, and there is no reason to push for more than your horse is physically, mentally, or emotionally able to give at any particular moment. You've done well so far — not just in your retraining techniques, but in the quality of your horsemanship.

Rescholing Ex-Racehorse for Dressage

Q I have recently purchased a rising five-year-old Thoroughbred mare. She retired from racing a little more than a year ago and has been going well. I hope to compete in dressage on her. She is exceptionally quiet and relaxed to handle, and we have been getting on great. I am 16 years old and have owned ponies for three years. Elle is my first real horse. I haven't had regular lessons for a while due to lack of a suitable trainer in my area, but I have dabbled in some "natural" horsemanship and read many books on dressage.

Now to my question: I am riding Elle in a simple eggbutt snaffle, and I try to ride with the softest contact possible. She has been getting the concept of dropping onto the bit, but as she does this she speeds up and usually comes above the bit again. In order to ask her to relax I reapply my hand and leg aids to ask her to come onto the bit. As she does this, however, she tends to rush more. I don't want to create an unresponsive mouth by checking her or using a stronger contact, but I find it difficult to slow her using my seat as I am light (approximately 90 pounds) and she is 16.3 hands and strongly built. How can I effectively slow her and keep her going soft without hanging onto her mouth, as one instructor has suggested?

Making the transition from racehorse to dressage horse means learning a new way of going.

A Your mare is tall and large, and she is also quite young and inexperienced. After unwinding from her brief racing career, she is now ready to be started from the ground up, like the green horse she is. It sounds as though you have the situation, and your mare, well in hand. You aren't dealing with a problem horse, nor do you have a problem. What you have are two different sets of expectations, and consequently, a minor breakdown in communication.

Your experience is limited, but your heart is obviously in the right place, and you are thinking in terms of dressage, lightness, and self-carriage. You would like to ride a balanced horse that is able to carry herself well with lowered hindquarters, lifted withers and neck, and a very light contact. That's an excellent goal, and one you will eventually reach if you take the time to teach your mare a new way of moving and a new way of responding to the rider.

The problem you are having is a simple one. Your mare's experience is also limited, and *her* expectations of you and of what you want from her are based on the only sort of riding she knows — going forward, speeding up when asked, extending her stride at the gallop, balancing on the forehand (as is natural at speed), and taking support from the bit. Horses trained for racing will go faster when the rider takes a stronger hold. Young, green, large horses will typically ask for a stronger contact while they are learning how to move comfortably under the rider. Even though your eventual goal is balance and lightness, remember that the *horse*, not the rider, must determine how much contact is appropriate at any given time. If your mare wants an amount of contact that you feel is excessive, don't worry — this is normal and she will lighten the contact herself, gradually, as her balance improves and as she learns to rely on your legs and seat for her security and signals.

Your mare is green, large, and accustomed to balancing on her forehand. If she gets "rushy" or quick when you apply the leg, it just means that she doesn't understand what you want.

Horses that become unbalanced, as young horses and green horses tend to do, typically speed up when they lose their balance. Picture yourself walking down a steep hill: If you go a little too quickly, you will find it difficult to slow down and balance yourself, but easy to lean forward and go into a sort of stumbling *run* down the hill. A horse that loses his balance and puts too much weight on his forehand does much the same thing.

The solution is also simple. Begin with work on the longe, using the longest longe line you can find, 35 feet or longer; if necessary, walk a circle of your own inside *her* circle, so as to allow her to work on the largest possible circle. Work for just a few minutes every other day; over several weeks, gradually increase the time until she is being worked on the longe for a maximum of 20 minutes. At that point, add a session every week until you are longeing her daily. Keep her moving steadily at walk and trot, with thousands of transitions, and focus on keeping her forward, attentive, and straight. As you know, "straight" in dressage terms means straight on straight lines, and conforming to the curve of the circle when on a circle.

As she progresses over the next three months, her body will become stronger, more supple, and more balanced, and she will find it increasingly easy to assume the posture of a riding horse rather than that of a racehorse. As the weeks go by and her balance and coordination improve, you will be able to reduce the size of the circles to 20 meters in diameter (nothing smaller, please).

When you work her under saddle, repeat the exercises she did on the longe. Practice large circles and ring figures, many transitions from walk to trot and back again, and so on. Toward the end of her three months on the longe, begin asking for halt-trot transitions. These are wonderful for helping a horse achieve balance. They are best done on the longe at first; then, when the horse has achieved good balance and is able to use her hindquarters more effectively, you can begin doing the same halt-trot transitions under saddle.

Once she is working under saddle, teach her to leg-yield. As soon as she can leg-yield easily in relaxation and good balance — without losing her impulsion, energy, straightness, and rhythm — teach her shoulder-fore. When she is similarly comfortable with shoulder-fore, begin shoulder-in. Shoulder-in is one of the most important tools in any rider's "toolbox." Focus on her balance, and on achieving good shoulder-in, again, while maintaining relaxation, impulsion, energy, straightness, and rhythm. (See chapter 9 for more information.)

Your mare is green, large, and accustomed to balancing on her forehand. If she gets "rushy" or quick when you apply your leg, it just means that she doesn't understand what you want. On the longe line, you can teach her that "forward" means *reach forward with your hind legs,* not *go faster!* Under saddle, you can teach her the same lesson, first at the walk, then at the trot. Leg-yielding, and eventually shoulder-in, will help her physically and mentally.

Use very little seat with her. She doesn't know what it means. Teach her that your (brief) leg pressure means "step forward from behind with more energy and reach" rather than "hurry along." While she is learning to listen to your leg and offer the response you want, sit lightly, allowing your thighs to carry weight. Putting all of your weight onto your seat bones, even if you don't weigh much, makes the horse's job harder and less pleasant. Your seat should never be used to push or grind into the horse's back.

Once your legs have indicated that you want an action from the horse, your seat either allows or blocks the action of the horse's back. You can allow the action by sitting softly or by rising into a half-seat; you can block the action by sitting still and keeping your own body from accompanying the horse's movement. But your seat's *most* important function is as a listening device. Your seat can tell you, at every moment, how your horse is moving, what each of her hind legs are doing, whether her hips are tight or flexible, to what degree she is bending each of the joints of her hind legs, whether her belly muscles are in use (allowing her back to lift and stretch) or whether her belly muscles are loose and her back is being held tight and rigid.

Over the next three months, her body will become stronger, more supple, and more balanced, and she will find it increasingly easy to assume the posture of a riding horse rather than that of a racehorse.

At first, she won't know what you mean. Racehorses aren't taught much about response to the legs and have no idea what is meant by the application of leg pressure low on the barrel. Even when she learns what you mean, it will still be hard work for her. It is easier for *any* horse to speed up than to take longer steps and use her hindquarters more actively. Remember that your *own* position is all-important. Don't compromise your own balance and position in an effort to "help" the horse — leaning forward to encourage her to reach with her head and neck, for instance, will just tip her onto her forehand. Maintain your good posture, be patient, and give your mare a lot of encouragement: Praise every tiny effort on her part.

Take all the time you need — in two years, people will be asking you where you found your wonderful mare, and whether they can buy her, but I promise you that not *one* person will ask, "Did it take four months or eight months or sixteen months to make her look that good?"

Jumping My Ex-Racehorse

Q I have recently purchased a six-year-old Thoroughbred mare. She is an ex-racehorse and has had little done with her since she was on the track. I am hoping to one day do some eventing or jumping. She is doing quite well, although she has started to get fit and can be a handful at times. I am doing lots of dressage work, and she comes down on the bit really well now. I am hesitant to do much canter work, because her only idea of canter is "run." She doesn't scare me, but I just want to be sure she knows "slow," too.

I have started her jumping, and she seems to love it. I have always been told that a "greenie" should be trotted to the fences in order to get a good spot. My mare insists on cantering to the jumps, and does quite well. I have tried trotting, but we wind up fighting and getting a bad approach because of it. Is it wrong for me to let her canter to the jumps since she is so green? She is very willing and doesn't stop. We have jumped cross-rails, verticals, and a gate, and nothing more than 2'3".

Take the time to teach your mare to approach jumps calmly, at a balanced trot.

Also, how will I know when to have her jump more height? I am entered in a starter combined test at the end of this month. How will I know when to start her in some mini horse trials? I don't know how she'll be cross-country, but I have faith that she'll be fine. My hesitation is with all the "scary" stadium jumps since my boarding barn has only run-of-the-mill jumps available. Any suggestions on getting her exposed to the new jumps we will face, and how slow should I take all this?

A To answer your last question first, *quite slowly*. In fact, I think that your interests will be best served if you back up a little. It takes time to recycle a racehorse, and you'll have more fun, and a better, safer horse, if you take the time to teach your mare to work well on the flat before you begin jumping her.

I suggest that you *not* jump her until you are entirely comfortable working her in the open at all three gaits. Ex-racehorses need to learn to canter without thinking that they're going to gallop; before you begin jumping your mare, you need to be able to shift her easily from trot to canter and back again, to ride figures at the canter, and to do simple lead changes. Make this your project through the autumn and winter, so that you can begin working over fences in the spring and perhaps go to a competition during the summer.

Look for a good instructor who can help you with your mare and who will be able to start both of you over fences when the time is right. If your mare is enthusiastic about jumping, proper training won't cause her to lose that enthusiasm. She needs to learn to jump when and where she is asked to jump; that includes trotting to jumps when asked.

Any reasonably healthy, sound horse can fling herself over low jumps at a canter, but that doesn't mean that it's a safe practice! You don't want your mare to develop a habit of running to the jumps or of taking control whenever she sees a jump. If you're going to jump her, you'll need some professional help to work out a systematic, progressive training and conditioning program for her. Here are some things you should consider.

Your mare is an ex-racehorse, which means that her legs, especially her front legs, have already been subjected to a good deal of punishment. If she is going to jump *and* remain sound, she needs to be built up physically so that she can jump quietly and safely and not ruin her legs in the process. Jumping and flat work make different physical demands on horses, but you can minimize the stress of jumping through good preparation on the flat.

There are very good reasons for trotting to fences. First, your mare needs to be balanced and listening to you, not fighting with you and running away over the jumps. Second, it *does* make a difference as to what "spot" she takes — it may not make a meaningful difference when the fences are tiny, but it will make an enormous difference later. Fighting as you approach a trappy fence on cross-country can have disastrous, even fatal, results.

If your horse can't balance well enough to jump from the trot at home in the ring, don't ask her to trot and canter up and down hills or jump up and down even small banks and drops. You may need to trot to individual cross-country jumps for various reasons: footing, light/dark changes, or just to get close enough to a wide fence to be able to jump across it safely. There may be times when you need to trot most of a course — for instance if there was a hard rain the night before, much of the course is waterlogged, and there are slippery patches in front of and after each jump. At the very lowest levels of eventing, one often sees riders racing around the course with the horse pulling his legs out of the way whenever he reaches a jump. This isn't good for the horse's body or mind; it doesn't teach lessons that will help you move up the levels. In fact, people who event this way usually *stay* at the lowest levels.

When you eventually begin jumping, your instructor should use small gymnastics to help your mare learn to use her body correctly and to develop the muscles and balance that will let her jump larger fences later. Speed isn't what takes you over a serious jump — *power* is what you need, and that power must be controlled. If you run your mare over small fences now, she will hurt herself on larger ones later because she won't have learned to jump well.

Take your time, and find a reliable professional to help you. It sounds as though you have a very nice mare — take care of her so that you can have many enjoyable years together.

Helping Ex-Racehorse to Bend and Stretch

Q I am in Germany, and my question concerns a horse that is not mine. I ride in a riding school where they board horses that are allocated for lessons. I know this is not ideal for the horses, but the owners work toward consistency to reduce the stress for the horses.

I am considering a half-lease of one of the horses, a seven-year-old former racehorse, now physically sound. However, he is problematic in several ways. He came to this place two years ago from a foreclosure, probably maltreated (certainly underfed). He constantly pretends to bite (he never does, but few students dare to enter his box) and is extremely sensitive to touch. He will stand for the owners and me when he is tied but acts up with most other people. You can use only soft brushes with long strokes and slow movements.

In riding, he is well behaved with small children and beginners who do not ask much of him. When I work him in lessons he sweats profusely after five minutes. He has difficulties bending and stretching toward the bit (no long and low). He does not pull on the bit but chews energetically or grinds his teeth. Recently I got him to snort, which I had thought a sign of releasing inner tension, but he doesn't feel that far yet. Sometimes he will spin and go backward or sideways instead of going forward. Canter is often interrupted by jumps landed on all fours. My feeling is that he is trying but sometimes does not know how to comply, and that he is lacking self-confidence. He does not run anymore, but it feels like he has lost his rhythm in the process of slowing down. He does this not just with me but with anybody asking him to work. If I lease him, how can I sort out the symptoms and help him to be a happier horse?

Developing longitudinal and lateral suppleness is essential for a stiff, tense horse.

A This is an interesting situation. The horse seems to have a good nature; I think that you are probably right in saying that he is confused and lacks confidence. You've already discovered that long, slow strokes and careful movements are correct, as indeed they are for most horses. I would suggest that you do some of the TTouch massage from Linda Tellington-Jones's earliest book.

Racehorses aren't taught to bend and stretch toward the bit. You'll have to teach him by riding him forward, encouraging him to reach, and praising him when he reaches forward even a little. Longitudinal suppling is the basis for all of his future work, so it's well worth the effort to help him develop this. Make rhythm and relaxation your two absolute priorities, as he will be physically and mentally unable to relax, reach forward with his hind legs, and stretch his topline unless he has *both*.

Offer him a definite but flexible, stretchy contact, and ride him consistently forward in that rhythm and relaxation. Working on straight lines is not always the best choice for an ex-racehorse; it may be better to keep him constantly changing direction. Soft changes, not sudden ones, are what you want; ride shallow serpentines around the arena and large ring figures inside. Leg-yielding is essential, as he needs to develop lateral suppleness, too; this will help him with his longitudinal suppling. Leg-yielding on a large circle, to enlarge and decrease the size of the circle, will help supple him laterally and encourage him to reach down and forward into your hand.

With any unfamiliar horse, it's best to assume that the horse knows nothing, treat him accordingly, and take him through the entire training process from the ground up. If the horse is well trained, this process may take only a few days. It will typically take longer, because you are almost guaranteed (with this horse, you *are* guaranteed) to find gaps in his training. Finding them in this way enables you to deal with them appropriately by teaching him what he doesn't know and enabling him to do what he can't yet do. It's much kinder and more effective than the all-too-common practice of assuming that the horse "should" do certain things and then fighting with or punishing the horse when he doesn't do them. Find out what he does *not* know, and then teach it to him.

If you want to work with this horse, a half-lease will certainly give you the opportunity to improve his life by improving his training and his comfort level; it will also give you the opportunity to improve your training skills and sensitivity to the horse.

Standardbred Stiff after Workout

Q I have a Standardbred that is 15 years old. She never learned the basics of walk, trot, and canter, so now we are retraining those muscles and learning balance. Leah is wonderful and willing, but I have noticed that after a good workout she is stiff for a couple of days. I turn her out every day the weather permits, usually for four and five hours at a time. I know what it is like to have sore muscles and obviously this is to be expected during her retraining phase. However, I don't want her muscles to become tight and knotted, as my own have in the past. I know that this leads to problems elsewhere in the body, and I was wondering what exercises, especially stretching ones, or massage I could do with her to keep these muscles from knotting up.

A You sound like a very thoughtful and sensible owner; Leah is lucky to have you. You're right; when a horse (or a human, for that matter) is learning to use her body in a completely different way, she is likely to be stiff and sore after workouts. This is especially true for the older horse. Here's how you can help your mare.

First, be sure that your warm-up is long and thorough. Start out with a good grooming, and then do at least 15 or 20 minutes of walking and 5 or 10 minutes of trotting on a long rein before you ask your mare to work.

When you work her, don't ask for too much at once. Focus on quality, ask for a little at a time, and don't forget to let your mare stretch her head and neck (all the way to the ground if she wants to) every three minutes. Talk to her, praise her efforts, and keep her as relaxed as possible.

After you work her, be sure that your cool-down (or "warm-down," as I prefer to call it) is equally thorough, and that she gets another chance to trot on a long rein and then walk until she is cool and dry.

On any day that you ask for more and come away thinking "Wow, that was a real workout!" plan an easy ride for the following day. If you alternate your days of intense schooling with days of hacking out, you'll both be happier and she'll make faster progress. Muscles need time to recover and rebuild after workouts. It's safe to go for long walks every day, but if you're also doing hard gymnastic work with Leah, use your good sense and treat your horse the way you would treat yourself if you began a bodybuilding program in the gym. You wouldn't work the same sets of muscles hard every day; you *would* do warming up, stretching, and suppling work every day.

You would also back off a little if you were sore *all* the time, because you would realize that you were trying to do too much, and that pushing your body too hard, too fast, was tearing it down instead of building it up. Keep your mare's work comfortable for her. You don't have to canter a mile: You can trot it, or walk it. It takes more time to walk a mile than to run the same distance, but the low-impact workout is safer and provides the same benefits. You could make Leah very supple and fit even if you did nothing but *walk* her for the next six months. I realize this might not appeal to you, but it's something to keep in mind.

As you suggest, it will help if you can also do some passive stretching after your warm-up (just before your workout) and before your warm-down (just after your workout). Massage is also very helpful. A good grooming is a nice massage for a horse, and there are a number of good books and videos on equine massage.

Use your powers of observation and your own good sense: Notice when your mare is moving with shorter strides or seems uncomfortable, and increase your warm-up time and grooming time accordingly. Notice whether she is most sore one day or two days after a big workout, and adjust your schooling and hacking schedule accordingly. Track weather changes: She'll need a longer warm-up and warm-down in colder weather. Track her activity level: She's likely to be more stiff and uncomfortable if she had limited or no turnout that day or the day before. Make it your habit to pay close attention to your horse's behavior and movement, draw the correct conclusions, and always act in your horse's best interest.

If your mare worked hard on Monday, she may be stiff and sore on Tuesday.

CHAPTER 18

Specific Situations

EVERY HORSE IS UNIQUE, but there are some principles you can apply across the board. Train your horse gently, carefully, correctly, systematically, and progressively, building understanding on understanding and skill on skill until he is physically, mentally, and emotionally prepared for the job you want him to do. Give the process the time it needs; the long, slow way may not be the only way, but it's the only way that will let you and your horse complete your journey as partners and arrive at a good destination. When you encounter a problem, look for its underlying cause. Identify the holes in your horse's training, understanding, and physical development, then go back and fill those holes before you proceed.

Symptoms, Causes, and Asking the Right Questions

Q My horse is a three-year-old Thoroughbred. We are going to train him for eventing, but we won't teach him to jump until he is five years old. My trainer says you are right that horses should not jump when they are too young. My problem with my horse is that he crosses his jaw. I know you say not to use tight nosebands, but you don't say what to do with a horse that crosses his jaw. If we put a flash noseband on him and tighten up both parts as tight as they will go, then he doesn't cross his jaw. I know that we are basically tying his mouth closed, which you say is bad, but he is still very green so my trainer and I think it is important for him not to learn resistances to the bit so early in his training. I want to do what is right for training and for the horse, but I can't do both at once because of this situation, so what can I do?

A There are really three problems here. First, your horse is trying to tell you something, but you aren't hearing him. Second, you and your instructor are asking the wrong question. You want to know "How can I keep him from crossing his jaw?" when what you should be asking is: "Why is he crossing his jaw?" Third, you think that the requirements of training are incompatible with the needs of your horse, and when training is correct, this is simply not true. Let's take these issues one at a time.

Horses don't "resist" in a vacuum. Something has to be bothering a horse before a resistance appears, and it's the rider's and the trainer's obligation, *always*, to ask, "Resistance to *what*? What are we doing that we shouldn't be doing, or what aren't we doing that we should be doing?" It's important that you listen to your horse, and that means doing your best to understand him.

Whenever a problem involves a horse's mouth, consider the possibility that something needs to change at *both* ends of the reins. Let's begin with his mouth and the bit. A horse won't cross his jaw for fun. It's usually a reaction to mouth pain; it's a fairly typical reaction to pain from a single-joint snaffle. Have your horse's mouth inspected by a good veterinarian or an equine dentist, preferably both. At his age, he could easily have a retained cap or wolf teeth coming in, in addition to all the usual issues of sharp edges and points.

If his teeth are fine, and he has no lacerations to his tongue or to the insides of his cheeks high up inside his mouth, you'll be able to turn your attention to his bit. Horses' mouths differ quite a lot. You'll find a lot of information about mouths and bits in the *Horse-Sense* archives, so all I'll say here is that you might want to try a French-link snaffle, adjusted so that it touches the corners of your horse's mouth but doesn't create wrinkles. If he is unhappy or the bit bangs against his teeth, take it up the least amount that will restore his comfort — you might end up with a tiny wrinkle, but beware of raising the bit until it creates multiple wrinkles. A too-high bit will make any horse uncomfortable.

Next, consider what's happening at the other end of the reins. You should be able to feel when your horse moves his tongue and when he swallows. If you are using the reins to help you keep your balance, pulling against your horse's mouth, or holding the reins with low hands and locked elbows, your horse will be unhappy with what he feels in his mouth. Ask yourself these questions:

- ▸ Are the reins long enough for my horse to carry his head and neck comfortably while staying in soft contact with my hands?
- ▸ Are my hands carried in front of me, at about belly-button level, so that my elbows are bent and my arms can move with the horse?
- ▸ Are my fingers closed on the reins, so that I can "talk" to my horse by tightening or relaxing my fingers very, very slightly?
- ▸ Can I feel my horse's mouth through the reins?

When any form of training is in opposition to the horse's welfare, *the training is wrong.* Training should educate a horse, improve his physique and his mind, and help him become more beautiful and responsive. But this works only when every aspect of the training puts the horse's interests first. If you learn to do this consistently, you'll be a good trainer, because you will never try to take shortcuts or employ gadgets to take the place of training. Knowing how to disguise or eliminate a *symptom*, such as the

Use your hands to "listen" to your horse's mouth.

horse's crossing of his jaw, isn't good training or horsemanship. Understanding why your horse would cross his jaw, and knowing how to make it unnecessary for him to try to protect himself in this way, is much more useful because it will help you deal with the *cause* of the problem.

One more thought: If your horse has been uncomfortable for quite some time, he may have muscular problems that won't disappear instantly even if his mouth is in perfect shape, you find the ideal bit, and your seat, hands, and understanding become perfect overnight. Ask your vet to recommend someone in your area who does equine massage. If you can find a truly qualified person, preferably someone who already had formal certification in human massage before extending his or her practice to include horses, your horse will be in good hands.

TRUE HORSEMANSHIP

Becoming a rider is quite different from becoming a horseman. Riding and horsemanship can overlap, but there are many riders who aren't horsemen, and quite a few horsemen who aren't riders. All true horsemen have three things in common:

1. They love horses and they know that "love" means "what you do and how you look after your horse," not "how you feel about your horse."
2. They take full responsibility; their first reaction to anything odd the horse does is to ask "Why?" followed quickly by "How did I cause that?" When they answer their own questions, they think in terms of communication and technique, not force and technology. They are quick to notice the symptoms of problems, and they always seek out the problems' causes, then adjust their horse-training and horse management practices to prevent such problems from arising again.
3. Their aim at all times is to have the horse develop his body and mind until he can do his job in beauty and with pleasure, trusting in the rider's sensitivity and interest in mutual communication.

My Horse Hates the Ring — What Can I Do?

Q I have been share-leasing a horse from a private owner for the past 18 months. His current owner rides very infrequently, so the horse is available to me every day. His owner is an experienced rider, and from what he says, he just likes to gallop the horse. I, on the other hand, want to learn and do everything. His owner has promised to sell him to me when he gets too busy to have a horse (soon, I hope). The horse is brave, well mannered on trail and on the ground, and very smart. I am told he has had experience in a little of everything Western: barrels, reining, team penning, and so on.

I love to trail ride, and so does the horse. The problem comes when I want to practice with him in the ring. He really dislikes the ring and makes it very hard on me. I can keep him on the rail at a walk, but when I ask him to trot, his ears go back (like "Oh, no, this stuff again!"), his head and neck crane to the inside of the ring, and he starts to pull to the center. He does not respond to leg pressure to keep him on the rail. So far, I have just given him a smack with my open hand on his neck, and this reprimand seems to insult him and make him behave for

Make the arena a pleasant place for your horse so that he won't be apprehensive about entering it.

the moment. A trainer has suggested that I twirl (i.e., threaten) a rope near his head when he pulls to the inside of the ring. He does go back on the rail when I do this, but I don't want to bully him, and I will not always have the rope. How do you think I should correct him?

Also, I was taking a lesson with a trainer, and she suggested I ride with no reins. She said I should ask for the trot, but give no directions. The idea is that first the horse would trot around, but eventually he would trot around the perimeter of the ring. Well, my horse went nuts. He trotted and trotted all over the place, went into corners and stopped short and then darted out again, ran to the exit gate, and threatened other horses (which he never does). He never slowed down to an easy trot and never did trot the perimeter. So, my second question is, what should I have done in this situation? I don't think he was being disobedient; I think he truly was nervous and upset because he wasn't getting the direction he was used to.

After that lesson, I have taken him to the ring, started trotting him while using the reins, and then quietly dropped the reins and tried to ride him with leg and seat only. He did a lot better when it wasn't obvious there was no rein direction. We have worked up to a full one-and-a-half revolutions of the ring. What is a good correction for a horse that is being difficult because he doesn't like the ring? And what about riding in the ring with no reins?

A Horses that have done barrel racing are often very nervous in and near riding arenas, sometimes refusing to go into them at all, sometimes jigging and sweating and tossing their heads until the rider takes them out, at which point they calm down. For many barrel horses, the ring is an unpleasant place. Your horse may be pulling to the center because as far as he knows, he needs to run toward the center, go around his barrels and out through the keyhole, and get the whole unpleasant experience over with as quickly as possible.

If a horse is being difficult because he doesn't like the ring, the best correction is to teach him that the ring can be a pleasant place. Slapping him and waving ropes at his head may work today, but in the long run they won't be effective, and they certainly won't make him like the ring any better than he does now. Also, as you point out, you ought not to become dependent on a rope that you may not always be carrying.

If he believes that the arena is a place where sooner or later he will be kicked to a gallop, jerked this way and that, whipped for more speed, and then finally jerked to a halt after the last run, he has no reason to expect good things when he enters the arena. He believes that he *knows* what will happen in there; it's only a question of *when*.

If you want him to be calm and accepting about the arena, you'll need to give him pleasant associations with the arena. That doesn't mean that you have to go in there with a bag of carrots and a box of sugar lumps — the best way to give him pleasant associations with the arena is to be sure that every time you go in there with him, you continually ask him to do things and you always praise him for trying, whether or not he does them well. Break everything you do down into tiny segments, ask one question at a time, pay attention to his response, and reward him with praise and a pat or a scratch if he gives any indication of trying to do what you've asked him to do. Instead of asking him to pick up a nice trot and go all the way around the perimeter, ask him to trot and praise him when he does, ask him to steady or slow the trot (do this by steadying or slowing your own posting) and praise him as soon as you notice even a tiny difference. Ask him to go in a certain direction, and praise him when he does; ask him to do a half-turn to change direction and praise him for that. If you look for every opportunity to communicate with him and praise him, you'll find there are thousands of such opportunities during even a half-hour ride.

> We all tend to rely on our reins for everything, even when we think that we're using our legs and seat more than our hands.

Your other question is actually quite intriguing. Riding without reins can be a very helpful exercise, and an illuminating one — for the rider. It doesn't have to be done in the arena; try riding him in a large field, at least 6 feet away from the fence line. Drop the reins on his neck and try to ride him in a straight line. If your experience is typical, you will quickly discover several facts that you will then (I hope) consider and ponder for some time and always keep in your mind thereafter.

Fact: Horses are not perfectly even; they tend to be one-sided and it takes years of careful training and riding to help them to develop both sides more equally.

Fact: Humans are not perfectly even; they tend to be one-sided, but most of them never realize this, even the ones who are very conscious of their *horses'* one-sidedness. Again, years of careful training and riding and exercise and conscious practice can make a great deal of difference, but even a superbly developed human athlete who has made every effort to become strong and supple and equally developed on both sides is going to be uncomfortable and uncoordinated if asked to twirl spaghetti or use chopsticks with the *other* hand.

Fact: Horses will generally choose to travel next to the fence, for security; riders will generally imagine that the horses are straight (hint: they aren't).

Fact: Horses are extremely sensitive and responsive to rider weight shifts, and the rider doesn't have to make an effort to shift her weight — just looking down or turning her head to look in a new direction will cause the horse to move differently and shift his direction as well. A rider on the rail who looks into the center of the ring may suddenly find herself there.

As for dropping the reins in the arena that first day, I think that was a bit silly since there were other people riding in the arena and you had no idea what your horse would do. What you've done since is very sensible and practical. When you're alone in the arena, try riding him across the arena, or even around the arena but on the inside track rather than on the rail. If you focus on your balance and position and on what your horse is doing, riding without reins can help you become aware of how sensitive your horse is to your position and actions, of how much influence your balance and position have on him, and — this is key — of how much we all tend to rely on our reins for everything, even when we think that we're using our legs and seat more than our hands. When you don't have the option of using your reins to straighten your horse, bend your horse, come off or onto the wall or make a change of direction, one of the first things you'll notice is that you will automatically try to use those reins, every time.

Sheath Noises = Tense Horse?

Q My new horse is a nervous type who dances around when he isn't feeling completely secure and safe. I never had to deal with this before, and unlike my *old* gelding (a completely relaxed type who never worried about anything except missing dinner), he makes that weird noise at the trot. I've heard that this noise is caused because of air being pulled into and out of the sheath, and I've also been told that it's a sign of tension in a horse. This is beginning to make sense to me because I've noticed that when my horse is relaxed, I don't hear that noise at all.

For example, I usually don't hear it after we have warmed up, but the other day some idiot brought her dogs to the barn and they ran through the arena barking at each other and trying to grab each other's ears. All the horses went ballistic, including mine. He quieted down quickly and started to listen to me again, but his sheath noise started up and he was still doing it when I put him away. I want to know if it's just a coincidence that he does it when he's nervous, and I want to know how to train him out of it.

A Somewhere there may be a gelding or stallion that can be relaxed and still make that noise, but in my experience, that sound *does* indicate tension. My theory is that muscle tension creates an air pocket in the horse's sheath, and movement then creates the strange, quacking sound. I don't think you can train a horse not to make the sound — the horse can't help it. What you can do, though, is to train your horse to relax. If my theory is correct, this would have the silencing effect that you want.

That "quacking" sound can make
you aware of your gelding's tension.

Rhythm and regularity are huge components of relaxation. When in doubt, trot a 20-meter circle on the aids: It's a comfortable, reassuring activity for a green horse. When he's less green you can make the circle somewhat smaller, but the principle will remain the same: To help a horse relax, give him something to do that he understands and can do easily and well; then give him constant feedback about how well he is doing it. Each time you mention what a nice circle it is, how forward he is, how responsive he is, he will relax a little more. Even a pat on the neck and a soft "thank you" once on each quarter of the circle will help him relax noticeably. When you're not directing his movement and praising his responses, check your own posture (straight and balanced) and your own breathing (deep and slow and regular). Any improvement you make to either will help your horse relax even more.

The next time your horse becomes nervous and anxious, check your own position and breathing *first* and then move smoothly into a trot on a circle. You'll be able to monitor your horse's level of tension by the noise — or lack thereof — down below. And if you consider the noise as an additional way for you to become aware of your horse's building tension, it may be more of a convenience than an annoyance.

By the way, if anyone ever tells you that sheath noise results from a dirty sheath, rest assured that it's not true.

Can I Train My Horse to Urinate?

Q I have a problem that sounds strange, but it's serious. My horse will not urinate when he is in the trailer, and I need to train him to urinate before we leave for a show. We live at least four hours from the shows I go to, so it's a long ride. Shadow gets frantic when he has to urinate, but he just won't go until he is in his stall or in a stall at the show. Do you have any secrets to tell me about how to get Shadow to urinate before we leave? Or is there some way to get him to urinate while we travel? Would it help if we stopped along the way and took him out of the trailer? My dad says there is no way to "command" a horse to urinate. Please help me.

A This isn't strange or even unusual; many horses are like Shadow, and there *are* some things you can do. At the racetrack, where horses must provide urine samples after races, grooms teach their horses to associate the act with the sound of whistling. You can do this — whistle whenever you see Shadow urinating, until he begins to associate the action with the sound. Traveling is tricky, but you can encourage Shadow to urinate during the trip. Don't take him off the trailer — it's really much safer to leave him in it. But stop for a little while somewhere, at a rest stop or roadside services, and park well away from everyone else so that the trailer won't rock when other vehicles pass it. This will give Shadow a chance to relax his muscles, and he may take advantage of the opportunity to urinate. Horses in moving trailers are constantly doing isometric exercise, tensing their muscles over and over just to keep themselves upright and balanced. Shadow can't possibly urinate until he relaxes. If the trailer stops moving and everything is still and quiet for 10 or 15 minutes, Shadow will be able to relax.

Bedding in the trailer may help, too. Horses dislike getting splashed with urine, and they will often avoid urinating until they get into their stalls, not just because the stall is a familiar place but because of the absorbent bedding. If you put several inches of shavings into the rear half of Shadow's trailer, it will provide an absorbent cushion that will encourage him to urinate.

In order to urinate, your horse must be relaxed.

Stiff Horse: Is It an Attitude?

Q I know that we should always look for a physical cause first when there's a problem with a horse. But I really think that my new horse may have an attitude problem, not a physical one, so I hope you can give me some advice for attitude problems. Devon is five years old and very muscular. He did not do much for the last year or two beyond living in a pasture because his owner was sick. Now he is in a stall with a big turnout (12' x 36') attached to it so he can move around, and he gets at least four hours in the pasture every day.

The reason I think he has an attitude problem is because of how he behaves when we longe him and ride him. The barn owner is very careful to always check saddle fit, and I have had the veterinarian out twice to check Devon's teeth and they are fine. He accepts his bit and doesn't open his mouth or pull or anything, so I am pretty sure he doesn't have a problem with his saddle or bit. I am a pretty good rider. You would probably call me an intermediate rider; my balance is good and I know I don't get in Devon's way a lot.

The problem is that he will be very cooperative and calm and nice for the first 20 minutes or so of the ride. I have tried longeing him before the ride and not longeing him before the ride. I have also tried riding him just after he comes in from his pasture time and at other times. It doesn't seem to make any difference how warmed up he is, so I think we are looking at an attitude problem. My friend Kellie does horse massage and she has checked him out three times and says he is not sore anywhere. It doesn't seem to matter what I do or don't do, he goes for about 20 minutes and then it's like his body starts to get stiff, he will twist his neck, and the barn owner says he is locking his jaw. His back feels all stiff too when he does that. He did the same thing when Kellie rode him and when we longed him in side reins — not too tight, just to keep him in a good frame the way I do with the reins when I ride, no different.

Now here is the last thing that convinced me. We go on trail rides every Sunday if the weather is good. Sometimes Devon is warmed up and sometimes he isn't warmed up, but he is always fine on the trail, and we are usually out for about four or five hours. He doesn't twist his neck or lock his jaw or get stiff on the trail. If it's a physical problem I would think he would get stiff after 20 minutes the same way he does in the arena and when we longe him, so I think stiffening up is something he does on purpose, although not on the trail because we are just mostly goofing around on a loose rein and talking, not working. I think he is making his body stiff when he doesn't want to work. I think I have considered everything. What can I do about it if it is attitude, and do you have any other ideas?

A Congratulations, you've been doing an excellent job of testing, investigating, and considering the various possibilities here. You've covered a lot of territory already, including the possibility that his stiffness could be caused by rider stiffness or imbalance. That would have been my checklist, too — look at the horse moving freely, look at him in pasture, look at him tacked up, check the tack, check the rider, try a different rider, check the horse for soreness, check him on the longe, observe everything, and take notes. You've done a good job.

However, I still don't think that you're dealing with an attitude problem. There's one more possibility you should consider, and that is simple fatigue. It's not that you're working Devon so hard that he becomes exhausted after 20 minutes, it's just that most often, when this sort of stiffness appears after the horse has been ridden or worked for a little while, the problem is simply that his muscles are tired.

I believe that Devon is probably a lovely, well-muscled young horse, and I'm sure that he is very strong. But the kind of strength that will let a horse carry himself and a rider easily, all day long, on the trails, which Devon obviously *can* do, is not the same kind of strength that will let a horse work *in a frame* for 15 or 20 minutes at a time without a break.

Devon is carrying you comfortably down those trails without becoming stiff because he's comfortable — this is something he can do. The stiffness he shows after 20 minutes of working in a frame under the rider or on the longe line is a reaction to tired muscles. All you need to do is back off a little when you're working. It's okay to ask him to carry himself in a frame for a few minutes, provided that the frame is appropriate to his level of training and fitness, and provided that after a few minutes, you give him a break and allow him to stretch.

Over time, Devon will develop the muscular strength he needs to carry himself the way you want him to, but you will always have to be attentive to his comfort level and his need to stretch and relax his muscles. Even a brief stretch will enable him to go back to work, only this time, he'll be physically and mentally comfortable. When you're in the arena, watch the clock. If watching the clock doesn't work, ask someone to help you. You need to make a habit of allowing — better still, *asking* — Devon to stretch whenever he's been working in a frame for a few minutes. Have your friend time you and remind you to encourage your horse to stretch every three minutes for the first week, then every five minutes.

This doesn't mean that you'll drop the reins and let Devon collapse into a slouch, it just means that you'll ask for and allow a stretch. You won't even need to change gait or direction. If you're working at the walk, do a free walk across the long diagonal. If you're working at the trot, do a "stretchy chewy circle." If you're working at canter, stay on your circle or continue around the arena — whatever you were already doing — but let him take the reins down until he's cantering with his nose on the ground. He'll just need a minute and then you can begin to bring him back into his working frame.

If you make it a habit to do this all the time, Devon won't ever become so uncomfortable that he *has* to twist and stiffen just to relieve the cramps in his muscles. And over time, he'll get stronger and stronger. *Never* ask him to go for 20 minutes without a stretch, though. Three minutes is quite a long time to hold some very large muscles (like those of the neck!) in one position.

To keep your horse comfortable, encourage him to stretch down with his head and neck every three minutes.

While you're thinking about giving your horse regular breaks, consider this: Working the same muscles every day, in the same way, will cause your horse to become tired and bored and will fatigue those muscles, but it won't help the horse develop those muscles. Workouts for horses are like workouts for humans — any good personal trainer would tell you not to work the same muscle groups every day.

Muscles build and strengthen by first being torn down and then rebuilding. If you're working your leg muscles and your back and chest muscles, say, you don't want to do the same leg-back-chest routine every day. You'll get sore, but you won't get stronger. Instead, your personal trainer would tell you to work the leg muscles on Monday, Wednesday, and Friday, and work your back and chest on Tuesday, Thursday, and Saturday. Your muscles would respond much better to the every-other-day workout, and you would begin to build strength and size.

It's the same for your horse, only a little bit better, perhaps, because your horse has something you probably don't — the personal trainer! *You* are your horse's personal trainer. *You* make the decisions about the schedule on which your horse's muscle groups are being worked and built up. Don't do the same thing every day. Vary the work, focusing on different muscle groups so that you aren't spending consecutive days working the same muscle groups over and over. And don't forget those trail rides — the rest of your workweek may be more deliberate and more focused, more "work," than the trail rides, but do everything in your power to make "work" days as pleasant and enjoyable for your horse as those trail rides. And *always* remember to ask the horse to stretch and then allow him to do it.

Here's an exercise you can try *off* the horse that will help remind you why those stretch breaks are so important. While you're sitting at the computer, straighten one of your arms and swing it back as far as you can, keeping it at shoulder level. Then *hold it there*. Look at the clock, and try to hold your arm in that position for five minutes. If you're like most people, you'll begin to get tired and uncomfortable in less than a minute; if you last for two minutes, your arm and shoulder will become painful. Soon you'll be unable to think about anything else. If you force yourself to keep the arm up and back, you'll begin twisting and stiffening your own neck and back to try to relieve the pain. (Does this sound familiar?)

If anyone tries to tell you anything or teach you anything, or if you try to remember anything you've read while you're doing this arm exercise, you'll quickly find that the "learning" part of your mind seems to shut down when you're trying to cope with extreme physical discomfort. You're unhappy, you're distracted, you can't think about anything but getting rid of the discomfort. It's the same for our horses.

Yes, a horse can develop an "attitude" about something, but if the cause of the "attitude" is pain, then the attitude, like the stiffness, is just a reaction to and a consequence of the problem. It's not the problem itself — for that, look all the way back to the physical cause.

If I asked you to try the arm exercise again right now, you would automatically stiffen and twist your back and shoulder — and your "attitude" would not be good. If I made you do it anyway, and said, "Oh, she's fine, she's not uncomfortable, she just has an *attitude*," you would be quite unhappy, and you would probably find some way of letting me know that you were unhappy. And I couldn't possibly blame you for reacting that way, because your "attitude" would be based on something real: your knowledge that what I'm asking you to do is going to *hurt*.

Devon sounds like a nice horse, and he clearly has a thoughtful rider — you very nearly figured out the answer for yourself. I think that you and Devon will do very well together. Let him stretch every few minutes, and always remember that whenever a horse is stiff or becomes stiff, the probability of the stiffness being caused by some kind of *physical* discomfort is very, very high. Think in terms of 99.999 percent, and you'll be there. And if you're ever absolutely certain that the horse you're riding is the .001 percent case, give him the benefit of the doubt anyway, because it's good for him and it's good for you.

Do everything in your power to make "work" days as pleasant and enjoyable for your horse as trail rides.

Talking during Dressage Test

Q I want to start taking my nervous Thoroughbred mare to dressage competitions. She is much calmer since I have been retraining her, but I am afraid that one reason she is calmer is that I talk to her all the time. I know I will not be allowed to do this during a dressage competition, but I don't know what Lady will do when she has to deal with a new environment and a test with me not saying anything to her.

A　Congratulations on making such progress with Lady! You've done a good job so far. Don't spoil it now by asking too much at once. I think that you are right to worry about the combined effect of a new environment and your silence, so here are my suggestions:

1. Go ahead and take Lady to a dressage show. Groom her, braid her, dress up, but don't go to compete, just go to hack her around the grounds. Talk to her, allow her to relax and be confident, let her eat grass and look around, and at the end of the day, take her home. That will help deal with the issue of "new environment".

2. When Lady is calm about trailering to a new venue, enter her in one or two low-level tests, no more. Then tell yourself that the first show day is a *teaching* day for Lady's benefit, not a *competing* day. Go right ahead and talk to her in the arena, whenever she needs it. You *can* talk, it's just that you'll be penalized for "use of voice." Since you're just there to school Lady, that will not be a problem.

If your nervous mare finds your voice reassuring, use it at your first show even if it costs you points.

You may even want to mention it to the judge, on your trip around the ring perhaps, or after your first test, so that he or she won't imagine that you don't *know* about the penalties. And then just have fun, and keep your focus where it should be: on Lady and her comfort level.

Don't worry about your scores; you're building your horse's security and confidence so that the next time out, you'll be able to show her instead of just school her. And, in fact, if you have to do this at several shows before Lady begins to relax enough to enjoy herself in the arena, that's fine — she is just at the beginning of her competitive career, and this is the time for you to create the attitude that she will have for the next 10 years or more. It's worth taking some time to allow her to relax and learn to enjoy her work.

Carrying the Flag

Q Call me a sentimental patriot, but every time I see a horse ride into an arena carrying our country's flag during the national anthem, I am totally covered with goose bumps. I always think, "I'd like to do that!" Through what progression of steps would I teach my six-year-old gelding how to carry a flag? He's a bit of a Nervous Nelson until he is used to something and then he is fine. I don't want to start out doing something that will scare him so badly that he will never want to do it. We just replaced our stable flag so I have one (4x6) that I can practice with. It's faded and a little frayed so I don't have to worry about damaging it. How do I begin?

A Your main job will be getting your horse and yourself used to the flag. A flag, to a horse, is just something large and flappy that eventually proves not to be a horse-eating, flying monster coming in for a landing.

Life can be quite interesting while your horse is learning about the flappy part, so don't start by trying to carry an actual flag. Use something little and soft, like a hand-towel or a stable rag, that you can drop immediately if your horse overreacts. You can flap it, wave it, shake it, drape it over your horse's neck, and eventually drape it over his head, first from the ground and then from the saddle. Don't tie it to a stick at first, because just in case the moving material creates a series of horse-bounces, you won't want to be holding or dropping anything hard and pointy.

Eventually your horse will be utterly bored by all of your flapping efforts, even if you cover both of his eyes and sing a song. When he accepts the small cloth, trade it in for a larger one; when he becomes bored with the larger one, trade it for one that's larger still. Keep increasing the size gradually until you're ready to deal with an actual, full-sized flag. This is all time well spent, by the way. If you do carry a flag, there will inevitably come a moment when your flag or someone else's flag comes loose or droops or blows across your horse's face, and when that happens, you'll want him to react by rolling his eyes and thinking "There she goes again with that boring old flapping stuff."

Somewhere during the process of letting your horse become bored with flappy things, introduce him to your real flag or at least a large piece of similar material. Different materials feel and sound different, so you'll want to accustom your horse to all kinds of material, including the slippery, rustly feeling of an actual flag.

If you're thinking about carrying a full-size flag in parade- or rodeo-style, then you'll need some help. There's a staff involved as well as a flag, and the staff is heavy. You'll need a stirrup attachment or a saddle attachment, depending on the size of the flag and length of the staff, on which to rest the base of the staff. If your local tack store fails you here, try one that caters to rodeo folk, or ask the nearest mounted police unit where they get their parade equipment. People who do cavalry reenactments are also very good at finding sources for things like this, so ask around, especially since what you really need at this point is something you can borrow, not buy.

When you've got one hand on the staff of a flag, it leaves (if you're a typical human) only one hand for your reins, so yes, you're going to have to teach your horse to neck-rein. It's not difficult — horses pick it up very quickly and easily. (See chapter 10 for information on how to train a horse to neck-rein).

Loping will be the most difficult, but if you're thinking in terms of carrying a flag in a show arena, you'll just need to be sure that your horse can pick up the correct lead. As a flag carrier, you're most likely to go two or three times around the arena; you won't have many changes of direction.

Make the whole process easier on yourself and your horse by using the right equipment. Begin with a snaffle, but make it a solid, heavy one and use the heaviest reins you can find. The same will apply when and if you shift to a curb: make it a heavy bit, and use heavy reins. It's much easier to be clear with substantial reins that don't flutter and flap. In any case, fluttering and flapping is the flag's job.

Philosophy of Problem Solving

Q After "starting over" with horses 10 years ago, I am now training my third horse and helping two friends with their horses. I'm not comfortable yet labeling myself a "horse trainer," but I hope to be one some day. At a lecture you gave some years ago, you said how important it was to trace problems back to their cause and solve *that* instead of just putting "training Band-Aids" over the symptoms of the problem. I didn't really understand this at the time, but it sounded right and it worked, and I have been trying to follow that advice ever since.

If I had learned this sooner my first horse would have had a much nicer life, and I might still have my second horse. He was the one I sold to my instructor because she knew enough to train him and I didn't. Right now I'm trying to improve my training and develop a philosophy of problem solving, or maybe I should say a philosophy of training and problem solving. The whole range of the subject is fascinating to me, and I'd like to know if I'm on the right path with the way I'm training now. Would you share some of your own training philosophy just to get me started?

A Over the years, you'll develop your own training philosophy — trust me, that's unavoidable! In the meantime, congratulations on your training activities and ambitions. I think that all thoughtful trainers must look back and think regretfully of how much better they could have trained their first horse, and how much happier that horse could have been. We've all been there. Look at it another way, though — you should *always* be able to look back, not just at your first horse but at the one you trained last year or last month, and tell yourself how much better his training would have been if you had known then everything you know now. That means that you're continuing to think and grow and improve as a trainer. The best trainers *always* strive to learn more and *never* imagine they know it all.

Problem solving is an important part of training, but truly, the first and most important way of dealing with problems is to prevent them from happening in the first place. Problem solving requires more time and energy than problem prevention. That said, we *need* problem-solving skills, because most of us will use them on a daily basis. Even if we don't create problems in the horses we breed and raise, we'll still have to deal with the problems of the horses that come to us for retraining.

You're probably already familiar with most of my training philosophy. It's fairly basic:

- ▸ Horses should enjoy being trained.
- ▸ Training should be based on trust and positive reinforcement.
- ▸ Training should develop the horse's body, mind, and spirit.
- ▸ Training should work with, not against, the horse's nature.
- ▸ It's the trainer's job to encourage desired behaviors and prevent problems from developing.
- ▸ It's the trainer's responsibility to understand horses in general *and* to learn to understand each specific individual horse.
- ▸ Problems must be solved appropriately; training problems must be solved with training, not with equipment or drugs.
- ▸ Problems should be solved in ways that don't create other problems.
- ▸ Trainers who are good problem solvers are the ones who are best equipped; not with ropes and leather straps but with knowledge, experience, compassion, patience, and — whenever possible — insight.
- ▸ Good trainers adapt to situations and circumstances by trying to anticipate all possible actions, reactions, and consequences, and by being ready to segue smoothly into "Plan B" (and, if necessary Plans C through Z).
- ▸ When something goes wrong, a good trainer's first question will not be "Why did my horse do that?" (or, worse, "Why did my horse do that *to me*?") but rather, "What am I doing that is causing my horse to respond in that way?"

Your responsibility as a trainer is to help your horse develop his musculature, movement, and mind while maintaining his soundness, dignity, and attitude of joyous cooperation. Whether you train one horse or one hundred horses in your lifetime, this should always hold true.

Problem solving is good; problem prevention is better.

GLOSSARY

Backing. The act of mounting a horse for the first time — a moment for which the horse should be fully prepared, and one that should not be traumatic or frightening for either the horse or the rider.

Balance. Horses balance both longitudinally (forelegs and hind legs) and laterally (left legs and right legs). Horses naturally carry more weight in front than behind. Achieving good balance under a rider means that the horse can maintain an even distribution of weight on each side, even when bent or turning; it also means that the horse is learning to carry a little more weight behind.

Bending lines. Jumping two or more jumps that require the horse's track to curve between one jump and the next. The horse landing from one does not face the next jump directly and has to follow a curving track in order to achieve a straight approach to the second jump.

Bradoon. A snaffle (usually a loose-ring, occasionally an eggbutt) used with a curb as one of the two bits in a double bridle. The bit rings are smaller in diameter than those of a conventional snaffle.

Bradoon carrier (bradoon hanger). A narrow leather strap added to an ordinary snaffle bridle (it passes through the browband) that allows the addition of a bradoon and another pair of reins, turning the snaffle bridle into a double bridle (the curb is attached to the bridle's cheek pieces).

Cavalletti. Poles attached at each end to Xs, adjustable (by turning) to three heights, used in schooling for both flat work and jumping.

Chambon. A training device designed to help a horse learn to stretch his topline and reach forward and down with his head and neck on the longe (never under saddle). It can be useful if employed briefly by an accomplished trainer. The chambon attaches to the girth between the horse's forelegs, then forks at chest level and each piece is taken straight up and passed through a small ring or pulley on a separate browband, then connected to the bit ring on that side. If it's adjusted loosely (as it should be) the horse feels its effect only if he lifts his head sharply, at which point the chambon tightens and puts pressure on the bit and on the poll at acupressure points that encourage the horse to drop his head. By reaching forward and down, even very slightly, the horse receives instant relief from those pressures.

Chip in (chip). A horse that adds an extra, short stride just in front of a jump is "chipping" or "chipping in."

Cold-backed. A horse that humps or drops his back or bucks when first saddled. This behavior generally indicates that the horse is experiencing pain somewhere; probably, but not necessarily, in his back.

Collection. A state in which the horse carries himself in balance by lowering his croup, flexing his leg joints, and placing more weight on his hindquarters, thus shortening his frame, lifting his shoulder, and lightening his front end. Collection requires great strength and it requires years of steady training before a horse can reliably collect at the rider's signal and maintain collection comfortably.

Contact. Although contact actually involves every part of the rider and the horse, the term is often used to describe only the contact between the rider's hands and the horse's mouth. The amount of rein contact that is appropriate must be determined by the horse, not the rider.

Counterbent. A horse that is bent away from the direction of movement; e.g. a horse trotting a circle to the left while bent to the right. Such a horse is not straight.

Crank cavesson. A cavesson (bridle noseband) designed to be fastened extremely tightly; most crank cavessons are heavily padded to avoid creating open wounds.

Draw reins. Draw reins can be attached to the sides of the girth, to the middle of the girth and running between the horse's forelegs, or to the ring of a breastplate, and then run through the bit rings and up to the rider's hands, allowing the rider to bring the horse's head in and (usually) down.

Engagement. The increased flexion of the horse's hind legs under his body during the weight-bearing phase of the stride; this creates energy for impulsion.

Eventing. An equestrian competition held over one, two, or three days and incorporating three disciplines: dressage, cross-country jumping, and show jumping.

Extension. The lengthening of a horse's frame and stride.

Flash attachment. Some flash nosebands are made in one piece; others have a simple cavesson with an attached loop and a narrow leather strap (the flash attachment) running through that loop.

Flat work. Work "on the flat" is work that does not involve jumps. This is the way in which jumping horses are prepared to learn to jump; it is also the way in which jumping horses are warmed up.

Gymnasticize. To work a horse over a combination of fences placed at relative distances to each other; used in the training of the jumping horse.

Half-pass. Dressage movement performed on two tracks in which the horse moves sideways and forward at the same time, bent in the direction of movement.

Impulsion. The horse's "pushing power," which uses the energy created by engagement to propel the horse more actively forward.

Lateral walk. A horse's walk should have a clear four-beat rhythm and propel a horse forward with energy. A lateral walk has a corrupted rhythm (two beats, pause, two beats) and is less energetic, stiffer, and less stable than a normal walk.

Lateral work. A series of exercises that send the horse both forward and sideways, improving his bend, flexion, and balance, thus helping him become more supple and responsive.

Leg-yield. Dressage movement in which the horse moves sideways and forward at the same time, with his body straight and head positioned away from the direction of movement.

On the aids. A horse that is alert and attentive to his rider's aids: legs, seat, weight, balance, voice, and reins.

On the bit. A horse is said to be "on the bit" when he is softly responsive to his rider's aids, carries his head in a near vertical position, and calmly accepts the rider's contact on the reins. This is the *result* of training and physical conditioning, not the starting point, and cannot be achieved by manipulating the reins. "On the bit" is an often-misunderstood term that puts too much emphasis on the bit, the reins, and the front end of the horse. A far more accurate and useful term is "on the aids."

Overbent. An overbent horse has his head behind the vertical; it may even be pulled or tucked in close to his chest. This interferes with the horse's breathing, vision, and forward movement.

Pressure and release. Correctly used, this method involves the application of pressure to elicit a desired reaction (e.g., putting pressure on the halter to ask the horse to stop or turn) and the immediate release of pressure as soon as the horse begins to comply or even thinks about complying.

"Rating" a horse. A rider who can control her horse's energy, stride length, and speed within a gait, sending him (for example) from a short, round canter to a longer-strided, flatter canter and bringing him back again is "rating" her horse. It's not so much a matter of acceleration and deceleration as it is a change of gear.

Rhythm. The beat — just as in music — of the gait and the horse's footfalls. The walk has a four-beat rhythm, the trot a two-beat rhythm, and the canter — this one is tricky — is a four-beat rhythm with three footfalls and one pause. Not to be confused with tempo, which refers to speed — a trot, for example, could be slow, fast, or somewhere in between.

Rising ("a rising five year old"). A horse that is not yet five years old, but is nearer five than four.

Round pen. A fenced or walled enclosure with no corners, generally between 35 and 90 feet in diameter, depending on its specific purpose.

Self-carriage. A horse is in self-carriage when he can move forward in balance without leaning on the rider's hand for support.

Shoulder-fore. A dressage movement in which the horse is evenly bent along the length of his spine away from the direction in which he is moving, with his inside hind foot tracking between his forefeet.

Shoulder-in. Two-track movement in which the horse is evenly bent along the length of his spine away from the direction in which he is moving, with his inside hind foot tracking his outside forefoot.

Starting. Beginning the horse's training. This can take place at any time during the horse's life. "Starting" is sometimes used to describe the process of teaching a horse a new skill, e.g. "starting a horse reining" or "starting a horse over fences." When talking to a trainer, always ask how that person defines the term. Some Western trainers use the word "starting" to mean "starting under saddle" (i.e., backing).

Straightness. Horses, like humans, have a dominant side. To make a horse straight means to develop his muscles and work his joints evenly on both sides so that he is in balance, always traveling with his right and left feet to the sides of an imaginary line from nose to tail. Whether moving in a straight line or along a curve, the horse's hind feet continue to track the front feet on the same side.

Suppleness. A supple horse is able to bend and flex his joints (including knees and hocks, shoulders and stifles, hips and poll and jaw) easily and evenly.

Tempi changes. Flying changes of lead performed in sequence; e.g. every four, three, or two strides, or even at every stride.

TTouch. A method of light massage, developed by Linda Tellington-Jones, that involves making circular movements of the fingers and hands all over the horse's body. The person doing the massage does not need any previous experience or even any understanding of equine anatomy.

Upside-down frame (inverted). Having a topline that resembles a "U" — a complete inversion of a well-developed topline. Instead of presenting a rounded silhouette with the back and the base of the neck lifted, an upside-down horse presents with a ewe neck, a hollow in front of the withers and a sway back with another hollow just behind the saddle. The upside-down horse's hind legs are typically out behind it, pushing the horse's body along instead of supporting and carrying it. Such a horse is uncomfortable to ride and experiences great discomfort when ridden.

RECOMMENDED READING

Aadland, Dan. *The Complete Trail Horse* (Lyons, 2004)

Allen, Linda. *101 Jumping Exercises for Horse and Rider* (Storey, 2002)

Bell, Jaki. *101 Schooling Exercises for Horse and Rider* (David & Charles, 2005)

Brainard, Jack. *Western Training* (Western Horseman, 1990)

Bucklin, Gincy Self. *What Your Horse Wants You to Know* (Howell, 2003)

Harris, Susan. *Horse Gaits, Balance and Movement* (Howell, 1993)

Harris, Susan. *USPC Guide to Longeing and Ground Training* (Howell, 1997)

Hart, Rhonda Massingham. *Trail Riding* (Storey, 2005)

Hill, Cherry. *101 Arena Exercises: A Ringside Guide for Horse and Rider* (Storey, 1995)

Karrasch, Shawna. *You Can Train Your Horse to Do Anything!* (Kenilworth Press, 2006)

Klimke, Reiner. *Basic Training of the Young Horse* (Lyons, 2000)

Kurland, Alexandra. *Clicker Training for Your Horse* (Ringpress Books, 2001)

Kurland, Alexandra. *The Click That Teaches* (Clicker Center, 2003)

Lilley, Claire. *Schooling with Ground Poles* (J.A. Allen, 2003)

Mailer, Carol. *Jumping Problems Solved* (Doubleday, 2006)

Rashid, Mark. *Considering the Horse* (Spring Creek Press, 1993)

Rashid, Mark. *Horses Never Lie* (Spring Creek Press, 2000)

Rashid, Mark. *A Good Horse Is Never a Bad Color* (Spring Creek Press, 1996)

Rashid, Mark. *Life Lessons from a Ranch Horse* (Johnson Books, 2003)

Savoie, Jane. *Cross-Train Your Horse* (Trafalgar, 1998)

Savoie, Jane. *More Cross-Training* (Trafalgar, 1998)

Shiers, Jessie. *101 Hunter/Jumper Tips* (Lyons, 2005)

Steiner, Betsy. *A Gymnastic Riding System Using Mind, Body & Spirit* (Trafalgar, 2003)

Strickland, Charlene. *Western Practice Lessons* (Storey, 2000)

Webb, Leslie. *Build a Better Athlete* (Primedia Equine, 2006)

Ziegler, Lee. *Easy-Gaited Horses* (Storey, 2005)

APPENDIX

Finding Suitable Longeing Equipment

At clinics, riders often complain that good longeing cavessons are ruinously expensive or that it's impossible to find 35-foot longe lines or longe lines without chains. If you've found those things to be true, here is some help:

Cottage Craft makes an affordable, correctly designed longeing cavesson from nylon. Your local tack shop should be able to order one for you.

Top Tack, Inc. (*www.toptackinc.com*) sells good quality 35-foot longe lines and gives customers the option of a single or double swivel snap. Some are conveniently marked at 10, 15, and 20 meters. The same firm also sells a 60-foot "pasture longe line" — very useful if you have the space to allow your horse to work on a truly big circle. Top Tack, Inc. also sells ReBalance sliding side reins (affordable, easy to adjust, and available in several sizes).

Building a Round Pen

Many people construct round pens using panels and gates made from strong metal tubing and designed to enclose cattle and horses. Better-quality metal panels are available in various heights, weights, lengths, and colors. These pens can be sturdy, attractive, and airy — excellent for views and ventilation. They also present two significant advantages. First, if you are on a budget and can initially afford only enough panels to set up a small pen, you can add panels and increase its size over time. Second, you can relocate the pen on your property if that becomes necessary (you need to erect an outbuilding in that spot) or even advantageous (you discover that another part of your farm has better drainage or catches more breezes in summer), and if you should move to another location, you'll be able to dismantle your pen and take it with you.

If you want to build a permanent round pen on your property, you'll find excellent "how to" instructions in *Horse Housing* by Cherry Hill and Richard Klimesh (Trafalgar Square, 2002).

Round Pen Math:
- Choose a diameter — how wide do you want the pen to be?
- Multiply that number by 3.14 to get the pen's circumference.
- Divide the circumference by the length of your panels.
- Don't forget to add a gate! If the gate is shorter than the panels, take that into account and tweak your calculations accordingly.

Remember that your horse will be traveling a foot or two inside the panels, which means that his circle will have a somewhat smaller diameter than the diameter of the pen itself. Take that into account when you make your plans, and be a little generous. If you want your horse to work comfortably on a 20-meter (66 foot) circle, you'll want your pen's diameter to be 70 feet or more. Usually you can achieve this by calculating the number of panels you'll need, rounding *up* to come out with an even number of panels, and *then* adding the gate.

INDEX

Page numbers in *italic* indicate illustrations.

equipment and tack, *continued*

"dressage headset," 246, 247

"fixed" vs. "pulling" hand, 258

flash noseband, 246, 252–54, *253,* 255

flexion at the poll, 256

forcing vs. teaching, 235, 242, 243, 247, 248, 249

forward motion, *257,* 257–59

grab straps, 246, 253, *253*

head-setting devices, 242–43, *243*

inverted (hollow) back, 242–43, *243,* 254, 255

learned behaviors vs. reactions, 245

leg aids or whip, *257,* 257–59

longeing, 84, 201, 277–78, 292, 302–3

longeing and side reins, 237–42, *239, 241,* 250

mental vs. physical issues, 264

neck stretchers, 242–43, *243*

physical vs. mental issues, 264

problems vs. equipment to solve problems, 208, 210, 211, 235

"pulling" vs. "fixed" hand, 258

reactions vs. learned behaviors, 245

resistance to bit and bend, 259–65, *260*

sliding (Vienna) vs. fixed side reins, *236,* 236–37

teaching vs. forcing, 235, 242, 243, 247, 248, 249

"training aids," 243

whip or leg aids, *257,* 257–59

young horse, longeing in side reins, 237, 239, 240–42, *241*

escalating aids, avoiding, 156, 258, 314, 343

eventer's forehand, lightening, 205–8, *206*

eventing for show hunters, 124–27, *125*

eventing saddle, 117

ex-racehorse, retraining, 346–65

angles of feet, correcting, 348

attitude (bad), 349–52, *350*

bending and stretching, 361–63, *362*

chain for leading, 347

contact, *355,* 355–58

dressage, reschooling for, *355,* 355–58

feet, cleaning, 347–48, *348*

footing, getting used to different, 348

head, handling, 349

jumping, *359,* 359–61

lateral work, 357, 363

leading, 347

leg aids, 355, 358

longeing, 357

patting horse caution, 349

rebuilding a horse's body, 352, 354, 357

"rushy," *355,* 355–58

soreness, 351

spoiling, 352–54, *353*

stiff after workout, 364–65, *365*

stretching and bending, 361–63, *362*

tips for dealing with, 347–49, *348*

trails, training for and on, 347, 348

trust, building, 354

turnout importance, 350, 351

warm-ups and warm-downs, 364, 365

extension and collection, 12–13, *13, 214,* 214–16

F

facing owner in stall, 27, 28

facing owner on longe line, 278–81, *279*

fatigue and stiffness, 378–81, *379*

feet, working with, 38–41, *39,* 347–48, *348*

first canter, 98–100, *99*

first solo ride, *64,* 64–65

first time mounting, 57–63, *59, 63*

fitness level of horse, 328

fitness of rider, 197, 198, *198*

five-stride line exercise, *128,* 128–30

"fixed" vs. "pulling" hand, 258

fixed vs. sliding (Vienna) side reins, *236,* 236–37

flag, carrying the, *383,* 383–84

flash noseband, 246, 252–54, *253,* 255

flat, fast canter, 111–13, *112*

flatwork and jumping, 117, 119, 120–21

flexion at the poll, 256

flight, fight, freeze, shut down responses, 25, 221–22, 331

flying changes, timing the aids, 113–15, *114*

foal, haltering and leading, 291, 292–95, *293. See also* young horse

footing, getting used to different, 348

footing and soundness, 181–84, *183*

force vs. conditioning, 209, 210

forcing vs. teaching, 235, 242, 243, 247, 248, 249

forehand, heavy on the, 203–8, *204, 206,* 337

forgetting training and time off, 231–34, *232,* 314–16

forgiving animals, horses as, 69, 71

"Forward"

age-appropriate training, 314

ground, training from the, 52

trails, training for and on, 183

transitions, 139–41, *140*

forward motion, equipment and tack, *257,* 257–59

forward-seat saddle, 117–18

four-beat rhythm (walk), 78–80, *79*

French-link snaffle, 252, 303–4, 340, 368

front-to-back riding, 211–13, *212*

frustration over collection, 208–11, *209*

full seat vs. half-seat (two-point), 117–18

tions; trot, training at the; walk, training at the; specific situations and solutions
training scale, 12–13, *13, 214,* 214–15
training vs. non-training interactions, 224–26, *225*
training your own horse vs. other options, 5–9, *6*
transitions, 134–49
  canter-canter transitions, *206,* 206–7
  canter-trot transitions, 137–38, *138,* 141–42, *142*
  connection, roundness, collection, 204
  "downward" vs. "upward" transition, 137–38
  "Forward," 139–41, *140*
  within gaits, transitions, 90, 148–49, *149,* 204
  half halts, 146, 147
  head tossing during, 141–42, *142*
  inverted (hollow) back, 141–42, *142,* 145–48, *147*
  leg aids, 139–41, *140*
  "long and low" position, 151
  longeing, 42–44
  rider position and aids, 135–38, *136, 138,* 141–42, 143, 144, *144,* 145, 146, 147, 148
  "Set it up and let it happen," 100, 145
  trot-canter transitions, 139–41, *140,* 145–48, *147,* 207
  trot-canter-trot transitions, 206
  trot-stop-walk transitions, 135–36, *136*
  "upward" vs. "downward" transition, 137–38
  walk-canter transitions, 142–45, *144*
traumatic week, fixing horse after, *320,* 320–24
trial-and-error training method, 8, 9
trot, training at the, 86–96
  age-appropriate training, 300, 301, *301*
  badly trained horse, rehabilitating, 93–95, *94*
  cantering before trotting, 87–88, *88*
  cavalletti poles for improving, 90
  circle, tossing head on, 91–92, *92*
  "draw reins trot," repairing, 93–95, *94*
  ground, training from the, 43
  head tossing on circle, 91–92, *92*
  inverted (hollow) back, 87–88
  jackhammer trot, 87–88, *88*
  longeing and warm-ups, 277, 282, 283
  metronome, 90, 96
  overbent horse, 93–95, *94*
  rhythm beads, 95–96, *96*
  saddle fit, 89, 90, *90,* 91
  spiral exercise, 92
  time required to improve rough trot, 89–91, *90*
  trot-canter transitions, 110, *112,* 112–13, 139–41, *140,* 145–48, *147,* 207
  trot-canter-trot transitions, 206
  trotting to fences, 359, 361

trot-walk transitions, 135–36, *136,* 300
turnout importance, 95
walk-trot transitions, 90, 92
warm-up, 87, 88, 270
young horse with jackhammer trot, 87–88, *88*
True Horsemanship, 369
trust, building, *320,* 320–24, 354
TTouch, 298, 330, 363
turning horse loose, observing, 11
turn on the forehand, 100, 157–59, *158,* 165, 172, *173*
turn on the haunches, 167–69, *168*
turnout importance
  abused or confused horse, retraining, 334–35
  age-appropriate training, 300
  connection, roundness, collection, 203
  ex-racehorse, retraining, 350, 351
  trot, training at the, 95
twisted wire snaffles, 246, 247
twitches, 18, 19, 244
two-point (half-seat) vs. full seat, 117–18
tying, 32–34, *33,* 292

U

understanding by horse, importance of, 58–60, 308
"unique noises" commands, 42, 44, 282, 284
"un-sticking" a horse and moving forward, 30–32, *31*
"upward" vs. "downward" transition, 137–38
urinating, training horse, 375–76, *376*

V

varying work for horse, 380
Vienna (sliding) vs. fixed side reins, *236,* 236–37
voice commands
  abused or confused horse, retraining, 322–23
  longeing, 42–44, 277, 282–84

W

walk, training at the, 72–85
  age-appropriate training, 298, 300, 301, *301*
  cavalletti poles for improving, 79, *79,* 80, 85
  contact, 73, 74, *74,* 79, 85
  developing the walk, 82–85, *83*
  dressage for improving, 78–80, *79*
  four-beat rhythm, 78–80, *79*
  ground, training from the, 29, 32, 43
  ground poles for improving, 79, *79,* 80, 85
  hurried (lateral) walk, 80–82, *81*
  leg aids, 74, 76
  longeing and warm-ups, 84, 277, 283
  metronome, 80
  ruined walks, 73–74, *74,* 212